ENCYCLOPAEDIA OF
RESEARCH METHODOLOGY IN APPLIED SCIENCES

ENCYCLOPAEDIA OF RESEARCH METHODOLOGY IN APPLIED SCIENCES

Volume 2

Harmeet Kour

ANMOL PUBLICATIONS PVT. LTD.
NEW DELHI - 110 002 (INDIA)

ANMOL PUBLICATIONS PVT. LTD.
H.O.: 4374/4B, Ansari Road, Daryaganj,
New Delhi-110 002 (India)
Ph.: 23278000, 23261597
B.O.: No. 1015, Ist Main Road, BSK IIIrd Stage
IIIrd Phase, IIIrd Block,
Bangalore - 560 085 (India)
Visit us at: www.anmolpublications.com

Encyclopaedia of Research Methodology in Applied Sciences

First Edition, 2009
ISBN 978-81-261-3814-2 (Set)

PRINTED IN INDIA

Printed at Mehra Offset Press, Delhi.

Contents

Preface

Applied Science is the application of knowledge from one or more natural scientific fields to solve practical problems. Fields of engineering are applied sciences. Applied science is important for technology development. Its use in industrial settings is usually referred to as research and development (R&D). No doubt, science is important to humans. It is a subject that teaches critical thinking and problem solving in a much applied way. Applied Science is all sciences "rolled" into some product or principle that we use every day. Applied science can be found in all sciences. Science and math is a reminder that two fields work together to solve problems. Physics explains the natural world into physical relationships with one another. Technology highlights examples of how science is used in society. Build Environment reflects technology's affect on the living planet. Major fields of applied sciences include: Artificial intelligence; Ceramic engineering; Computing technology; Electronics; Energy; Energy storage; Engineering physics; Environmental technology; Fisheries science; Materials science and engineering; Microtechnology; Nanotechnology; Nuclear technology; Optics; Zoography. In other words, Applied Science & Technology covers a wide range of specialties, such as: Acoustics; Aeronautics; Applied Mathematics; Artificial Intelligence; Atmospheric Sciences; Automatic Control; Automotive Engineering; Chemical Engineering; Chemistry; Civil Engineering; Communication & Information Technology; Computer Databases & Software; Construction; Electrical & Electronic Engineering; Engineering & Biomedical Materials; Energy Resources & Research; Environmental Engineering; Fire & Fire Prevention; Food & Food Industry; Geology; Industrial Engineering; Machine Learning; Machinery; Marine Technology; Mechanical Engineering; Metallurgy; Mineralogy; Mining Engineering; Neural Networks; Nuclear Engineering; Oceanography; Optical & Neural Computing; Petroleum & Gas; Physics; Plastics; Robotics; Solid State Technology; Space Science; Textile Industry & Fabrics; Transportation; Waste Management; Other Industrial & Mechanical Arts. Applied Sciences also have following major subcategories:

- Administration
- Agriculture
- Applied mathematics
- Aquariology

- Audiology
- Cybernetics
- Earth sciences
- Education
- Engineering
- Ergonomics
- Forestry science
- Health sciences
- Industrial design
- Management
- Military science
- Pyrotechnics
- Simulation
- Tactics

Majors and Programs in Applied Science Research Guides include:

- Architecture and Construction Management
- Business and Commerce
- Chemical Technology
- Culinary Arts and Science
- Electrical and Computer Engineering
- Fire Science
- Horticulture
- Information Technology
- Mathematics and Physics
- Mechanical Engineering
- Professional Practice
- Sport Administration

Methodology is defined as the analysis of the principles of methods, rules, and postulates employed by a discipline or the systematic study of methods that are, can be, or have been applied within a discipline or a particular procedure or set of procedures. The primary aim for research is discovering, interpreting, and the development of methods and systems for the advancement of human knowledge on a wide variety of scientific matters of our world and the universe. Research can use the scientific method, but need not do so. A hypothesis can never be proven, but rather only supported by surviving rounds of scientific testing and, eventually, becoming widely thought of as true, but this is not the same as it has been proven. A useful hypothesis allows prediction and within the accuracy of observation of the time, the prediction will be verified. The goal of the research process is to produce new knowledge. Research methodology applies exploratory approach. All of the techniques, methods and procedures adopted in terminology work to carry out terminology research. M*ethodology* includes the concepts as they relate to a particular discipline or field of inquiry, such as; a collection of theories, concepts or ideas; comparative study of different approaches; and critique of the individual methods. Methodology refers to more than a simple set of methods; rather it refers to the rationale and the philosophical assumptions that underlie a particular study. This is why scholarly literature often includes a section on the methodology of the researchers.

This publication titled "Encyclopaedia of Research Methodology in Applied Sciences" provides readers with an introduction to research methodology in applied sciences. It also focuses on the research methodology in engineering physics and optical science. The research methodology in materials science is also covered. Similarly, the research methodology in energy science and electronic science are dealt in detail. The research methodology in nuclear science has received considerable coverage. The research methodology in artificial intelligence science, computer science nanoscience and micro sciences have been discussed in detail. The research methodology applied in ceramic science has been also highlighted. Glossary of terms, acronyms/ abbreviations, bibliography and links have been provided to make this publication more users' friendly.

—Editors

7

Research Methodology in Energy Science

INTRODUCTION

In physics and other sciences, energy is a scalar physical quantity that is a property of objects and systems which is conserved by nature. Energy is often defined as the ability to do work. Several different forms of energy, including kinetic, potential, thermal, gravitational, sound energy, light energy, elastic, electromagnetic, chemical, nuclear, and mass have been defined to explain all known natural phenomena. While one form of energy may be transformed to another, the total energy remains the same. This principle, the conservation of energy, was first postulated in the early 19th century, and applies to any isolated system. According to Noether's theorem, the conservation of energy is a consequence of the fact that the laws of physics do not change over time. Although the total energy of a system does not change with time, its value may depend on the frame of reference. For example, a seated passenger in a moving airplane has zero kinetic energy relative to the airplane, but non-zero kinetic energy relative to the earth. The concept of energy emerged out of the idea of vis viva, which Leibniz defined as the product of the mass of an object and its velocity squared; he believed that total vis viva was conserved. To account for slowing due to friction, Leibniz claimed that heat consisted of the random motion of the constituent parts of matter — a view shared by Isaac Newton, although it would be more than a century until this was generally accepted. In 1807, Thomas Young was the first to use the term "energy", instead of vis viva, in its modern sense. Gustave-Gaspard Coriolis described "kinetic energy" in 1829 in its modern sense, and in 1853, William Rankine coined the term "potential energy." It was argued for some years whether energy was a substance (the caloric) or merely a physical quantity, such as momentum. Energy has many different forms. Kinetic, Sound, Nuclear and Electrical are the main forms.

He amalgamated all of these laws into the laws of thermodynamics, which aided in the rapid development of explanations of chemical processes using the concept of energy by Rudolf Clausius, Josiah Willard Gibbs and Walther Nernst. It also led to a mathematical formulation of the concept of entropy by Clausius, and to the introduction of laws of radiant energy by Jo•ef Stefan.

The concept of energy and its transformations is useful in explaining and predicting most natural phenomena. The *direction* of transformations in energy (what kind of energy is transformed to what other kind) is often described by entropy (equal energy spread among all available degrees of freedom) considerations, since in practice all energy transformations are permitted on a small scale, but certain larger transformations are not permitted because it is statistically unlikely that energy or matter will randomly move into more concentrated forms or smaller spaces. The concept of energy is used often in all fields of science.

- In chemistry, the energy differences between substances determine whether, and to what extent, they can be converted into other substances or react with other substances.
- In biology, chemical bonds are broken and made during metabolic processes, and the associated changes in available energy are studied in the subfield of bioenergetics. Energy is often stored by cells in the form of substances such as carbohydrate molecules (including sugars) and lipids, which release energy when reacted with oxygen.
- In geology and meteorology, continental drift, mountain ranges, volcanos, and earthquakes are phenomena that can be explained in terms of energy transformations in the Earth's interior. While meteorological phenomena like wind, rain, hail, snow, lightning, tornadoes and hurricanes, are all a result of energy transformations brought about by solar energy on the planet Earth.
- In cosmology and astronomy the phenomena of stars, nova, supernova, quasars and gamma ray bursts are the universe's highest-output energy transformations of matter. All stellar phenomena (including solar activity) are driven by various kinds of energy transformations. Energy in such transformations is either from gravitational collapse of matter (usually molecular hydrogen) into various classes of astronomical objects (stars, black holes, etc.), or from nuclear fusion (of lighter elements, primarily hydrogen).

Energy transformations in the universe over time are characterized by various kinds of potential energy which has been available since the Big Bang, later being "released" (transformed to more active types of energy such as kinetic or radiant energy), when a triggering mechanism is available. Familiar examples of such processes include nuclear decay, in which energy is released which was originally "stored" in heavy isotopes (such as uranium and thorium), by nucleosynthesis, a process which ultimately uses the gravitational potential energy released from the gravitational collapse of supernovae, to store energy in the creation of these heavy elements before they were incorporated into the solar system and the Earth. This energy is triggered and released in nuclear fission bombs.

In a slower process, heat from nuclear decay of these atoms in the core of the Earth releases heat, which in turn may lift mountains, via orogenesis. This slow lifting represents a kind of gravitational potential energy storage of the heat energy, which may be released to active kinetic energy in landslides, after a triggering event. Earthquakes also release stored elastic potential energy in rocks, a store which has been produced ultimately from the same radioactive heat sources. Thus, according to present understanding, familiar events such as landslides and earthquakes release energy which has been stored as potential energy in the Earth's gravitational field or elastic strain (mechanical potential energy) in rocks; but prior to this, represents energy that has been stored in heavy atoms since the collapse of long-destroyed stars created these atoms. In another similar chain of transformations beginning at the dawn of the universe, nuclear fusion of hydrogen in the Sun releases another store of potential energy which was created at the time of the Big Bang. At that time, according to theory, space expanded and the universe cooled too rapidly for hydrogen to completely fuse into heavier elements. This meant that hydrogen represents a store of potential energy which can be released by fusion. Such a fusion process is triggered by heat and pressure generated from gravitational collapse of hydrogen clouds when they produce stars, and some of the fusion energy is then transformed into sunlight.

Such sunlight from our Sun may again be stored as gravitational potential energy after it strikes the Earth, as (for example) water evaporates from oceans and is deposited upon mountains (where, after being released at a hydroelectric dam, it can be used to drive turbine/generators to produce electricity). Sunlight also drives many weather phenomena, save those generated by volcanic events. An example of a solar-mediated weather event is a hurricane, which occurs when large unstable areas of warm ocean, heated over months, give up some of their thermal energy suddenly to power a few days of violent air movement. Sunlight is also is captured by plants as *chemical potential energy*, when carbon dioxide and water are converted into a combustible combination of carbohydrates, lipids, and oxygen. Release of this energy as heat and light may be triggered suddenly by a spark, in a forest fire; or it may be available more slowly for animal or human metabolism, when these molecules are ingested, and catabolism is triggered by enzyme action. Through all of these transformation chains, potential energy stored at the time of the Big Bang is later released by intermediate events, sometimes being stored in a number of ways over time between releases, as more active energy. In all these events, one kind of energy is converted to other types of energy, including heat.

Regarding Applications of the Concept of Energy

Energy is subject to a strict global conservation law; that is, whenever one measures (or calculates) the total energy of a system of particles whose interactions do not depend explicitly on time, it is found that the total energy of the system always remains constant.

- The total energy of a system can be subdivided and classified in various ways. For example, it is sometimes convenient to distinguish potential energy (which is a function of coordinates only) from kinetic energy (which is a function of coordinate time derivatives only). It may also be convenient to distinguish gravitational energy, electric energy, thermal energy, and other forms. These classifications overlap; for instance thermal energy usually consists partly of kinetic and partly of potential energy.
- The *transfer* of energy can take various forms; familiar examples include work, heat flow, and advection, as discussed below.
- The word "energy" is also used outside of physics in many ways, which can lead to ambiguity and inconsistency. The vernacular terminology is not consistent with technical terminology. For example, the important public-service announcement, "Please conserve energy" uses vernacular notions of "conservation" and "energy" which make sense in their own context but are utterly incompatible with the technical notions of "conservation" and "energy" (such as are used in the law of conservation of energy).

In classical physics energy is considered a scalar quantity, the canonical conjugate to time. In special relativity energy is also a scalar (although not a Lorentz scalar but a time component of the energy-momentum 4-vector). In other words, energy is invariant with respect to rotations of space, but not invariant with respect to rotations of space-time (= boosts).

Main Energy Topics

The main energy topics which identifies articles and categories that relate to energy. In general, the energy refers to "the potential for causing changes". The word is used in several different contexts. The engineering use has a precise, well-defined meaning, whilst many non-technical uses often do not. In science and physics, it's a physical system's capacity to do work and this page contains items that are related to that defintion.

Energy Infrastructure

See especially Category:Electric power and Category:Fuels for a large number of conventional energy related topics.

- Energy storage
- Electricity generation
- Electricity retailing
- Grid energy storage
- Liquified natural gas
- Microwave power transmission
- Power plant
- Power supply
- Power transmission
- Underground power station

General Energy Topics

- Energy
- Electricity
- Exergy
- Blue energy
- Green electricity
- Orders of magnitude (energy) — list describes various energy levels between 10^{-31} joules and 10^{70} joules
- Thermodynamics
- Perpetual motion
- Heat

Energy Culture

- Biosphere
- Ecology
- Energy balance
- Energy conservation — includes tips for consumers
- Conservation of energy

- Earth Day
- Energy development — Ongoing effort to provide abundant and accessible energy, through knowledge, skills and constructions.
- Energy economics
- Energy resources
- Free energy suppression
- Future energy development — Provides a general overview of future energy development.
- History of perpetual motion machines
- Hubbert peak theory, also known as peak oil — the theory that world oil production will peak (or has peaked), and will then rapidly decline, with a corresponding rapid increase in prices.
- Power harvesting
- Renewable energy development

Energy Applications

- Distributed generation
- Electric vehicle
- Hybrid vehicle
- Hydrogen vehicle
- Steam engine
- Biofuel

Energy Technology Inventors

- Alessandro Volta
- Charles Kettering
- Farrington Daniels — solar energy
- Georges Leclanché — battery
- John Frederic Daniell —
- Rudolf Diesel -compression ignition internal combustion engine
- Georges Imbert — Wood gas
- Leonardo da Vinci

- Moritz von Jacobi
- Nikolaus Otto — internal combustion engine
- Viktor Schauberger — Implosion
- Robert Stirling — Stirling engine (external combustion)
- Nikola Tesla
- James Watt — steam engine with separate condensor

Issues

- Environmental concerns with electricity generation
- Soft energy path — an energy use and development strategy delineated and promoted by some energy experts and activists
- Nuclear power phase-out — a policy to abandon nuclear power.
- Fuel poverty
- Peak Oil
- Low-carbon economy

Energy Policies and Use — National and International

International

- Energy policy — Introductory article
- Eesi — Energy & Environmental Security Initiative

Regional and National

- *Energy Tax Act* — United States energy-related legislation. See also: Category:United States federal energy legislation
- Proposed oil phase-out in Sweden
- United Kingdom:
- Energy policy of the United Kingdom
- Energy use and conservation in the United Kingdom

Energy Unit Terms

- Barrel of oil equivalent
- British thermal unit
- Calorie

- Current solar income — the amount of solar energy that falls as sunlight
- Direct current
- Electronvolt — (symbol: eV) is the amount of energy gained by a single unbound electron when it falls through an electrostatic potential difference of one volt
- Energy conversion — process of converting energy from one form to another
- Enthalpy
- Erg — (symbol "erg") unit of energy and mechanical work in the centimetre-gram-second (CGS) system of units
- EU energy label
- Fill factor — defined as the ratio of the maximum power (Vmp x Jmp) divided by the short-circuit current (Isc) and open-circuit voltage (Voc) in light current density — voltage (J-V) characteristics of solar cells.
- Foot-pound — (symbol ft·lbf or ft·lbf) is an Imperial and U.S. customary unit of mechanical work, or energy, although in scientific fields one commonly uses the equivalent metric unit of the joule (J). There are approximately 1.356 J/(ft·lb_f).
- Francis turbine the most common water turbine in use today
- Gibbs free energy
- Gigaton — Metric Unit of mass, equal to 1,000,000,000 (1 billion) Metric tons, 1,000,000,000,000 (1 trillion) kilograms
- Gray (unit) — (symbol: Gy), is the SI unit of energy for the absorbed dose of radiation. One gray is the absorption of one joule of radiation energy by one kilogram of matter. One gray equals 100 rad, an older unit.
- Heat
- Internal energy — (abbreviated E or U) is the total kinetic energy due to the motion of molecules (translational, rotational, vibrational) and the total potential energy associated with the vibrational and electric energy of atoms within molecules.
- Ionization energy — the (IE) of an atom is the energy required to strip it of an electron.
- Joule — (symbol J, also called newton meter, watt second, or coulomb volt)
- Kilowatt-hour — (symbol: kW·h) corresponds to one kilowatt (kW) of power being used over a period of one hour.

- Kinetic energy
- Mass driver
- Mass-energy equivalence — where mass has an energy equivalence, and energy has a mass equivalence
- Megawatt
- Net energy gain
- Photoelectric effect
- Potential energy — (U, or E_p), a kind of scalar potential, is energy by virtue of matter being able to move to a lower-energy state, releasing energy in some form
- Power beaming — Moving energy from one place to another through "empty space" (or air)
- Power factor — of an AC electric power system is defined as the ratio of the real power to the apparent power.
- Primary energy — Energy contained in raw fuels and any other forms of energy received by a system as input to the system.
- Radiant energy — energy that is transported by waves
- Rotational energy — An object's rotational energy or angular kinetic energy is part of its total kinetic energy
- Solar radiation — radiant energy emitted by the sun, particularly electromagnetic energy
- Therm — (symbol thm) a non-SI unit of heat energy. It is approximately the heat equivalent of burning 100 cubic feet of natural gas. In the US gas industry it is defined as exactly 100,000 $BTU_{59°F}$ or 105.4804 megajoules.
- Ton of oil equivalent
- TPE — Ton Petroleum Equivalent, 45.217 GJ, see ton of oil equivalent

Energy Companies

- Exxon Mobil
- Enercon GmbH — Company based in Germany that operates in the wind turbine industry. One of the biggest producers in the world.
- Saudi Aramco
- Sasol

- United States Enrichment Corporation — contracts with the United States Department of Energy to produce enriched uranium.

Non-profit Organisations

- Musicians United for Safe Energy
- IEA Energy Conservation in Buildings and Community Systems (ECBCS) Programme
- ASPO — Association for the Study of Peak Oil & Gas

Energy-Related Categories

- Electric power
- Emergy
- Energy
- Energy development
- Energy economics
- Energy storage
- Fuels
- Fuel cell vehicles
- Green vehicles
- Low-energy building
- Nuclear fusion
- Orders of magnitude (energy)
- Peak oil
- Petroleum production
- Renewable energy
- Sustainability
- Turbines
- Units of energy
- Waste management

Industry Associations

- OPEC — Organisation of Petroleum-exporting Countries
- IEA — International Energy Agency
- CAPP — Canadian Association of Petroleum Producers

Liquefied Natural Gas (LNG)

Liquefied natural gas or LNG is natural gas that has been processed to remove either valuable components e.g. helium, or those impurities that could cause difficulty downstream, e.g. water, and heavy hydrocarbons and then condensed into a liquid at almost atmospheric pressure (Maximum Transport Pressure set around 25 kPa) by cooling it to approximately -163 degrees Celsius. LNG is transported by specially designed cryogenic sea vessels and cryogenic road tankers; and stored in specially designed tanks. LNG is about 1/614th the volume of natural gas at standard temperature and pressure (STP), making it much more cost-efficient to transport over long distances where pipelines do not exist. Where moving natural gas by pipelines is not possible or economical, it can be transported by LNG vessels, where the most common tank types are membrane(prismatic), Moss Rosenberg (spheres) or Self-Supporting Prismathic Type.

Basic Facts on LNG

LNG offers an energy density comparable to petrol and diesel fuels and produces less pollution, but its relatively high cost of production and the need to store it in expensive cryogenic tanks have prevented its widespread use in commercial applications.

Conditions required to condense natural gas depend on its precise composition, the market that it will be sold to and the process being used, but typically involve temperatures between -120 and -170 degrees Celsius (pure methane liquefies at -161.6 °C) and pressures of between 101 and 6000 kPa (14.7 and 870 lbf/in^2). High pressure natural gas that is condensed is then reduced in pressure for storage and shipping.

The density of LNG is roughly 0.41 to 0.5 kg/L, depending on temperature, pressure and composition. In comparison water has a density of 1.0 kg/L.

LNG does not have a specific heat value as it is made from natural gas, which is a mixture of different gases. The heat value depends on the source of gas that is used and the process that is used to liquefy the gas. The higher heating value of LNG is estimated to be 24 MJ/L at -164 degrees Celsius. This corresponds to a lower heating value of 21 MJ/L.

The natural gas fed into the LNG plant will be treated to remove water, hydrogen sulfide, carbon dioxide and other components that will freeze (e.g., benzene) under the low temperatures needed for storage or be destructive to the liquefaction facility. Purified LNG typically contains more than 90% methane. It also contains small amounts of ethane, propane, butane and some heavier alkanes. The purification process can be designed to give almost 100% methane.

The most important infrastructure needed for LNG production and transportation is an LNG plant consisting of one or more LNG trains, each of which is an independent unit for gas liquefaction. The largest LNG train is the SEGAS Plant in Egypt with a capacity of 5 million ton per annum (mtpa). Exxon Mobil operating Qatargas stage 2, of which one train has a production ability of 5 mtpa. Other facilities needed are load-out terminals for loading the LNG onto vehicles, LNG vessels for transportation, and a receiving terminal at the destination for discharge and regasification, where the LNG is reheated and turned into gas. Regasification terminals are usually connected to a storage and pipeline distribution network to distribute natural gas to local distribution companies (LDCs) or Independent Power Plants (IPPs).

In 1964 the UK and France were the LNG buyers under the world's first LNG trade from Algeria, witnessing a new era of energy. As most LNG plants are located in "stranded" areas not served by pipelines, the costs of LNG treatment and transportation were so huge that development has been slow during the past half century. The construction of an LNG plant costs USD 1-3 billion, a receiving terminal costs USD 0.5-1 billion, and LNG vessels cost USD 0.2-0.3 billion. Compared with the crude oil, the natural gas market is small but mature. The commercial development of LNG is a style called value chain, which means LNG suppliers first confirm the downstream buyers and then sign 20-25 year contracts with strict terms and structures for gas pricing. Only when the customers were confirmed and the development of a greenfield project deemed economically feasible could the sponsors of an LNG project invest in their development and operation. Thus, the LNG business has been regarded as a game of the rich, where only players with strong financial and political resources could get involved. Major international oil companies (IOCs) such as BP, ExxonMobil, Royal Dutch Shell; and national oil companies (NOCs) such as Pertamina, Petronas are active players. Japan, South Korea and Taiwan import large sums of LNG due to their shortage of energy. In 2002 Japan imported 54 million tons of LNG, representing 48% of the LNG trade around the world that year. Also in 2002, South Korea imported 17.7 million tons and Taiwan 5.33 million tons. These three major buyers purchase approximately 70% of the world's LNG demand.

In recent years, as more players take part in investment, both in downstream and upstream, and new technologies are adopted, the prices for construction of LNG plants, receiving terminals and vessels have fallen, making LNG a more competitive means of energy distribution. The standard price for a 125,000-cubic-meter LNG vessel built in European and Japanese shipyards used to be USD 250 million. When Korean and Chinese shipyards entered the race, increased competition reduced profit margins and improved efficiency, reducing costs 60%. The per-ton construction cost of a LNG liquefaction plant fell steadily from the 1970s through the 1990s, with the cost reduced approximately 35%.

Due to energy shortage concerns, many new LNG terminals are being contemplated in the United States. Concerns over the safety of such facilities has created extensive controversy in the regions where plans have been created to build such facilities. One such location is in the Long Island Sound between Connecticut and Long Island. Broadwater Energy, an effort between TransCanada Corp. and Shell (A British-Dutch Corporation) wishes to build a LNG terminal in the sound on the New York side. Local politicians including the Suffolk County Executive have raised questions about the terminal. New York Senators Chuck Schumer and Hillary Clinton have both announced their opposition to the project. Several terminal proposals along the coast of Maine have also been met with high levels of resistance and questions. The Government Accountability Office (GAO) is planning to deliver its report to Congress on LNG in the spring of 2006.

Trade in LNG

LNG is shipped around the world in specially constructed seagoing vessels. The trade of LNG is completed by signing a sale and purchase agreement (SPA) between a supplier and receiving terminal, and by signing a gas sale agreement (GSA) between a receiving terminal and end-users. Most of the contract terms used to be DES or Ex Ship, which meant the seller was responsible for the transportation. But with low shipbuilding costs, and the buyer preferring to ensure reliable and stable supply, there are more and more contract terms of FOB, under which the buyer is responsible for the transportation, which is realised by the buyer owning the vessel or signing a long-term charter agreement with independent carriers.

The agreements for LNG trade used to be long-term portfolios that were relatively inflexible both in price and volume. If the annual contract quantity is confirmed, the buyer is obliged to take and pay for the product, or pay for it even if not taken, which is called the obligation of take or pay (TOP).

In contrast to LNG imported to North America, where the price is pegged to Henry Hub, most of the LNG imported to Asia is pegged to crude oil prices by a formula consisting of indexation called the Japan Crude Cocktail (JCC).

The pricing structure that has been widely used in Asian LNG SPAs is as follows: PLNG = A+B×Pcrude oil, where A refers to a term that represents various non-oil factors, but usually a constant determined by negotiation at a level that can prevent LNG prices from falling below a certain level. It thus varies regardless of oil price fluctuation. Typical figures of ex-ship contracts range from USD 0.7 to 0.9. B is a degree of indexation to oil prices; typical figures are 0.1485 or 0.1558, and Pcrude oil usually denominated in JCC. PLNG and Pcrude oil stand for price of oil in USD per million British Thermal Unit (MMBTU (in the fuel industry, M stands for 1000 and MM for 1 000 000)). With the demand of LNG moving up and down, the price of

LNG moves in a "S" curve. With new demand from China, India and US increasing dramatically, and crude oil price skyrocketing, the LNG price is on the rise too.

In the mid 1990s LNG was a buyer's market. At the request of buyers, the SPAs began to adopt some flexibilities on volume and price. The buyers had more upward and downward flexibilities in TOP, and short-term SPAs less than 15 years came into effect. At the same time, alternative destinations for cargo and arbitrage were also allowed. By the turn of the 21st century, the market was again in favor of sellers. Sellers now propose rigid SPAs and would like an association similar to OPEC to be established to protect their interests. It is certain that the competition between sellers and buyers will go on.

Receiving terminals exist in several countries (see the list of *importing* countries in table below; China is expected to move onto the list by 2006), allowing gas imports from other areas (see list of *exporting* countries in table below).

The United States Department of Energy's Energy Information Administration provides estimates of LNG trade in 2002 as follows:

Country	*Export volume*		*Country*	*Import volume*	
	(10^9 ft^3)	*(10^6 t)*		*(10^9 ft^3)*	*(10^6 t)*
Indonesia	1,100	23.0	Japan	9,200	188.3
Algeria	935	19.6	South Korea	2,000	40.7
Malaysia	741	15.6	France	511	10.7
Qatar	726	14.9	Taiwan	363	7.5
Nigeria	394	8.2	United Kingdom	356	7.3
Australia	367	7.7	United States	229	4.8
Oman	356	7.3	Turkey	224	4.6
Brunei Darussalam	351	7.2	Portugal	146	3.3
United Arab Emirates	278	5.7	Spain	131	2.7
Russia	234	4.8	Italy	130	2.6
Trinidad and Tobago	189	4.0	Belgium	124	2.7
United States	68	1.4	India	122	2.5

In 2005, Egyptian NG production outpaced consumption and it joined the LNG exporting countries.

LNG Safety and Accidents

In its liquid state, LNG is not explosive. For an explosion to occur with LNG, it must first vaporize and then mix with air in the proper proportions (the explosive range is 5% to 15%), and then be ignited afterwards. Serious accidents involving LNG to date are listed below:

- 1944, 20 October. The East Ohio Natural Gas Company experienced a failure of an LNG tank in Cleveland, Ohio. 128 people perished in the explosion and fire. The tank did not have a dike retaining wall, and it was made during World War II, when metal rationing was very strict. The steel of the tank was made with an extremely low amount of nickel, which made the tank brittle when exposed to the extreme cold of LNG, and the tank ruptured, spilling LNG into the city sewer system.
- 1973, February,Staten Island, New York. While repairing the interior of an empty storage tank, a fire started. The pressure increased inside the tank so fast the concrete dome on the tank lifted and then collapsed falling inside the tank and killing the 37 construction workers below.
- 1979, Lusby, Maryland, at the Cove Point LNG facility a pump seal failed, releasing gas vapors, which entered and settled in an electrical conduit. A worker switched off a circuit breaker, igniting the gas vapors, killing a worker and causing heavy damage to the building. National fire codes were changed as a result of the accident.
- 2004, 19 January, Skikda, Algeria. Explosion at Sonatrach LNG liquefaction facility. 27 killed, 80 injured, three LNG trains destroyed, 2004 production was down 76% for the year. A cold hydrocarbon leak occurred introducing the high-pressure steam boiler with gases via a combustion air fan. The explosion inside the boiler fire box precipitated a larger explosion of vapors outside the box.

Seaborne LNG transport tankers (including their loading terminals) have not had a "major" (term undefined) accident in over 33,000 voyages since maritime inception in 1959. There have, however, been several significant incidents with LNG ships, but with no spills. In addition to accidents, terrorism experts are concerned that intentional sabotage could lead to unprecedented releases, resulting in massive fires and other damaging effects. The latter may include detonations (producing large blast waves) and deflagration-to-detonation transition phenomena. As the Department of Energy notes in its December 2004 report (Sandia National Labs, SAND2004-6258), the available testing data on LNG spills are based on releases of very small size in comparison to releases expected from intentional attacks. Despite intense local opposition, the Federal Energy Regulatory Commission has approved a site permit for an LNG terminal in Fall River, Massachusetts in a densely populated harbor area

LNG Storage

LNG above-ground tanks are mainly of double-wall, high-nickel steel construction with extremely efficient insulation between the walls. Large tanks are low aspect ratio (height to width) and cylindrical in design with a domed roof. Storage pressures in these tanks are very low, less than 5 psig. Sometimes more expensive frozen-earth, underground storage is used. Pre-stressed concrete backed up with suitable thermal insulation, are designed to be both under and above ground to suit sites conditions and local safety regulations and requirements. Smaller quantities, 190,000 US gallons (700 m^3) and less, are stored in horizontal or vertical, vacuum-jacketed, pressure vessels. These tanks may be at pressures anywhere from less than 5 psig to over 250 psig (35 to 1700 kPa gauge pressure).

LNG must be maintained cold (at least below -117 °F or -83 °C) to remain a liquid, independent of pressure. There will inevitably be some degree of boil-off as a result of heat gained from the outside ambient atmosphere. This gas is may be returned to storage by recompression and reliquefaction, or used in the liquefaction process.

LNG Refrigeration

The insulation, as efficient as it is, will not keep the temperature of LNG cold by itself. LNG is stored as a "boiling cryogen," that is, it is a very cold liquid at its boiling point for the pressure it is being stored. Stored LNG is analogous to boiling water, only 470 °F (260 °C) colder. The temperature of boiling water (212 °F or 100 °C) does not change, even with increased heat, as it is cooled by evaporation (steam generation). In much the same way, LNG will stay at near constant temperature if kept at constant pressure. This phenomenon is called "autorefrigeration". As long as the steam (LNG vapor boil off) is allowed to leave the tea kettle (tank), the temperature will remain constant.

If the vapor is not drawn off, then the pressure and temperature inside the vessel will rise. However, even at 1000 psig (7 MPa), the LNG temperature will still be only about -200 °F (-130 °C).

LNG, LPG, and CNG

Liquefied Natural Gas (LNG)

When natural gas is cooled to a temperature of approximately -260 °F (-160 °C) at atmospheric pressure it condenses to a liquid called liquefied natural gas (LNG). One volume of this liquid takes up about 1/600th the volume of natural gas at a stove burner tip. LNG is only about 45% the density of water. LNG is odorless, colorless, non-corrosive, and non-toxic. When vaporized it burns only in concentrations of 5% to 15% when mixed with air. Neither LNG, nor its vapor, can explode in an unconfined environment.

Natural gas is composed primarily of methane (typically, at least 90%), but may also contain ethane, propane and heavier hydrocarbons. Small quantities of nitrogen, oxygen, carbon dioxide, sulfur compounds, and water may also be found in "pipeline" natural gas. The liquefaction process removes the oxygen, carbon dioxide, sulfur compounds, and water. The process can also be designed to purify the LNG to almost 100% methane.

Compressed Natural Gas (CNG)

Compressed natural gas (CNG) is natural gas pressurized and stored in welding bottle-like tanks at pressures up to 3,600 psig (25 MPa). Typically, it is same composition of the local "pipeline" gas, with some of the water removed. CNG and LNG are both delivered to the engines as low pressure vapor (ozf/in² to 300 psig, up to 2.1 MPa). CNG is often misrepresented as the only form natural gas can be used as vehicle fuel. LNG can be used to make CNG. This process requires much less capital intensive equipment and about 15% of the operating and maintenance costs.

Liquid Petroleum Gas (LPG)

Liquid petroleum gas (LPG, and sometimes called propane) is often confused with LNG and vice versa. They are not the same and the differences are significant. Varieties of LPG bought and sold include mixes that are primarily propane, mixes that are primarily butane, and mixes including propane, propylene, n-butane, butylene and iso-butane. Depending on the season—in winter more propane, in summer more butane. Vapor pressures, at 30 °C, are for commercial propane in the range 10-12 barg (1 to 1.2 MPa), for commercial butane, 2-4 barg (0.2 to 0.4 MPa). In some countries LPG is composed primarily of propane (upwards to 95%) and smaller quantities of butane. The vapor pressure of commercial butane is generally too low to release it from the top vapor space. Pumps and (hot water, steam, electricity or direct-fired) vaporizers are frequently used. An alternative to using neat butane vapor which overcomes the need for pipework heating, is to use a gas-air mixture (well outside flammability limits). Air depresses the vapor dew-point temperature. Another advantage is that the mixture can be made to "simulate" natural gas or town gas to produce the same heat release through a burner under equal supply pressures, characterized by a term known as Wobbe number or Wobbe index.

LPG compared to natural gas has a significantly higher heating value and requires a different air-to-gas mixture (propane: 24:1, butane: 30:1) for good combustion.

LPG can be stored as a liquid in tanks by applying pressure alone. While the distribution of LNG requires heavy infrastructure investments (pipelines, etc.), LPG is portable. This fact makes LPG very interesting for developing countries and rural areas. LPG (sometimes called autogas) has also been used as fuel in light duty vehicles for many years. An increasing number of petrol stations around the world offers LPG pumps as well. A final example that should not be forgotten is that the "bottled gas" can often be found under barbecue grills.

LNG Terminals

The following LNG off-loading and processing terminals are located in the United States:

- Massachusetts
- Maryland
- Louisiana
- Elba Island
- Cove Point

Proposed LNG Terminals

- Crown Landing LNG Terminal — A proposed LNG terminal to be located along the Delaware River in Gloucester County, New Jersey. BP is working for regulatory approval and the LNG terminal is expected to be online by the end of 2008.
- Gate Terminal — LNG terminal to be located on the Maasvlakte, part of the harbour and industrial area of the city of Rotterdam in the Netherlands, proposed by the gas company N.V. Nederlandse Gasunie and logistics company Vopak. Regulatory approval has been granted for the largest part. See the website of the Gate Terminal.
- LionGas Terminal — LNG terminal to be located in the harbour and industrial area of the city of Rotterdam in the Netherlands, which will be developed by 4Gas. All necessary regulatory approval has been granted. Construction is expected to start mid-2007 and operation to start in the second half of 2010. See the website of LionGas
- Eemshaven LNG Terminal — LNG terminal to be located in the seaport of Eemshaven in the Dutch province of Groningen, proposed by Dutch energy company Essent and American oil company ConocoPhillips.
- Fos Cavaou LNG Terminal — LNG terminal under construction in the south of France. Operation is expected to start operation mid-2007.

Important FAQs

What is LNG?

Liquefied natural gas, or LNG, is natural gas in its liquid form. When natural gas is cooled to minus 259 degrees Fahrenheit (-161 degrees Celsius), it becomes a clear, colorless, odorless liquid. LNG is neither corrosive nor toxic. Natural gas is primarily methane, with low concentrations of other hydrocarbons, water, carbon dioxide, nitrogen, oxygen and some sulfur compounds. During the process known as liquefaction, natural gas is cooled below its boiling point, removing most of these compounds. The remaining natural gas is primarily methane with only small amounts of other hydrocarbons. LNG weighs less than half the weight of water so it will float if spilled on water.

How is LNG stored?

When LNG is received at most terminals, it is transferred to insulated storage tanks that are built to specifically hold LNG. These tanks can be found above or below ground and keep the liquid at a low temperature to minimize the amount of evaporation. If LNG vapors are not released, the pressure and temperature within the tank will continue to rise. LNG is characterized as a cryogen, a liquefied gas kept in its liquid state at very low temperatures. The temperature within the tank will remain constant if the pressure is kept constant by allowing the boil off gas to escape from the tank. This is known as auto-refrigeration. The boil-off gas is collected and used as a fuel source in the facility or on the tanker transporting it. When natural gas is needed, the LNG is warmed to a point where it converts back to its gaseous state. This is accomplished using a regasification process involving heat exchangers.

How is natural gas stored?

Natural gas may be stored in a number of different ways. It is most commonly stored underground under pressure in three types of facilities. The most commonly used in California are depleted reservoirs in oil and/or gas fields because they are more available. Aquifers and salt cavern formations are also used under certain conditions. The characteristics and economics of each type of storage site will dictate its suitability for use. Two of the most important characteristics of an underground storage reservoir are its capability to hold natural gas for future use and its deliverability rate. The deliverability rate is determined by the withdrawal capacity of the associated valves and compressors and the total amount of gas in the reservoir. In other states, natural gas is also stored as LNG after the natural gas has been liquefied and placed in above-ground storage tanks. (Source: U.S. Department of Energy, Energy Information Administration.)

Where does LNG come from?

A majority of the world's LNG supply comes from countries with large natural gas reserves. These countries include Algeria, Australia, Brunei, Indonesia, Libya, Malaysia, Nigeria, Oman, Qatar, and Trinidad and Tobago.

What countries import LNG?

There are 40 LNG receiving terminals located worldwide. Japan, South Korea, the United State and a number of European Counties import LNG.

How is LNG transported?

LNG is transported in double-hulled ships specifically designed to handle the low temperature of LNG. These carriers are insulated to limit the amount of LNG that boils off or evaporates. This boil off gas is sometimes used to supplement fuel for the carriers. LNG carriers are up to 1000 feet long, and require a minimum water depth of 40 feet when fully loaded. There are currently 136 ships which transport more than 120 million metric tons of LNG every year. (Source: University of Houston IELE, *Introduction to LNG.*)

How is LNG used?

LNG is normally warmed to make natural gas to be used in heating and cooking as well as electricity generation and other industrial uses. LNG can also be kept as a liquid to be used as an alternative transportation fuel.

Is LNG flammable?

When cold LNG comes in contact with warmer air, it becomes a visible vapor cloud. As it continues to get warmer, the vapor cloud becomes lighter than air and rises. When LNG vapor mixes with air it is only flammable if it's within 5%-15% natural gas in air. If it's less than five percent natural gas in air, there is not enough natural gas in the air to burn. If it's more than 15 percent natural gas in air, there is too much gas in the air and not enough oxygen for it to burn.

Is LNG explosive?

As a liquid, LNG is not explosive. LNG vapor will only explode if in an enclosed space. LNG vapor is only explosive if within the flammable range of 5%-15% when mixed with air.

What is a "peak-shaving" facility?

LNG peak-shaving facilities are used for storing surplus natural gas that is to be used to meet the requirements of peak consumption later during winter or summer. Each peak-shaving facility has a regasification unit attached but may or may not have a liquefaction unit. These facilities without a liquefaction unit depend upon tank trucks to bring LNG from other nearby sources to them. Of the approximate 113 active LNG facilities in the United States, 57 are peak-shaving facilities. The other LNG facilities include marine terminals, storage facilities, and operations involved in niche markets such as LNG vehicular fuel.

Why use LNG?

Natural gas is the cleanest burning fossil fuel. It produces less emissions and pollutants than either coal or oil. The North American supply basins are maturing and as demand for natural gas increases in California and throughout the United States, alternative sources of natural gas are being investigated. Natural gas is available outside of North America, but this gas is not accessible by pipelines. Natural gas can be imported to the United States from distant sources in the form of LNG. Since LNG occupies only a fraction (1/600) of the volume of natural gas, and takes up less space, it is more economical to transport across large distances and can be stored in larger quantities. LNG is a price-competitive source of energy that could help meet future economic needs in the United States.

What is a Rapid Phase Transition?

When enough LNG is spilled on water at a very fast rate, a Rapid Phase Transition, or RPT, occurs. Heat is transferred from the water to the LNG, causing the LNG to instantly convert from its liquid phase to its gaseous phase. A large amount of energy is released during this rapid transition between phases and a physical explosion can occur. While there is no combustion, this physical explosion can be hazardous to any nearby person or buildings.

What about security?

All LNG ships must comply with all pertinent local and international regulatory requirements, which include regulations and codes set forth by the International Maritime Organisation (IMO), the U.S. Maritime Administration (MARAD), the U.S. Coast Guard (USCG), and the U.S. Department of Transportation (DOT), as well as the hosting Port Authority.

DOT regulations must be followed at onshore LNG facilities and marine terminals. The Research and Special Programmes Administration, DOT, regulations include 49 CFR Part 193 - Liquefied Natural Gas Facilities: Federal Safety Standards. These standards specify siting, design, construction, equipment, and fire protection requirements that apply to new LNG facilities and to existing facilities that have been replaced, relocated, or significantly altered.

Offshore marine terminals must follow regulations set by the USCG. The USCG monitors the safety of coastal waters around the U.S. and ensures the safety of ships while in U.S. waters and in port by preventing other ships from getting near LNG tankers. The USCG works with local harbor authorities and LNG facility personnel to ensure that proper procedures are followed. The USCG and MARAD are the federal agencies responsible for siting off-shore LNG facilities and are currently developing regulations.

What is Biodiesel?

With some modification it is possible to run many diesel engined vehicles on unrefined vegetable oil. The oil needs to be pre-heated, and well filtered before use to prevent coagulation. Biodiesel is the name given to any diesel equivalent biofuel which can be used in an unmodified diesel engined vehicle. In general biodiesel is most commonly made with a mixture of vegetable oil and methanol. With a flash point of 160 degrees C it is classified as non-flammable, and it is also biodegradable and non-toxic. On its own biodiesel has much lower emissions than petrodiesel, and it can also be mixed with petrodiesel to reduce emissions. B20 for example is a fuel containing 20% biodiesel and 80% petrodiesel. Pure biodiesel is B100. Biodiesel reduces carbon dioxide emissions by 78%, and carbon monoxide emissions by 50%. It also completely eliminates sulphur emissions. For vehicles made before the early 1990's there is a problem with the use of biodiesel. The rubber hoses and gaskets used before that time can degrade in the presence of biodiesel. Newer cars have synthetic hoses and gaskets, and of course older cars can have their hoses and gaskets replaced before biodiesel is introduced. Biodiesel is also more solvent than petrodiesel and so it will rapidly break down any deposits of old residue in a vehicle's fuel lines and fuel tank and clog the fuel filter. Therefore, after making the transition to biodiesel it is important to change the fuel filter around 1000 miles after switching.

Biodiesel could be an important renewable substitute for fossil fuels. And, in certain parts of the world, governments and some corporations consider the jatropha plant, common in hot climates, one of the most promising sources of biodiesel. The plant can grow in wastelands, and it yields more than four times as much fuel per hectare as soybean, and more than ten times that of corn. But the commercial-scale cultivation of jatropha, which has not previously been grown as a crop, raises several significant challenges.

This year, the Energy and Resources Institute (TERI), an Indian research group, launched a 10-year, $9.4 million project to research issues involved in taking jatropha from seed to filling station. One challenge is growing the plant in poor soil.

The first crops of jatropha, planted in what was wasteland, have now flowered, says Alok Adholeya, director of TERI's Biotechnology and Management of Bioresources division. "It proves that we can do this," he says. He and other researchers at TERI spent five years testing different mycorrhiza microorganisms, symbiotic fungi that improve the ability of many plants to grow in poor soil. Adholeya's team found that the most effective was a fungus in the glomus species (he is not currently disclosing the exact fungus), which improves jatropha yields by 15 percent.

The TERI project is working in rural Andra Pradesh, a state in southeast India, collaborating with local financial institutions to develop loan guarantees to fund seed purchases; it's also collaborating with insurers to back the farmers against potential losses. In addition, it had to educate the farmers on how to cultivate the plant.

So far, the project has signed up 5,000 farmers representing 1,000 hectares of land. The goal is to have 8,000 hectares under cultivation by March 2008, and Adholeya says that the success of the first crops has drawn interest from many more farmers. By the end of 2008, TERI plans to have a production facility producing biodiesel from jatropha. Eventually, it aims to produce 90 million liters of biodiesel annually.

We live in an age where oil acquisition and consumption dominate the course of national and international politics throughout much of the world. Nowhere is this more apparent than in the United States, where 25% of the world's oil is consumed. Our dependence on oil is responsible for many of the greatest environmental, health, and security problems our nation faces today. Finding a way to ease our dependence on oil presents one of the paramount challenges of our age.

In 2002, diesel fuel accounted for almost 23% of total vehicle fuel consumption in the United States. Biodiesel, which is a relatively clean-burning, renewable fuel produced from new and used animal and vegetable oil, could be used to replace at least a portion of the diesel fuel consumed in this country. Biodiesel has three large advantages over regular petroleum diesel. First, it is not a petroleum-based fuel, which means that using biodiesel would reduce our dependency on foreign oil. Second, biodiesel is produced domestically, which means that using biodiesel will create jobs and contribute to local economies. The third major advantage of biodiesel is that it is cleaner than conventional diesel; biodiesel produces significantly less harmful emissions than regular petroleum diesel when burned in a combustion engine.(Note, however, that biodiesel emits higher levels of nitrogen oxides (NOx); pure biodiesel emits up to 13% more NOx than conventional diesel.) Best of all, we already have much of the equipment and infrastructure necessary to use biodiesel. Biodiesel can be used in almost any normal internal combustion diesel engine with few special modifications, and it can be stored and transported using existing infrastructure and equipment already used for regular petroleum diesel. Biodiesel can be burned in its neat form (100% biodiesel) or it can be mixed with regular petroleum diesel in any concentration, such as 2% (B2), 5% (B5) and 20% (B20).

Studies have shown that biodiesel blends actually increase the performance of diesel engines. It is estimated that the current production capabilities of biodiesel in the U.S. is around 3 billion gallons annually.m. This is sufficient to replace about 8 percent of the diesel fuel used on our highways. The model bill presented in this packet calls for statewide use of a low-blend biodiesel fuel, providing numerous economic, security, and environmental benefits for states. This web site offers the tools necessary to introduce and pass legislation to encourage the use of biodiesel in your state. These tools include a sample bill, talking points, press clips, a fact pack, links, and other background information.

Benefits

There are many benefits of using biodiesel rather than petroleum based diesel. Biodiesel is a renewable resource that is significantly cleaner that traditional diesel.

- Biodiesel emits up to 100% less sulfur dioxide, a major component of acid rain, and 80-100% less carbon dioxide than traditional diesel.
- Biodiesel reduces smoke particulates up to 75%.
- Pure biodiesel is biodegradable and breaks down as fast as sugar.
- Biodiesel breaks down four times faster than regular diesel, with most of a spill broken down only 28 days later.
- Biodiesel is safer to transport and handle, because it has a higher flash point (greater than 150C) than traditional petroleum (77C), and is non-corrosive for human contact.
- Since biodiesel can be used in conventional diesel engines, the renewable fuel can directly replace petroleum products; reducing the country's dependence on imported oil.

Fortunately the transition between biodiesel and traditional diesel can be easy. Transport, storage, distribution of biodiesel can be facilitated through traditional means of tankers and gas pumps. Because biodiesel and traditional diesel can be mixed without issue no major cleaning would be required of existing infrastructure. In fact, biodiesel and diesel are currently sold as mixtures, for example, a 20% biodiesel blend is B20, and a 5% blend is B5. Biodiesel is also more lubricating than diesel fuel, which increases the life of your engine, and replaces the sulfur that can produce acid rain. Engines running on biodiesel normally have similar fuel mileage, and auto ignition, fuel consumption, power output, and engine torque are also similar to traditional diesel. Most websites agree that no modification of engines is required to run pure biodiesel, but the U.S. Department of Energy suggests consulting your engine warranty or dealership when running blends or pure biodiesel.

Drawbacks

The advantages of biodiesel far outweigh the disadvantages, although there are a few. Biodiesel has long been shown to reduce all regulated emissions, with the exception of nitrous oxide (NOx) emissions, which are a contributing factor in the localized formation of smog and ozone. Few studies suggest a slight increase in nitrous oxide emissions and varies widely based upon the type engine and type of biodiesel used. Older engines may be responsible for higher nitrous oxide emissions as well. Nitrous oxide emission increases range from 1-15%. Nitrous oxide emissions can be reduces through additives in the biodiesel. Emissions can also be reduced through the lowering of the combustion temperature of the fuel, which will decrease nitrous oxide emissions to, or below, the current level of NOx emissions. One other issue with biodiesel is the effect extreme cold weather. Cold temperatures cause wax crystals to form which plug fuel filters. Different derivatives of biodiesel result in different temperature thresholds, for example, soy biodiesel can be used down to –1C, and cooking oil biodiesel can be used down to –9-12C. Standard diesel goes down to –29C. In order to compensate for cold temperatures a number of things can be done. Additives can be added to the fuel, electrical elements that heat the fuel back up, fuel tank heaters, and coolant operated fuel heaters. Beside these minor issues biodiesel is easily compatible with current fuel infrastructure and diesel engines.

Biodiesel Processors

Building your own processor isn't difficult. You don't need any special skills or special tools. There are plans, designs and ideas here to help you. There's also good advice to be had from experienced users at the Biofuel mailing list, and lots of good information in the searchable list archives: We don't recommend buying the ready-made cone-bottomed plastic biodiesel processors on sale at quite a few websites. They're overpriced, most of them don't work well, some of them have been prone to catching fire, most of them are not properly sealed, and they can't withstand much or any pressure, which is something that might happen. One major plastic-reactor vendor, rather than including the separate washing tank that's a standard basic requirement, claims instead that washing biodiesel isn't necessary. Or you can pay another $1,000 for a plastic washing tank. Some of these reactors don't even have a way of heating the oil (an ineffective electric heating band comes as an expensive extra). In most cases you'd be paying a high price for a poor-quality processor that won't make good fuel and might not even be safe. You can build a better processor yourself, for much less money: "You could make an excellent processor plus more than 8,000 gallons of high-quality biodiesel for that price."

The processor designs that follow are for different types of processors and cover the range of the different methods used. They can be used with any of the main processing methods and all of them will produce high-quality biodiesel efficiently. These are the original designs, they're not copies of other processors, and many thousands of people have built them using the information presented here. They can be adapted to suit a wide range of circumstances — there's a solution for everyone. Only one design comes with precise blueprints, the others are built from recycled oil drums, water heaters, propane gas bottles, kerosene tanks (in our case), or whatever works — they're designed to be adaptible to whatever is easily available. Study all the designs before making your choices. The text will also give you essential information about processing techniques and how to make good biodiesel.

Scaling up

Scaling up from small test-batches to a full-sized processor might not be straightforward. Your processing procedure might need some adjusting. Processing instructions have to use averages and approximations to some extent because processors vary so widely. Short fat reactor tanks need more and better agitation than tall thin ones, using a pump for agitation might not be the same as using a stirrer and baffle plates, stirrer speeds and blade configurations vary widely, while insulated tanks maintain better temperature control than uninsulated ones, some reactors don't have in-tank heaters and rely on insulation, etc etc.

All the processing methods provided at this website will get you close to the mark, then use the fuel quality tests to fine-tune your process to your particular processor. You'll also need larger-scale measuring equipment. You'll be measuring methanol by the litre, not millilitres. We use a postal scale measuring up to 1,000 grams to weigh the lye for full-scale batches. Mark out volume measures on the side of your processor or pre-heating tank for the oil. Heat some oil to the processing temperature (usually 55 deg C, 131 deg F), decant one litre into an accurate measuring flask, let it cool to room temperature and check the volume again. Multiply by 10. Using an accurate measuring flask, measure out this amount into a large bucket, mark the level in the bucket. Fill the pre-heating tank or the processor one bucket at a time, marking the tank as you go until you reach the right level for your processor's batch capacity of heated oil, whether 60, 80, 140 or however many litres. If you prefer, use 15 or 20-litre increments instead of 10 litres. For future batches, just fill the tank straight up to the level you've marked. The main variables you'll face in scaling up are agitation, duration of the process and temperature. Adjust the rate of agitation and the process duration, stay with the usual process temperature and with your usual proportions of methanol and lye catalyst.

Agitation is usually via a pump or a stirrer. With stirrers you can vary the speed and the shape and configuration of the paddles, and you can use baffle plates in the tank. With pumps, there's not much you can do to adjust the action. To take one example, the 1" Clear Water Pump from Northern Tools is a popular choice. Usually it's stepped down to 3/4", but then the maximum-sized batch it will handle efficiently in the usual processing time is about 80 litres (21 US gal), 100 litres at the most (26 US gal). Nonetheless these pumps are often used with 200-litre reactors (55 US gal), which can result in the washing problems that come with poor process completion. Using full-sized 1" fittings would help, but it might still need extended processing times. For 100 litres and more use a more powerful pump.

For temperature control, once the oil is hot it doesn't need much reheating during the processing, especially if the reactor tank is insulated. But accurate temperature control helps and a thermostat is useful (though not essential). It's also a safety factor — if you forget to switch off the heating element the oil will get too hot and boil off the methanol, upsetting the process and building up pressure in the reactor. With a thermostat you can just set it to the required temperature, switch it on and leave it until the processing is finished. Next to oil quality, the main variable in making biodiesel is user error, especially inaccurate measurements. Starting off with small 1-litre test batches not only means your mistakes are small ones, it's also the best way to learn good technique. Learn the process first, building the processor comes later. Start here.

Hazards

The main hazards of making biodiesel are poisonous fumes, dangerous chemicals, and fires. For poisonous fumes, the best advice is not to expose yourself to the fumes in the first place. Don't use "open" reactors — biodiesel processors should be closed, with no fumes escaping. See Safety. In practice this either means having a reactor vessel built to withstand some pressure, or having a vent leading to the open air outside. Here we'll deal with fires. There have been several fires, one operator lost all his equipment, his shed, and was slightly burned. But don't be put off, as he admitted it was due to his own carelessness and shouldn't have happened. It's easy to avoid fires. Most important, again, use closed processors.

Good advice from Todd Swearingen of Appal Energy:

The main fire hazard is using an open reactor and poor ventilation in the presence of an ignition source.

Ignition sources can be:

- Using combustible fuels as a direct heat source (propane, natural gas, wvo, wood, etc.), rather than a heat exchange system where the flame is far removed from the processing area.
- Open electric motor housings, rather than TEFC motors (Totally Enclosed Fan Cooled AC Motors) or explosion proof motors.

- Disconnecting an electrical appliance by pulling out the wall plug, rather than using enclosed switches.
- Any open flame.

Other fire sources can be over-taxed pumps and motors in close proximity to combustible materials (garage walls, plastic barrels, almost anything), oversized breakers and/or fuses, undersized wiring (such as 14/2 Romex) for higher amperage draws. And there always is the danger of spontaneous combustion in the presence of oily rags, especially when working with drying oils such as hemp and linseed. This threat decreases as the saturation of the oil/fat increases.

Touchless Processor

Dale Scroggins's processor is "a 100-liter processor made from mostly salvaged materials that almost fills itself, mixes everything, recovers the unused methanol, washes and dries the ester pretty much by flipping a few switches... I first evacuate the tank, then suck in the oil through a filter setup that I'm a little too proud of. I mix the methoxide separately, using a drill press with a paint mixer installed, then suck it into the tank. Then flip the pump switch. Watch through the clear hoses and marvel. Later, flip off the pump switch, check for separation, flip on the vacuum pump (which is connected to a liquid trap which is connected to the condenser which is connected to the tank) and watch the excess methanol collect in the trap. When no more collects, I open the tank to the atmosphere and drain off the glycerine. Then suck in water and begin the wash."

Joe Street's Processor

This processor is made from a 36-liter hot water tank and employs vacuum which has many advantages in safety and reduced processing time. Vacuum drying the oil ensures low water content, which is critical to a good process, and vacuum allows catalysts to be drawn into the processor rather than forcing them in by pressurizing containers, which eliminates a potential serious hazard. Vacuum allows fuel to be quickly dried at the end of the process as well giving considerable time savings from start to end of process. The reactor employs forced exhaust for safety and is designed to allow methanol recovery and two-stage processing without any additional modifications. The design is easily scaled up in size, and makes quality fuel consistently, and easily. From the start, the purpose of this project was to offer something easily reproducible so that as many people as possible can have a chance to make their own fuel.

Mechanical, Maritime and Materials Engineering

The Energy Technology Section is part of the Process and Energy Department of the Mechanical, Maritime and Materials Engineering (3mE) faculty at the Delft

University of Technology. Our director is temporarily Associate Prof. P. Colonna. Research and education focus on the design and modeling of energy conversion systems. Our lively group consists of around 30 scientific staff members (professors, research associates and PhD students) and 40 M.Sc. students. We also manage excellent research facilities. This slide show gives an overview of the prospects and trends of the technologies and research demand in the field of Energy Technology, as well as an overview of the activities and plans of the Energy Technology Section. Our Laboratories behind the P&E building. Almost a Zeppelin with inside quite some advanced experimental equipment! The central heating station of the TU Delft campus is also close by. Maybe one day it will be replaced by a biomass fuelled power station employing an hybrid high-temperature fuell cell, cogenerating electricty and heat for the whole campus. The efficiency of such system would be extraordinary and the environmental impact reduced to a minimum!

Microwave Power Transmission

Microwave power transmission (MPT) is the use of microwaves to transmit power through outer space or the atmosphere without the need for wires. Following World War II, which saw the development of high-power microwave emitters known as cavity magnetrons, the idea of using microwaves to transmit power was researched. In 1964, William C. Brown demonstrated a miniature helicopter equipped with a combination antenna (electronics) and rectifier device called a rectenna. The rectenna converted microwave power into electricity, allowing the helicopter to fly. In principle, the rectenna is capable of very high conversion efficiencies — over 90% in optimal circumstances.

Most proposed MPT systems now usually include a phased array microwave transmitter. While these have lower efficiency levels they have the advantage of being electrically steered using no moving parts, and are easier to scale to the necessary levels that a practical MPT system requires.

Using microwave power transmission to deliver electricity to communities without having to build cable-based infrastructure is being studied at Grand Bassin on Reunion Island in the Indian Ocean.

Common Safety Concerns

The common reaction to microwave transmission is one of concern, as microwaves are generally perceived by the public as dangerous forms of radiation — stemming from the fact that they are used in microwave ovens. While high power microwaves can be painful and dangerous as in the United States Military's Active Denial System, MPT systems are generally proposed to have only low intensity at the rectenna.

Though this would be extremely safe as the power levels would be about equal to the leakage from a microwave oven, and only slightly more than a cell phone, the relatively diffuse microwave beam necessitates a large rectenna area for a significant amount of energy to be transmitted.

Research has involved exposing multiple generations of animals to microwave radiation of this or higher intensity, and no health issues have been found.

Proposed Uses

- MPT is the most commonly proposed method for transferring energy to the surface of the Earth from solar power satellites or other in-orbit power sources.
- MPT is occasionally proposed for the power source in beamed energy orbital space ships. Although lasers are more commonly proposed, their low efficiently in light generation and reception has led some designers to opt for a microwave based systems. Although microwave are more easily scaled to high powers and suffer from less atmospheric distortion, the engineering hurdles in building a man capable craft to reach orbit involving the beaming of megawatts of power have prevented the realisation of such plans.

Electricity Generation

Electricity generation is the first process in the delivery of electricity to consumers. The other processes are electric power transmission and electricity distribution. The importance of dependable electricity generation, transmission and distribution was revealed when it became apparent that electricity was useful for providing heat, light and power for human needs. Centralised power generation became possible when it was recognised that alternating current electric power lines can transport electricity at low costs across great distances by taking advantage of the ability to transform the voltage using power transformers. Electricity has been generated for the purpose of powering human technologies for at least 120 years from various sources of energy. The first power plants were run on wood, while today we rely mainly on petroleum, natural gas, coal, hydroelectric and nuclear power and a small amount from hydrogen, solar energy, tidal harnesses, wind generators, and geothermal sources.

Electricity Demand

The demand for electricity can be met in two different ways. The primary method thus far has been for public or private utilities to construct large scale centralized projects to generate and transmit the electricity required to fuel economies. Many of these projects have caused unpleasant environmental effects such as air or radiation pollution and the flooding of large areas of land. Distributed generation creates power on a smaller scale at locations throughout the electricity network. Often these sites generate electricity as a byproduct of other industrial processes such as using gas from landfills to drive turbines.

Main Generating Electricity Methods

Turbines

Rotating turbines attached to electrical generators produce most commercially available electricity. Turbines are driven by a fluid which acts as an intermediate energy carrier. The fluids typically used are:

- steam — Water is boiled by nuclear fission or the burning of fossil fuels (coal, natural gas, or petroleum). Some newer plants use the sun as the heat source: solar parabolic troughs and solar power towers concentrate sunlight to heat a heat transfer fluid, which is then used to produce steam.
- water — Turbine blades are acted upon by flowing water, produced by hydroelectric dams or tidal forces,
- wind — Most wind turbines generate electricity from naturally occurring wind. Solar updraft towers use wind that is artificially produced inside the chimney by heating it with sunlight.
- hot gases — Turbines are driven directly by gases produced by the combustion of natural gas or oil.

Combined cycle gas turbine plants are driven by both steam and gas. They generate power by burning natural gas in a gas turbine and use residual heat to generate additional electricity from steam. These plants offer efficiencies of up to 60%.

Photovoltaic Panels

Unlike the solar heat concentrators mentioned above, photovoltaic panels convert sunlight directly to electricity. Although sunlight is free and abundant, solar panels are expensive to produce and have only a 10-20% conversion efficiency. Until recently, photovoltaics were most commonly used in remote sites where there is no access to a commercial power grid, or as a supplemental electricity source for individual homes and businesses. Recent advances in manufacturing efficiency and photovoltaic technology, combined with subsidies driven by environmental concerns, have dramatically accelerated the deployment of solar panels. Installed solar capacity is growing by 30% per year in several regions including Germany, Japan, California and New Jersey.

Reciprocating Engines

Small electricity generators are often powered by reciprocating engines burning diesel, biogas or natural gas. Diesel engines are often used for back up generation, usually at low voltages. Biogas is often combusted where it is produced, such as a landfill or wastewater treatment plant, with a reciprocating engine or a microturbine, which is a small gas turbine.

Power Station

A power station or power plant is a facility for the generation of electric power. 'Power plant' is also used to refer to the engine in ships, aircraft and other large vehicles. Some prefer to use the term *energy center* because it more accurately describes what the plants do, which is the conversion of other forms of energy, like chemical energy, gravitational potential energy or heat energy into electrical energy. However, *power plant* is the most common term in the U.S., while elsewhere *power station* and *power plant* are both widely used, *power station* prevailing in many Commonwealth countries and especially in the United Kingdom. At the center of nearly all power stations is a generator, a rotating machine that converts mechanical energy into electrical energy by creating relative motion between a magnetic field and a conductor. The energy source harnessed to turn the generator varies widely. It depends chiefly on what fuels are easily available and the types of technology that the power company has access to.

Thermal Power Stations

In thermal power stations, mechanical power is produced by a heat engine, which transforms thermal energy, often from combustion of a fuel, into rotational energy. Most thermal power plants produce steam, and these are sometimes called steam power plants. Not all thermal energy can be transformed to mechanical power, according to the second law of thermodynamics. Therefore, is found for the heat, it is lost to the environment. If reject heat is employed as useful heat, for industrial processes or district heating, the power plant is referred to as a cogeneration power plant or CHP (combined heat-and-power) plant. In countries where district heating is common, there are dedicated heat plants called heat-only boiler stations. An important class of power stations in the Middle East uses byproduct heat for desalination of water.

Classification

Thermal power plants are classified by the type of fuel and the type of prime mover installed.

By Fuel

- Nuclear power plants use a nuclear reactor's heat to operate a steam turbine generator.
- Fossil fuel powered plants may also use a steam turbine generator or in the case of Natural gas fired plants may use a combustion turbine.
- Geothermal power plants use steam extracted from hot underground rocks.
- Renewable energy plants may be fuelled by waste from sugar cane, municipal solid waste, landfill methane, or other forms of biomass.
- In integrated steel mills, blast furnace exhaust gas is a low-cost, although low-energy-density, fuel.
- Waste heat from industrial processes is occasionally concentrated enough to use for power generation, usually in a steam boiler and turbine.

By Prime Mover

- Steam turbine plants use the pressure generated by expanding steam to turn the blades of a turbine.
- Gas turbine plants use the heat from gases to directly operate the turbine. Natural-gas fuelled turbine plants can start rapidly and so are used to supply "peak" energy during periods of high demand, though at higher cost than base-loaded plants.
- Combined cycle plants have both a gas turbine fired by natural gas, and a steam boiler and steam turbine which use the exhaust gas from the gas turbine to produce electricity. This greatly increases the overall efficiency of the plant, and most new baseload power plants are combined cycle plants fired by natural gas.
- Internal combustion Reciprocating engines are used to provide power for isolated communities and are frequently used for small cogeneration plants. Hospitals, office buildings, industrial plants, and other critical facilities also use them to provide backup power in case of a power outage. These are usually fuelled by diesel oil, heavy oil, natural gas and landfill gas.
- Microturbines, Stirling engine and internal combustion reciprocating engines are low cost solutions for using opportunity fuels, such as landfill gas, digester gas from water treatment plants and waste gas from oil production.

Cooling Towers and Waste Heat

Because of the fundamental limits to thermodynamic efficiency of any heat engine, all thermal power plants produce waste heat as a byproduct of the useful electrical energy produced. Natural draft wet cooling towers at nuclear power plants and at some large thermal power plants are large hyperbolic chimney-like structures (as seen in the image at the left) that release the waste heat to the ambient atmosphere by the evaporation of water (lower left image). However, the mechanical induced-draft or forced-draft wet cooling towers (as seen in the image to the right) in many large thermal power plants, petroleum refineries, petrochemical plants, geothermal, biomass and waste to energy plants use fans to provide air movement upward through downcoming water and are not hyperbolic chimney-like structures. The induced or forced-draft cooling towers are rectangular, box-like structures filled with a material that enhances the contacting of the upflowing air and the downflowing water. In desert areas a dry cooling tower or radiator may be necessary, since the cost of make-up water for evaporative cooling would be prohibitive. These have lower efficiency and higher energy consumption in fans than a wet, evaporative cooling tower. Where economically and environmentally possible, electric companies prefer to use cooling water from the ocean, or a lake or river, or a cooling pond, instead of a cooling tower. This type of cooling can save the cost of a cooling tower and may have lower energy costs for pumping cooling water through the plant's heat exchangers. However, the waste heat can cause the temperature of the water to rise detectably. Power plants using natural bodies of water for cooling must be designed to prevent intake of organisms into the cooling cycle. A further environmental impact would be organisms that adapt to the warmer temperature of water when the plant is operating that may be injured if the plant shuts down in cold weather.

Other Sources of Energy

Other power stations use the energy from wave or tidal motion, wind, sunlight or the energy of falling water, hydroelectricity. These types of energy sources are called renewable energy.

Pumped Storage

A pumped storage hydroelectric power plant is a net consumer of energy but decreases the price of electricity. Water is pumped to a high reservoir during the night when the demand, and price, for electricity is low. During hours of peak demand, when the price of electricity is high, the stored water is released to produce electric power. Some pumped storage plants are actually not net consumers of electricity because they release some of the water from the lower reservoir downstream, either continuously or in bursts.

Hydroelectricity

Hydroelectric dams impound a reservoir of water and release it through one or more water turbines to generate electricity.

Wind

Wind turbines can be used to generate electricity in areas with strong, steady winds. Many different designs have been used in the past, but almost all modern turbines being produced today use the Dutch three-bladed, upwind design. Grid-connected wind turbines now being built are much larger than the units installed during the 1970's, and so produce power more cheaply and reliably than earlier models. With larger turbines (greater than 100 kW), the blades move more slowly than older, smaller (less than 100 kW) units, which makes them less visually distracting and safer for airborne animals. However, the old turbines can still be seen at some wind farms, particularly at Altamont Pass and Tehachapi Pass.

Solar

A solar photovoltaic power plant converts sunlight directly into electrical energy, which may need conversion to alternating current for transmission to users. This type of plant does not use rotating machines for energy conversion. Solar thermal electric plants are another type of solar power plant. They direct sunlight using either parabolic troughs or heliostats. Parabolic troughs direct sunlight onto a pipe containing a heat transfer fluid, such as oil, which is then used to boil water, which turns the generator. The central tower type of power plant uses hundreds or thousands of mirrors, depending on size, to direct sunlight onto a receiver on top of a tower. Again, the heat is used to produce steam to turn turbines. There is yet another type of solar thermal electric plant. The sunlight strikes the bottom of the pond, warming the lowest layer which is prevented from rising by a salt gradient. A Rankine cycle engine exploits the temperature difference in the layers to produce electricity. Not many solar thermal electric plants have been built. Most of them can be found in the Mojave Desert, although Sandia National Laboratory, Israel and Spain have also built a few plants.

Break Junction

A break junction is an electrical junction between two wires formed by pulling the wires apart to produce electrodes separated by a few atomic distances. In this technique a metal wire is bent or pulled, often using a piezoelectric crystal to apply the necessary force. The bending or pulling causes the metal wire to break in a controlled manner since piezoelectric elongation can be controlled to a precision of angstroms or less (such crystals are used for motion control in scanning tunneling microscopy). As the wire breaks, the separation between the electrodes can be indirectly controlled by monitoring the electrical current through the junction. A typical conductance versus time trace during the breaking process (conductance is simply current divided by applied voltage bias) shows two regimes. First is a regime where the break junction comprises a quantum point contact. In this regime conductance decreases in steps equal to the conductance quantum $G_Q = 2e^2 / h$ which is expressed through the electron charge e and Planck's constant h. The conductance quantum has a value of 7.74×10^{-5} Siemens, corresponding to a resistance increase of roughly 12.9 KÙ. These step decreases are interpreted as the result of a decrease, as the electrodes are pulled apart, in the number of single-atom-wide metal strands bridging between the two electrodes, each strand having a conductance equal to the quantum of conductance. As the wire is pulled, the neck becomes thinner with fewer atomic strands in it. Each time the neck reconfigures, which happens abruptly, a step-like decrease of the conductance can be observed. This picture inferred from the current measurement has been confirmed by "in-situ" TEM imaging of the breaking process combined with current measurement. In a second regime, when the wire is pulled further apart, the conductance collapses to values less than the quantum of conductance. This is the tunneling regime where electrons tunnel through vacuum between the electrodes. Digging into the literature on break junctions and quantum point contacts reveals that the above conceptual description is somewhat oversimplified, but the description is a good first approach to understanding the topic.

Use

This method has been developed to study the conductance of few-atom constrictions of varied metals. The conductance of these constrictions has been compared with theoretical predictions for both the stability and the conductance of possible few-atom configurations. More recently it has been used to study molecules which are inserted in the junction in the liquid phase and binds to them (dithiols) or in the gas phase. This method has several advantages. It is clean, since the junctions can be made in a controlled atmosphere (high vacuum). It is fast and thus enables many independent measurements to be done in a few hours. It is then possible to study the statistical occurrence of a particular type of contact, and build conductance histograms. Lately this method has enabled the more accurate determination of the conduction of a single molecule. The disadvantage of this technique is that it is a two-terminal technique (that is, it uses only two wires and can be considered an electrical diode), whereas complete determination of electronic properties requires using a three-terminal configuration similar to the source, drain and gate of an MOS transistor.

Energy Storage

Energy storage as a natural process is billions of years old — the energy produced in the initial creation of the Universe has been stored in stars such as our Sun, and is now being used by humans directly (e.g. through solar cells) or indirectly (e.g. by growing crops). As a purposeful activity, energy storage has certainly existed since pre-history, though it was often not recognized as such. An example would be the use of logs or boulders as defensive measures in ancient forts — the logs or boulders would be collected at the top of a hill, and the energy thus stored would be released as a defense against invaders. A more recent application was the control of waterways to power water mills for processing grain or powering machinery. Often complex systems of reservoirs and dams were constructed to store and release water (and the potential energy it contained) when required. Energy storage only became a major concern, however, with the introduction of electricity. Unlike the other common power sources at the time, such as natural gas, electricity had to be used as it was generated. This meant that changes in demand were difficult to cater for without either cutting supplies at times, or having expensive excess capacity. An early solution was the battery, but this is of limited use both due to its small capacity and relatively high cost. A similar solution with the same type of problems is the capacitor. Some areas of the world (Washington and Oregon in the USA, and Wales in the United Kingdom are examples) have used geographic features to store large quantities of water in reservoirs at the top of hills, using excess electricity at times of low demand to pump water into the reservoirs, then letting the water fall through generators to retrieve the energy when demand peaks. A number of other technologies have been investigated, but to date no widely available, affordable solution to the challenge of mass energy storage has been found.

Reasons

Load Levelling

The demand for electricity from consumers and industry is constantly changing, broadly within the following categories:

- Seasonal (during dark winters more electric lighting and heating is required, while in other climates hot weather boosts the requirement for air conditioning)
- Weekly (most industry closes at the weekend, lowering demand)
- Daily (such as the peak as everyone arrives home and switches the television on)
- Hourly (one method for estimating television viewing figures in the United Kingdom is to measure the power spike when advertisements are shown and everyone goes to switch the kettle on)
- Transient (fluctuations due to individual's actions, differences in power transmission efficiency and other small factors that need to be accounted for)

There are currently three main methods for dealing with changing demand:

- Electrical devices generally having a working voltage range that they require, commonly 110-120V or 220-240V. Minor variations in load are automatically smoothed by slight variations in the voltage available across the system.
- Power plants can be run below their normal output, with the facility to increase the amount they generate almost instantaneously. This is termed 'Spinning Reserve'.
- Additional power plants can be brought online to provide a larger generating capacity. Typically, these would be combustion gas turbines, which can be started in a matter of minutes.

The problem with relying on these last two methods in particular is that they are expensive, because they leave expensive generating equipment unused much of the time, and because plants running below maximum output usually produce at less than their best efficiency. Grid energy storage is used to shift load from peak to off-peak hours. Power plants are able to run closer to their peak efficiency for much of the year.

Portability

This is the area of greatest success for current energy storage technologies. Single-use and rechargeable batteries are ubiquitous, and provide power for devices with demands as varied as digital watches and cars. Advances in battery technology have generally been slow, however, with much of the advance in battery life that consumers see being attributable to efficient power management rather than increased storage capacity. This has become an issue as pressure grows for alternatives to the internal combustion engine in cars and other means of transport. These uses require far more energy density (the amount of energy stored in a given volume or weight) than current battery technology can deliver. Liquid hydrocarbon fuel (such as gasoline,ethar ol/petrol and diesel) have much higher energy densities.

Reliability

Virtually all devices that operate on electricity are adversely affected by the sudden removal of their power supply. Solutions such as UPS (uninterruptible power supplies) or backup generators are available, but these are expensive. Efficient methods of power storage would allow for devices to have a built-in backup for power cuts, and also reduce the impact of a failure in a generating station. Examples of this are currently available using fuel cells and flywheels.

Storage Methods

The upper reservoir (Llyn Stwlan) and dam of the Ffestiniog Pumped Storage Scheme in north Wales. The lower power station has four water turbines which generate 360 MW of electricity within 60 seconds of the need arising. The size of the dam can be judged from the car parked below.

- Electrochemical
- Batteries
- Flow batteries
- Fuel cells
- Electrical
- Capacitor
- Superconducting magnetic energy storage (SMES)
- Mechanical
- Compressed air energy storage (CAES)
- Flywheel energy storage
- Hydraulic accumulator
- Hydroelectric energy storage
- Spring
- Potential (Gravity)
- Hydroelectric
- Thermal
- Molten salt
- Cryogenic liquid air or nitrogen
- Seasonal thermal storeage
- Solar pond
- Hot bricks
- Steam

Energy Storage: An Introduction

What is Energy Storage?

- Energy storage is needed to store electricity, heat and cold, which is produced at times of low demand and low generation cost and from intermittent energy sources such as wind and solar power. It is released at times of high demand and high generation cost or when there is no more generation capacity available.

- Reliable and affordable energy storage is a prerequisite for using renewable energy in remote locations, for integration into the energy system and the development in a future decentralised energy supply system. Energy storage therefore has a pivotal role to play in the effort to combine a future, sustainable energy supply with the standard of technical services and products that we are accustomed to and need.
- Energy storage is the most promising technology currently available to reduce fuel consumption in the transport sector.

What are the Current Uses of Energy Storage Technologies?

- The decision to use an energy storage system depends both on the requirements of the application and the cost of competing solutions. In renewable energy systems, for instance, the use of fossil fuel based back-up generation and grid connection are competing solutions
- Power stations, compressors, heating systems, etc. all have different performance characteristics as regards their response time to changing demand, their lead times for starting up or shutting down, and their most efficient points of operation.
- Energy storage systems can usually be replaced by conventional energy generation. However, this can lead to an inefficient use of fossil fuels and a demand for investment in additional energy generators with high power output and fast response time.
- The time required for energy generation from renewable sources, be they electricity or heat, cannot always be matched to the time of demand.
- Energy storage systems are therefore an integral part of any renewable energy sources (RES) system.
- Even when fuel-powered generation is used to cover periods of low RES generation, energy storage is required for economic reasons, as it is cheaper than the frequent use of a motor-driven generator.
- Also, the stability of the electricity system and quality of the voltage supplied will be considerably higher when an energy storage system is used. The technical and economic optimum concerning the size of an electricity storage system needs to be defined in each case individually.

- Conventional, commercially available lead-acid batteries have a very high-energy efficiency and all other technologies have to compete with this.
- Batteries are the most expensive item in RES systems when the system's total lifetime costs are considered; and there are big variations in battery lifetime in different installations.
- Excess electricity can always be stored cheaply in the form of heat and for a long time. However, the value of heat energy is much lower than the value of electricity.
- In solar thermal systems for heating and cooling it is also necessary to store energy because heat generation depends on solar radiation for energy production. Overview of the technology

What are the Different Energy Storage Technologies?

- Batteries
- Flywheels
- Reversible fuel cells
- Electromagnetic
- Compressed air
- Super-capacitor
- Pumped hydro storage

Different energy storage technologies coexist because their characteristics make them attractive to different applications. From a user point of view there are both technical and commercial criteria for selecting the most suitable technology.

What are the Various Aspects of the Technologies and their Applications?

Batteries and Advanced Batteries

Rechargeable batteries or accumulators are the oldest form of electricity storage and widely used. Batteries store electric energy in a chemical form. Their performance is linked in a complex manner to the materials used, the manufacturing processes and the operating conditions. Consequently, progress in battery technology is slow and the transfer of laboratory results into commercial applications is sometimes risky. Lithium ion and nickel-metal-hydride (NiMH) batteries are the only new battery technologies which have achieved significant market penetration in the last decade. Batteries can respond to changes in power demand within microseconds.

Only super-capacitors equal such a response time. Batteries usually have very low standby losses and can have high energy efficiency, depending on the application and the details of the operation. Most batteries contain toxic materials, hence the ecological impact from uncontrolled disposal of batteries must always be considered.

Super-capacitors

Super-capacitors store electrical energy in the electric field between two electrodes. Ultra-capacitor, super-capacitor and electric double layer capacitor (EDLC) are also called electro-chemical capacitors working with chemical reactions or not like true capacitors. The fundamental design and electrical properties are those of conventional capacitors used throughout the electrical and electronics industry. EDLC uses electric double layer capacitance on both positive and negative electrodes.

Reversible Fuel Cell Systems and Redox Flow Batteries

Fuel cells convert hydrogen from a storage tank and oxygen from the air to water and generate a current from the electrochemical process. The electrochemical reaction itself is reversible. The fuel cells' energy capacity is determined by the size of the storage tanks for the active materials, and the power by the area of the electrodes and design of the reactor. Standby losses are low because the active materials are kept physically separate. Redox flow batteries are systems using materials other than hydrogen and oxygen. Their energy efficiency is higher than those of reversible fuel cells, but still below the energy efficiency of most batteries.

SMES (Super-conducting Magnetic Energy Storage Systems)

SMES store energy in the magnetic field of a coil made from special alloys. By cooling the conducting wires to — 269°C the resistance of the material to electrical current disappears, allowing it to conduct very high currents without electrical losses. When looking at the complete system, however, it is clear that there is considerable energy requirement for refrigeration. Also, the current has to flow through non-super-conducting components and solid-state switches, which cause resistive losses. Despite this, the overall efficiency in commercial applications is very high.

Flywheels

The energy is stored as kinetic energy in a rotating mass. The amount of energy stored increases with the square of the rotational speed, which is limited by the tensile strength of the material used.

Thermal Storage (Heat and Cold)

Conventional heat and cold storage systems simply store excess energy in a large tank using the working medium at the temperature required for later use. Virtually every cooling and heating system has such storage tanks.

Compressed Gas Storage

Compressed air tanks are widely used in industry to provide a constant source of compressed air with uniform pressure in the range of 8-10 bar. There is renewed interest in compressed air storage for covering the demand of peak electricity or for small wind/hybrid applications, where the energy-to-power ratio of batteries is unsuitable, either because the energy content is very high but the power requirement low, or the energy through-put is very high compared to the energy content.

Pumped Hydro Storage

Pumped hydro storage is a conventional energy storage technology utilised by the electrical industry. Water in a basin at the top of a mountain is used to drive a generator in a reservoir at a lower level. When surplus energy is available, the water is pumped back up again. The power output and the cost efficiency of pumped hydro storage depends on the difference in height.

Early High Field Magnets and Energy Storage

In 1961, Kunzler et al. prepared the first practical powder-in-tube conductor by drawing a Nb tube filled with Nb_3Sn powder into a wire. Following reaction, the wire showed very high critical current density of ~10^5 A/cm^2 at magnetic fields up to 8.8 T (Kunzler et al. 1961). This result provided the first real demonstration of high field superconductivity and at that time produced as much excitement for superconductivity as seen in the subsequent HTS discovery. Kunzler and his coworkers eventually wound the Nb_3Sn wire into a solenoid, which produced a field of 7 T (Kunzler 1987).

Kunzler's experiment thus opened the "Type II" superconductor era, offering enormous potential for superconducting magnets and electric power applications. Shortly after Kunzler's work was reported, Hulm and Blaugher (1961) published their studies on the Group IV, V, and VI transition metal alloys, four of which included the body-centered-cubic Nb-Ti and Nb-Zr systems. The Nb-Ti alloys were eventually shown to be the most technologically important because researchers were able to inexpensively fabricate long lengths of conductor with high current properties in useful magnetic fields of 3 to 5 T. Within a year, these hard superconducting Nb-Ti and Nb-Zr alloys were fabricated into wires and wound into solenoids, producing magnetic fields up to 7 T. The successful test of these solenoids quickly established that the transition alloy conductors could be easily applied to a wide range of electric power-related applications.

The idea of using large SC magnets for energy storage follows directly from the expression describing the energy stored in an inductor, which is simply, $E(J) = \frac{1}{2}LI^2$, where L is in henrys (H) and I, the current, is in amps. Following the demonstration of high field magnets in the early 1960s, magnetic energy storage was immediately considered, but advances in fabricating a high-current cabled conductor were necessary before serious programmes could be pursued. The major programmes on superconducting magnetic energy storage (SMES) started in the early 1970s and continued through the mid-1980s. The primary interest for these magnets was directed at diurnal storage for load leveling and the need for a pulsed power source for current induction in the plasma of Tokamak fusion power devices. The SMES effort during this period, which was concentrated in the United States and Japan, is reviewed by John Rogers (1981).

The energy storage effort in Japan from the 1970s to the 1980s was fairly widespread, with most of the major electric power companies such as Hitachi, Mitsubishi, and Toshiba participating. In addition, storage programmes were pursued by some of the utilities, in particular, by Chubu Electric Power and Kansai Electric Power companies. Hitachi built and tested a 5 MJ SMES system in 1986, which was connected to the 6.6 kV power line of the Hitachi Works to evaluate transmission line stability (Ishigaki, Shirahama, and Kuroda n.d.). In 1989, Chubu Electric and Hitachi jointly studied and developed a 1 MJ SMES system to evaluate how a SMES could provide power system stability (Fujita 1989). It is important to note that a "SMES system" includes the SC magnet, a fairly sophisticated solid state ac-dc power conditioning system (PCS), and usually a closed-cycle refrigerator. The SMES system is also interfaced with additional circuit breakers and a control system for protection and isolation. The response time for the SMES is thus dependent on the PCS and associated switch gear, which is fairly fast, allowing a SMES system to demonstrate high-speed response with capability of the order of one cycle or approximately 10 ms.

The major achievement in the United States during this period was the construction and successful test on the Bonneville Power System of a 30 MJ SMES to damp low frequency oscillations between two ac transmission lines running from the Pacific Northwest to Southern California (Rogers et al. 1985).

Ac Rotating Machine Efforts in the United States

Reliance Electric DOE SPI Motor Programme

Reliance Electric and EPRI started a joint programme in 1987 to evaluate the use of HTS materials in electric motors. This initial programme, which included American Superconductor Corporation (ASC) as the HTS conductor supplier and coil manufacturer, eventually was transitioned in 1994 into a DOE SPI programme. The SPI programme also included Centerior Energy, representing the end user, and Sandia National Laboratory, which assisted Reliance on the cryogenic cooling system analysis. The focus for the SPI programme was demonstration of a "model" air core ac synchronous motor using HTS windings that would be scaleable to larger ratings suitable for commercialization. In February 1996, Reliance Electric successfully tested a four-pole, 1,800 rpm synchronous motor using HTS windings operating at 27 K, at a continuous output of 200 hp. The HTS coils, manufactured by ASC using Bi-2223 tape, achieved currents of 100 A at 27 K, which is 25% over the initial goal of 80 A. The current density at 100 A corresponds to 7,500 A/cm^2 in the SC wire. The peak field achieved by the four coils during the testing was nearly 1 T. Rotating tests at different speeds (600 rpm and 1,800 rpm) did not show any change in performance for the coil current, indicating good mechanical capability and acceptable cryogenic cooling for the HTS coils (Schiferl, Zhang, Shoykhet et al. 1996; Schiferl, Zhang, Driscoll et al. 1996).

A shows a cross-section of the rotor, illustrating the placement of the HTS coils and cryogenic cooling scheme. This rotor design, which is similar to designs used for SC ac generators, has a number of key features (Schiferl, Zhang, Shoykhet et al. 1996; Schiferl, Zhang, Driscoll et al. 1996):

- the vacuum jacket includes the use of multilayer insulation
- torque tubes (at each end of the rotor) are cooled by exhaust gas
- the rotor drive end is demountable, allowing access to cryogenic space to allow coil changing and modification of the cooling scheme
- the rotor cold space and coils are fully instrumented to measure stresses, temperature, pressure, and magnetic field levels

In August 1996, the Reliance-led programme was extended by DOE into a Phase II SPI effort to develop a precommercial prototype of an HTS motor at a 5,000 hp rating. This motor, combined with an adjustable speed drive (ASD), will offer marked efficiency improvement and more operational flexibility than conventional induction motors; specifically, the motor speed will be able to be properly matched to changing load requirements. Superior performance and energy efficiency compared to conventional induction motors should make the HTS motor extremely attractive to customers and highly competitive in the marketplace for large motors above 1,000 hp. The 5,000 hp motor with ASD should show an efficiency exceeding 98%, which is ~2% higher than that of a conventional motor (Schiferl, Zhang, Shoykhet et al. 1996; Schiferl, Zhang, Driscoll et al. 1996). The impact of refrigerator power on the efficiency for a motor operating at 77 K should result in a ~0.1% decrease, giving an efficiency improvement of 1.9%. Reducing the operating temperature from 77 K to 30 K would decrease efficiency by an additional ~0.2%, yielding a net efficiency improvement of 1.7%. This estimate is based on an assumption of ~100 W heat removal at 30 K and a refrigerator coefficient of performance of 0.10, which would require approximately 10 kW of refrigerator power. In contrast to ac generators at ratings of 200-400 MW, which are much larger than the 5,000 hp motor, which corresponds to 3.73 MW, the refrigerator power has a much more severe impact on motor efficiency, giving an incentive to operate at higher temperatures near 77 K. If operation at 30 K is necessary due to HTS characteristics, the net efficiency improvement of ~1.7 % is still likely to be attractive for the 5,000 hp application.

General Electric DOE SPI Generator Development Programme

Another DOE SPI programme on SC ac synchronous generators was carried out by a General Electric Company (GE) team made up of engineers from its research laboratory and its Power Generation Engineering and Power Systems Engineering departments. The GE programme, initiated in 1994, was directed at the conceptual design and assessment of a 100 MVA generator and the development of an HTS racetrack coil suitable for use in a full-scale generator. A more complete description of the GE programme is presented in an article in *Advances in Cryogenic Engineering* (Lay, Herd, and King 1996), which discusses the programme objectives; the early coil development; related HTS conductor activities on Bi-2223 tape; and the alternate HTS conductor, Tl-1223 tape, which offers improved temperature and field capability over Bi-2223. The complete details on the development, fabrication, and final testing of the racetrack coil is reported in two companion papers presented at the August 1996 Applied Superconductivity Conference (Herd et al. 1996; Salasso et al. 1996). The Bi-2223 tape for this racetrack was supplied by Intermagnetics General Corporation (IGC), which was the HTS wire and tape manufacturer and key partner on the GE/SPI programme. The Argonne National Laboratory also supported IGC on the Bi-2223 development. Additional national laboratory support to GE and IGC was provided by Oak Ridge National Laboratory and Los Alamos National Laboratory.

The racetrack coil, is cooled by a heat exchanger wound external to the surface of the epoxy-impregnated coil. Helium gas, from a closed-cycle helium refrigerator, is circulated through the heat exchanger to maintain steady state temperature control at test temperatures of 20 K and 25 K. The racetrack coil achieved 34 A at 25 K, corresponding to ~40,000 ampere-turns, which is sufficient for consideration in a generator application (Herd et al. 1996). Additional tests on the ac loss and transient capability were also extremely positive. GE researchers concluded, "The capability of 20 K (and higher temperature) HTS coils to deal with current overdrive transients of many times I_c is remarkable. Much higher transient heating can be tolerated than for the previous generation of LTS generator designs" (Salasso et al. 1996). This conclusion supports the earlier comments in the section Advantages of HTS Coils (p. 12) on the use and limitations of liquid-helium-cooled LTS coils for generator applications.

Magnetic Energy Storage Efforts in Germany

The most active effort on SMES in Germany is carried out at the Forschungszentrum Karlsruhe (FZK) laboratory. FZK is constructing a 250 kJ SMES with a toroidal field design to address a power quality problem due to the frequent startup of large motors at a saw mill. Analysis indicates that a SMES system would be ideal for reducing this flicker problem, due to its fast response capability. FZK is also investigating at the Deutsches Electronen Synchrotron (DESY) laboratory the use of SMES as a pulse power source for providing ~10 GW pulses with a 1.0 ms duration at a 10 Hz rate in order to power rf klystrons (Jungst 1995).

At Siemens, the interest in SMES over the last few years has been mostly restricted to design and evaluation studies, with no current plans for development or construction. Siemens, RWE (the largest German utility), and Preussen Electra have completed an evaluation and conceptual design of a 2 MWh/50 MW SMES for use in providing frequency stabilization to the electric system. SMES continues to be of interest, but recent economic studies by Siemens indicate it may be too expensive compared to other storage technologies (Prescher et al. 1995).

It is thought that using HTS current leads, which are now a commercial reality and can be purchased from a number of manufacturers including Hoechst and ASC, would reduce the heat leak in a SMES system. Use of HTS conductors to fabricate a high field, high density SMES is also projected as an interesting future application. An HTS SMES operating at 10-20 T would require a "composite conductor" capable of carrying thousands of amperes with low ac losses to minimize heating for multiple charge and discharge cycles.

The Technical University of Munich has also been conducting research on SMES and is constructing a 1.4 MJ toroidal field system using LTS conductors (Lorenzen et al. 1995). At ABB in Switzerland, energy storage is considered an important area; it has been strongly followed in the past, and an LTS SMES has been constructed for experimental evaluation. At the time of the WTEC visit, a major SMES system using LTS conductor under development for the Swiss railroad had been terminated due to realisation of an alternative, less costly, solution. ABB had no plans for SC storage but did plan to continue to evaluate the technology.

Conceptual Design of Energy Storage

There are several commercially viable energy storage systems that are being developed for hybrid electric vehicles (HEV's) on the market today. Major advances are being made on almost a daily basis. The reason for this is the fact that the government is subsidizing a large amount of the alternative fuel research going on by many U.S. as well as foreign manufacturers. The types of devices that hold the most promise to solve the energy storage problems are batteries, flywheels, and ultracapacitors. As shown in the plot below, both gasoline and hydrogen have a higher specific energy than the rest of these electrical storage devices. An advantage of HEV's is that they can use the high specific energy of liquid or gaseous fuels to provide the vehicle with long-range capabilities. Conversely, the HEV can use the high specific power of electrical energy storage to provide the peak power requirements.

Flywheels store energy mechanically by turning a heavy rotor at ever-increasing revolutions per minute to store kinetic energy. Flywheels are good because they store energy very efficiently but their specific energy is very low at this point in their development compared to other available products making them an inappropriate choice for energy storage in our HEV.

The next possible technology is ultracapacitors. These devices work by accumulating and separating unlike charges. The positive abilities lie in the fact that they have no moving parts as well as the number of times that they can be cycled through their charge discharge cycle is very high. Ultracapacitors shortfalls arise from the fact that they have low specific energy in addition to the fact that those commercial versions that are available are in their infant development stages. In the future they may be a very viable solution but just like flywheels they still need to improve in both price and availability.

The final device is one that the team has chosen to store energy in our car. That device is the battery. Batteries are the maturest technology to date that is available right now for implementation into our HEV. Our choice of batteries in itself is a diverse subject. There are hundreds of types of batteries that are available to do the job, but finding the right battery that matches all of our systems requirements is a daunting task. The best method for speeding the process to a final decision on battery choice is to limit that choice as much as possible. There any many battery technologies currently being developed, with Lead-Acid (Pb-acid), nickel cadmium (NiCd), and nickel metal hydride (NiMH) being the most promising technologies. For the purposes of comparing the advantages and disadvantages of the three different types of batteries, the graphs below show the specific energy, specific power, specific cost, and cycle life.

Specific energy is important because it affects the number of batteries, and hence the mass of batteries, that a vehicle needs to carry on-board to attain a certain electric-only range. It is perhaps the most important factor for EV's because it determines their total range, but not as critical for HEV's which carry most of their energy in the concentrated form of a liquid or gaseous fuel. As you can see in the above charts, NiMH batteries currently have the most energy per unit mass of the three battery chemistries, and in the future, are projected to maintain a specific energy of over 2 times that of lead-acid and NiCd batteries.

Specific power, in the units of power per mass, is perhaps the most important parameter measuring electrical energy storage requirements for hybrid electric vehicles. Because hybrids normally depend on the electric energy stored on-board to provide the power for accelerations and hill climbs, the higher the power for less mass, the better. Hence, high specific power batteries are critical to the success of HEV's. The graph shows how today's lead-acid batteries have a slight edge in specific power, but in the future, it is anticipated that bi-polar lead-acid batteries will have twice the specific power of both NiCd and NiMH batteries.

The definition of specific cost is the cost per unit energy (kWh) of battery, and is related to the economic viability of the battery. Because lead-acid have been around the longest and are the most fully developed battery technology, they have the lowest specific cost at around $125/kWh currently, with projections to decrease that cost to $75 with bi-polar batteries. Currently, NiCd and NiMH batteries are 4 to 5 times more expensive than lead-acid. Once these batteries become standard, their price will decrease. However, due to the raw material cost, they will never be as low as lead-acid.

The cycle life is a measure of how long the battery will last before it needs to be replaced. Historically lead-acid batteries have had a relatively short life when deep-cycled, while NiCd and NiMH have 3-4 times the cycle life of lead-acid.

Final Design Decision

Limiting our choice of batteries has enabled us to save considerable time. In talking to our team members as well as EV owners and distributors we discovered several limiting factors to our choices. These factors include weight, geometry, availability, and system integration ability. When we started the initial design of this HEV we decided to study currently running vehicles that we could benchmark our own decisions from. After seeing AC Propulsions T-Zero EV we realised that our car was going to be very similar and therefore it would be a good benchmark for us to work from. Collecting a large amount of energy storage data from this car proved to be invaluable. We interviewed the creators of the car extensively and were able to find out the advantages and disadvantages of their energy storage design approach. We found out what they would change if they could start over again and used this powerful information for the foundation of our own design and component choices.

Our first major constraint was the weight of the battery pack. The AC Propulsions car used 28 Optima sealed lead acid Batteries for a combined weight of 1200 pounds. With standard engineering practice in mind the team decided a new battery weight goal of 25% less (900 pounds) would be acceptable for a new upper end goal. In addition to that 900-pound weight we decided that a 700-pound weight would be our ideal number to achieve with 500 pounds or less being the "dream number".

The second major constraint was geometry. Being that the body of our HEV is a production kit car body it confines us to the usable geometry possibilities. Initially we saw that the Riot had two excellent possibilities for battery placement. Those positions were along the side rails of the car on both sides as well as the possibility of a battery storage compartment located under the driver and passenger compartment. We believed that retrofitting these two areas with the proper mounting brackets and wiring distribution capabilities would yield excellent load distribution as well as impact protection. The answer for our battery storage eventually did come with the design of a battery storage box that would fit under the car.

The third constraint to our battery choice was system integration ability. This factor includes things like the battery bus voltage of 336 volts that is necessary to run the AC Propulsion motor and controller unit. It turned out that after extensive dialogue with the engineers at AC Propulsion that we discovered their entire battery charging system and algorithms are based all on lead acid technology. This ended up being an extremely valuable constraint for our project. It limits our choices but simplifies them as well. Researching batteries led us to other important factors to be aware of in order to make a sound choice in battery selection; these being specific energy, specific power, and Ampere-Hour ratings of the batteries. These factors when combined and rated for importance will confine the final battery choice. The cost and cycle life are not in our list of accountable characteristic parameters for battery choice. This is because upon discussion with Dr. Burns early on in the project we decided that due to the fact that this is a prototype vehicle and test bed we would not have to worry about cost of ownership of the vehicle thereby dismissing the cost and cycle life from our decision tree. We did keep in mind the fact that we needed to choose a battery that was not so hyper exotic in its technology that we did not have the knowledge, capability, or time to make our choice successful. Our final battery choice did reflect this fact in making a safer choice to facilitate our success in this project. The fourth design constraint for the project is component availability. Initially we knew all battery types were available in at least some small production quantities. This ended up being a frivolous belief. We kept finding very attractive choices that met nearly all of our constraints like Lithium Polymer batteries as well as advanced Nickel Cadmium batteries. Unfortunately, we found out they are not commercially available to us and would be impossible to get in accordance with our production timeline. We found an excellent possibility with the Nickel Metal Hydride Batteries from the Ovonics corporation. At this point we are still investigating their possibility for implementation at a later date. Ovonics offers the only batteries in the industry that were specifically designed and built for Hybrid Electric Vehicles. This is important because HEV batteries due not have to be capable of the same characteristics that purely EV batteries do. There are three major differences in EV and HEV design, which are:

1. HEV's require higher power capabilities on both the discharge and the charge
2. HEV's need to be able to withstand many more charge discharge cycles than EV's.
3. HEV's have a much smaller need of energy storage because of the onboard APU and alternator.

Battery Choices

The culmination of battery research to this point has resulted in two primary battery choices. Although both choices come from the same manufacturer it is important to note that each serves a different purpose in HEV design. The battery that will be used for a preliminary testing in the vehicle will be the Genesis sealed lead acid batteries from Hawker Energy.

Each cell is 12 volts and is rated at 26 amp-hours, which at this point seems adequate but this could change. This possible change in amp-hour size represents the reasoning behind our secondary battery choice. The secondary choice for our vehicle is the Genesis 42Ah cells. This battery offers the same flexibility as its smaller brother except for its increased size of energy storage. It is appropriate at this point to comment on some of the characteristics that make Genesis batteries a superior choice than that of other manufacturers. One of the details that we noticed about the Genesis series batteries is that the internal heat generation within the cells as a function of recharge rate is nearly linear. The data in figure 6 was supplied by the Hawker Corporation shows their test results that prove this fact. In general high heat within and around a cell kills battery life very quickly. What that this means is that proper temperature control within the battery compartment needs to be applied. This becomes far easier to control with charts like these so can predict upcoming spikes in battery temperature and begin pre-cooling the compartment. In addition to the easily traceable temperature range of the Genesis batteries there is a long list of reasons to use these batteries in electric and hybrid electric vehicles. The following is a shortened list of the positive reasons to put Hawker Genesis batteries in our vehicle. In an attempt to explain these points as clearly as possible the following was taken from literature provided by Hawker Energy.

- Greater Cycle Life: Genesis achievable cycle life is remarkable for sealed-lead batteries. A conventional sealed-lead battery would typically have a cycle life to 80% DOD (degree of discharge) anywhere between 200 and 300 cycles. By using Genesis batteries the EV manufacturer is able to experience almost double life expectancy, very high power density, plus all the advantages of a sealed chemistry.

- Existing Product: Genesis batteries are in commercial production today as an ooff the shelf Product. The EV manufacturer does not have to wait for prototypes to be converted to production status and to demonstrate consistent reliability.
- Low Internal Resistance: The extremely low fully charged internal resistance of 4.5 to 10 milli-ohms per 12V module is another distinct advantage of Genesis batteries. The resistances are 33% to 50% of those of conventional sealed-lead batteries. Because of such low internal resistance, the voltage profile of the Genesis battery under discharge is relatively flat, enabling the voltage to stay at a higher level longer during discharge.
- Opportunistic charging: Another advantage of having very low internal resistance is that it allows high currents to be pumped back into the battery pack during regenerative braking. Such a scheme enables "opportunistic" charging of the battery.
- Fast recharge capability: Provided that the charger has the requisite power handling capability, Genesis batteries may be recharged to better than 95% state of charge (SOC) in under (1) hour using only a single level constant voltage charge.
- Dry cell classification: Hawker Energy Products Inc.'s batteries are dry by design and have been tested and determined to be in compliance with section 173.159 (d) of the United States Department of Transportation regulation. They are exempted from other shipping requirements of 49CFR, sub-chapter 173.159, and as of September 30, 1995 all shipments are classified as unregulated "nonspillable wet electric storage batteries." All batteries and shipping containers are marked "NONSPILLABLE" or "NONSPILLABLE BATTERY"
- Mounting flexibility: The Genesis battery may be installed in any orientation without sacrificing any performance attributes.
- High-rate discharge capability: Genesis batteries are ideally suited for HEV's due to their extraordinary short duration, high rate discharge capability. This helps greatly in providing brief bursts or pulses of very high magnitudes to accelerate the HEV without causing the battery voltage to drop too low.

In addition to these incredible characteristics that the Genesis batteries offer we need to look at how well they fulfill the design constraints listed in the beginning of this section. It is important to review how these two battery choices have done in meeting our set requirements.

The first and most important constraint was the weight of the battery pack. Our first choice ended up meeting this goal rather well. With 28 cells at 12V apiece needed to achieve the 336V necessary to run the motor and controller and a single 26Ah battery weighing 22.3 pounds it equated to 624.4 pounds. Our secondary choice of the 42Ah size equated to a battery pack that weighs 938 pounds. Although it is above our *acceptable* upper limit it must be noted that this choice is in place only to be available in a wworst casescenario situation. What I mean by worst case scenario is that the 26Ah-size battery is already considered to be a conservative choice from initial calculations. Initial testing of the vehicle will hopefully show that we can actually shrink the battery size to the 13Ah range if possible.

The second given constraint was that of geometry. The ability for our battery choices to fit within the battery box under the driving compartment is obviously an important detail. The dimensions of our battery box are roughly 63 x 50 x 7 inches. With seven inches being the critical dimension it is important that at least one dimension of both our battery choices was under that. As you can see from the geometry specifications from both the 26 and 42Ah batteries, it will not be a problem getting all of our batteries for both choices inside the battery box. The open ended in housef this battery box allows for future types of batteries that are not available yet to be implemented in our car in the future.

The third constraint was based on availability of the product. Luckily for us, as stated above, both choices are commercially available. We currently have the 26Ah batteries in-house and are awaiting continued fabrication so we can integrate them soon. As a note it took only two weeks to order and receive our batteries at a total cost of less than $2,500. In comparison the Ovonics Nickel-Metal Hydride batteries cost $60,000 that includes a 60% educational discount and a six month lead time. Lastly came the importance of system integration ability. This means how easily these batteries will interface with the rest of the car and the separate subsystems. Overall they should both work rather well. The 28 batteries will supply the needed 336V to the AC controller at a fairly constant rate. The recharge controller system was designed specifically for lead-acid batteries making that whole area of battery control fairly complete. Aside from the AC motor controller package LabVIEW also has all the necessary current, voltage, and temperature sensors too properly track and adjust the entire energy storage system. The information listed above as well as the fact that they come recommended to us by both AC Propulsion for use on there charging system and other EV manufacturers has proven there potential satisfactory for use in our HEV.

Optimisation

The optimization section for the vehicle energy storage is difficult in a sense because we have no way of changing a manufactured battery. We can however do our best to make sure that the control system takes care of our batteries as well as possible. This proper care begins with an accurate description of common driving cycles, the power (kW) loads associated with these driving cycles and finally the effect these road loads have on the energy storage system at a real time level. From this information we can build and more importantly test an efficient energy control scheme.

The beginning of this description began with a need for standardized driving cycles so we could accurately represent realistic road loads that our vehicle might see. We thought it would be an easy thing to do, but finding assorted federal driving cycles in a usable numeric array proved to be time consuming. Driving cycles are just standardized descriptions of velocity versus time for an average driving cycle. The two that we use the most being the aggressive driving cycle and the federal urban driving cycle are shown below respectively.

Having just these driving cycles as well as many others like them is just the first step in shaping a viable energy storage plan. The next step was to develop a cycle digester that would take the given information from the cycles and supply a point by point description of the loads that the car was experiencing. The software that we used was LabVIEW. It was used for two reasons. The first was that we all had a fairly good understanding of how to use it and, secondly, whatever we developed could be directly integrated with the cars future control system. Developing the necessary virtual instruments in LabVIEW was slow going at first in an effort to collect the most accurate equations to describe road loads. The equations that we ended up using came from the NASA Lewis Research Center hybrid electric bus project. These equations were the only readily available that had been developed for design purposes and then tested for accuracy against the final product. The equations used as well as a brief description is listed below.

Inputting the given variables into the road load equations gives a separated description of individual loads. To combine the separated loads we inputted them into the equation below.

The output of this equation gives the kW road load at a given interval of time. When used in conjunction with LabVIEW the completed programme that contains these equations combined with the imputed times and velocities of a driving cycle delivers an instantaneous description of the power required to move the vehicle. This piece of information is very valuable because it is the foundation for the entire vehicle dynamics programme that we are developing for this project. The vehicle dynamics programme also contains the state-of-charge (SOC) calculator for the batteries that was the final piece to the puzzle for accurately describing the power losses and gains within the vehicle.

The SOC calculator makes use of another equation used by the NASA hybrid bus programme. We contacted Dale Stalnaker at NASA who was responsible for writing their original programme that uses these equations. He said that the equations all worked well giving consistently reasonable results. The SOC equation used in our programme is:

The equation makes use of the instantaneous power in kilowatts needed for the *t instant in time. These two inputs are directly linked to the cycle analysis programme that was described above allowing for accurate comparison between velocity and power demand. The C term in the denominator is the Ampere-hour capacity of the individual batteries (in our case 26) and the V is the system bus voltage 336V. To accurately represent the internal resistance changes and efficiencies within our batteries at any point in their state-of-charge, calibration tables have been integrated. These tables actively account for the difficulty a battery has in accepting a charge as it nears 100% state of charge. This phenomenon occurs mostly in charge because the internal resistance of our batteries during discharge is both low and constant. The value of V in our equation has been reduced by 5% from the true 336V to 322V in order to take into account the efficiency of the pulse width modulator (PWM) and controller that drives the AC motor. The theory behind this is that because the PWM creates a square wave that cannot be railed out to 336V all the time. They have to return to the zero point even when maxed out in order to create the square wave. This instantaneous dip to zero results in roughly a 5% reduction of bus voltage. At this point in time the vehicle dynamics programme as well as the state-of-charge calculator are all up and running. Initial results of the programme look very promising in supporting our choice in batteries as well as proving our energy storage schemes. For the federal urban driving cycle the Genesis 26Ah batteries are showing a 78.4% SOC upon the completion of the roughly thirty minute cycle. The aggressive driving cycle left the vehicle with a 92% SOC for that ten minute cycle. These initial SOC numbers left us very excited because it proves that the batteries that we chose are very well sized to the energy pool that they need to accept and deliver. If the batteries were completely discharged after three minutes or if the SOC never left 100% over these driving cycles then there would be reason for concern, but fortunately the results look very promising. In fact, as the vehicle dynamics programme continues to mature and become more accurate the numbers keep getting better.

Grid Energy Storage

Grid energy storage is the use of various energy storage techniques to complement electric power generation plants on the transmission grid. By using the grid connection, intermittent sources of power need not be installed with their own energy storage facilities. For example, a solar panel installed on a home may produce more energy than is required during the day, which can be exported to the grid; at night, the needs of the home can be met from the grid, as if the solar energy had been stored locally. In the most narrow sense, grid energy storage does not involve actual storage. Instead, a grid-connected user/producer provides power to the grid that is surplus to its own requirements, and draws power from the grid when needed, netting out its usage. This is called grid energy storage as the user uses the grid instead of arranging storage, which may not be economically feasible. In this sense, grid energy storage is related to distributed generation. For it to function correctly, technical and economic arrangements (such as net metering) will be required, and often requires regulatory support. Use of the term in this limited sense is not consistent, however.

Demand for electricity from the world's various grids varies over the course of the day and from season to season. For the most part, variation in electric demand is met by varying the amount of electrical energy supplied from primary sources, usually hydroelectric power plants and natural gas-fired turbines. Increasingly, however, operators are storing lower-cost energy produced at night, then releasing it to the grid during the peak periods of the day when it is more valuable. In areas where hydroelectric dams exist, release can be delayed until demand is greater; this form of storage is common and can make use of existing reservoirs. This is not storing "surplus" energy produced elsewhere, but the net effect is the same — although without the efficiency losses. Energy storage is economical when the marginal cost of electricity varies more than the energy losses of storing and retrieving it.

For instance, 1.2 gigawatt-hours might be stored at night in a pumped-storage reservoir, at a cost of 1.5 cents/kilowatt-hour. The next day, 1 gigawatt-hour might be recovered, and 200 megawatt-hours lost, and sold at 4.0 cents/kilowatt-hour, for a profit of $22,000. If this profit can be realised on most days, and if the storage facility cost less than perhaps $100M, the operator makes a profit. The marginal cost of electricity varies because of the varying economics of different kinds of generators.

At one extreme, water from a dam can go through a turbine for little more than the cost of releasing it down a spillway, so the marginal cost of generation is nearly zero. Coal-fired and nuclear power plants are also low marginal cost generators, as they have high capital and maintenance costs but low fuel costs. At the other extreme, most peaking generators burn natural gas, which is expensive. Operators prefer cheaper electricity, so they run the low-marginal-cost generators most of the time, and run the more expensive ones only when necessary. This is called "economic dispatch". Renewable supplies with prediction errors and variable production, like wind and solar power, tend to increase the net variation in electric load. Because they are not dispatchable and must run when available, power from these supplies is generally sold to grid operators for less than power available on demand. As renewable supplies become increasingly popular, this difference in price opens an increasingly large economic opportunity for grid energy storage.

Pumped Water Storage

In many places, pumped storage is used to even out the daily generating load, by pumping water to a high storage reservoir during off-peak hours and weekends, using the excess base-load capacity from coal or nuclear sources. During peak hours, this water can be used for hydroelectric generation, often as a high value rapid-response reserve to cover transient peaks in demand. There is over 90 GW of pumped storage in operation, which is about 3% of global generation capacity. Pumped storage recovers about 75% of the energy consumed. The chief problem with pumped storage is that it usually requires two nearby reservoirs at considerably different heights, for which there are few suitable locations, and often requires considerable capital expenditure. Traditional hydroelectric dam configurations are far more common than pumped storage; in this configuration, release is delayed until needed. The net effect is the same as pumped storage, but without the round-trip efficiency loss.

Energy Demand Management

The easiest way to deal with varying electrical loads is to decrease the difference between varying generation and demand. This is referred to as demand side management (DSM). For decades, utilities have sold off-peak power to large consumers at lower rates, to encourage these users to shift their loads to off-peak hours, in the same way that telephone companies do with individual customers. Usually, these time-dependent prices are negotiated ahead of time. In an attempt to save more money, some utilities are experimenting with selling electricity at minute-by-minute spot prices, which allow those users with monitoring equipment to detect demand peaks as they happen, and shift demand to save both the user and the utility money. Demand side management can be manual or automatic and is not limited to large industrial customers. In residential and small business applications, for example, appliance control modules can reduce energy usage of water heaters, air conditioning units, and other devices during these periods by turning them off for some portion of the peak demand time or by reducing the power that they draw. Energy demand management includes more than reducing overall energy use or shifting loads to off-peak hours. A particularly effective method of energy demand management is the installation of more energy efficient equipment. For example, many utilities give rebates for the purchase of insulation, weatherstripping, and appliances and light bulbs that are energy efficient. Companies with factories and large buildings can also install such products, but they can also buy energy efficient industrial equipment, like boilers, or use more efficient processes to produce products. Companies may get incentives like rebates or low interest loans from utilities or the government for the installation of energy efficient industrial equipment.

Compressed Air Storage

Another grid energy storage method is to use off-peak electricity to compress air, which is usually stored in an old mine or some other kind of geological feature. When electricity demand is high, the compressed air is burned with natural gas to run a turbine and generate electricity.

Thermal Energy Storage

Off-peak electricity can be used to make ice from water, and the ice can be stored until the next day, when it is used to cool either the air in a large building, thereby shifting that demand off-peak, or the intake air of a gas turbine generator, thereby increasing the on-peak generation capacity.

Flywheel Storage

Mechanical inertia is the basis of this storage method. A heavy rotating disc is accelerated by an electric motor, which acts as a generator on reversal, slowing down the disc and producing electricity. Electricity is stored as the kinetic energy of the disc. Friction must be kept to a minimum to prolong the storage time. This is often achieved by placing the flywheel in a vacuum and using magnetic bearings, tending to make the method expensive. Larger flywheel speeds allow greater storage capacity but require strong materials such as steel or composite materials to resist the centrifugal forces (or rather, to provide centripetal forces). The use of carbon nanotubes as a flywheel material is being researched. The ranges of power and energy storage technically and economically achievable, however, tend to make flywheels unsuitable for general power system application; they are probably best suited to load-levelling applications on railway power systems.

Superconducting Magnetic Energy Storage

Superconducting magnetic energy storage (SMES) systems store energy in the magnetic field created by the flow of direct current in a superconducting coil which has been cryogenically cooled to a temperature below its superconducting critical temperature. A typical SMES system includes three parts: superconducting coil, power conditioning system and cryogenically cooled refrigerator. Once the superconducting coil is charged, the current will not decay and the magnetic energy can be stored indefinitely. The stored energy can be released back to the network by discharging the coil. The power conditioning system uses an inverter/rectifier to transform alternating current (AC) power to direct current or convert DC back to

AC power. The inverter/rectifier accounts for about 2-3% energy loss in each direction. SMES loses the least amount of electricity in the energy storage process compared to other methods of storing energy. SMES systems are highly efficient; the round-trip efficiency is greater than 95%. Due to the energy requirements of refrigeration, and the limits in the total energy able to be stored, SMES is currently used for short duration energy storage. Therefore, SMES is most commonly devoted to improving power quality. If SMES were to be used for utilities it would be a diurnal storage device, charged from baseload power at night and meeting peak loads during the day. The high cost of superconductors is the primary limitation for commercial use of this energy storage method.

Battery Storage

Battery storage was used in the very early days of electric power networks, but is no longer common. Many "off-the-grid" domestic systems rely on battery storage, but means of storing large amounts of electricity as such in giant batteries or by other means have not yet been put to general use. Batteries are generally expensive, have maintenance problems, and have limited lifespans. One possible technology for large-scale storage are large-scale flow batteries. Sodium-sulfur batteries could also be inexpensive to implement on a large scale and have been used for grid storage in Japan and in the United States. Vanadium redox batteries and other types of flow batteries are also beginning to be used for energy storage including the averaging of generation from wind turbines. If battery electric vehicles were in wide use with modern high cycle batteries, such mobile energy sinks could be utilized for their energy storage capabilities. Vehicle to Grid technology could be employed, turning each vehicle with its 20 to 50 kWh battery pack into a load-balancing device or emergency power source. This represents 2 to 5 days per vehicle of average household requirements of 10 kWh per day, assuming annual consumption of 3650 kWh. This quantity of energy is equivalent to between 40 and 300 miles of range in such vehicles consuming 0.5 to 0.16 kWh per mile. These figures can be achieved even in home-made electric vehicle conversions.

Hydrogen Fuel Cells

Hydrogen is not a primary energy source, but a portable energy storage method, because it must first be manufactured by other energy sources in order to be used. However, as a storage medium, it may be a significant factor in using renewable energies, See Hydrogen storage. It is widely seen as a possible fuel for hydrogen cars, if the problem of energy return on energy invested can be overcome. It may be used in conventional internal combustion engines, or in fuel cells which convert chemical energy directly to electricity without flames, similar to the way the human body burns fuel. Making hydrogen requires either reforming natural gas with steam, or, for a possibly renewable and more ecologic source, *the electrolysis of water* into hydrogen and oxygen. The former process has carbon dioxide as a by-product, which exacerbates greenhouse gas emissions relative to present technology. With electrolysis, the greenhouse burden depends on the source of the power, and both intermittent renewables and nuclear energy are considered here.

With intermittent renewables such as solar and wind, the output may be fed directly into an electricity grid, with a penalty due to the requirement to operate some extra conventional plant on part load to allow for the fluctuations in power output. At penetrations below 20% of the grid demand, this penalty is usually small and does not severely change the economics; but beyond about 20% of the total demand, the penalty may make the generated power uneconomic. If these sources are used for electricity to make hydrogen, then they can be utilized fully whenever they are available, opportunistically. Broadly speaking, it does not matter when they cut in or out, the hydrogen is simply stored and used as required.

Nuclear advocates note that using nuclear power to manufacture hydrogen would help solve plant inefficiencies. Here the plant would be run continuously at full capacity, with perhaps all the output being supplied to the grid in peak periods, and any not needed to meet demand being used to make hydrogen at other times. This would mean far better efficiency for the nuclear power plants. High temperature (950-1000°C) gas cooled nuclear reactors have the potential to split hydrogen from water by thermochemical means using nuclear heat (i.e. without using electrolysis).

About 50 kWh (180 MJ) is required to produce a kilogram of hydrogen by electrolysis, so the cost of the electricity clearly is crucial. At $0.03/kWh, common off-peak high-voltage line rate in the U.S., this means hydrogen costs $1.50 a kilogram for the electricity, equivalent to $1.50 a US gallon for gasoline if used in a fuel cell vehicle. Other costs would include the electrolyzer plant, compressors, liquifaction, storage and transportation, which will be significant.

Electricity Retailing

Electricity retailing is the final process in the delivery of electricity from generation to the consumer. The other main processes are transmission and distribution.

Beginnings

Electricity retailing began at the end of the 19th century when the bodies who generated electricity for their own use made supply available to third parties. In the beginning, electricity was primarily used for street lighting and trams. The general public were allowed to purchase electricity only after large scale electric companies were started. The provision of these services was generally the responsibility of electric compaines or municipal authorities who either set up their own departments or contracted the services from private entrepreneurs. Residential, commercial and industrial use of electricity was confined, initially, to lighting but this changed dramatically with the development of electric motors, heaters and communication devices. The basic principle of supply has not changed much over time. The amount of energy used by the domestic consumer, and thus the amount charged for, is measured through an electricity meter that is usually placed near the input of a home to provide easy access to the meter reader. Customers are usually charged a monthly service fee and additional charges based on the electrical energy (in kWh) consumed by the household or business during the month. Commercial and industrial consumers normally have more complex pricing schemes. These require meters that measure the energy usage in time intervals (such as a half hour) to impose charges based on both the amount of energy consumed and the maximum rate of consumption, i.e. the maximum demand, which is measured in kW.

Creating a Market

In 1990 there was a significant development in the way electricity was bought and sold. In many countries, the electricity market was deregulated to open up the supply of electricity to competition. In the United Kingdom the Electricity Supply Industry was radically reformed to establish competition. This trend continued in other countries and the role of electricity retailing changed from what was essentially an administrative function within an integrated utility to a become a risk management function within a competitive electricity market. Electricity retailers now provide fixed prices for electricity to their customers and manage the risk involved in purchasing electricity from spot markets or electricity pools. This development has not been without casualties. The most notable example of poor risk management (coupled with poor market regulation) occurred in California in the summer of 2001, when Pacific Gas and Electric and Southern California Edison were driven into bankruptcy by having to purchase electricity at high spot prices and sell at low fixed rates. Customers may choose from a number of competing suppliers. They may also opt to pay more for "green" power, i.e. electricity sourced from renewable energy generation such as wind power or solar power. An electricity provider is often known as "the electric company" or "the power company".

Monopoly Supply

The rapid growth in electric appliances in the early part of the 20th century contributed to an explosive growth in electrification around the world. The supply of electricity to homes, offices, shops, factories, farms, and mines became the responsibility of public utilities, which were either private organisation subject to monopoly regulation or public authorities owned by local, state or national bodies. In some countries a statutory or government-granted monopoly was created, which was controlled by legislation (for example Eskom in South Africa). Electricity retailing in the period from approximately 1890 to 1990 consisted of managing the connection, disconnection and billing of electricity consumers by the local monopoly supplier. In many utilities there was a marketing function which encouraged electricity usage when there was excess capacity to supply and encouraged conservation when supply was tight.

Rates

The rates charged for electricity vary between countries, regions and states. The reason for the variation is primarily regulation and the way it is generated. For example, some states in the US have large hydroelectric generation facilities that are largely subsidized and relatively efficient, and rates are as low as $0.06 per kWh, as in Idaho. In other states, such as California, which has to import electricity from neighboring states, the rates can be as high as $0.38 per kWh during peak hours for high-use residential customers that pay based on time of use (.xls file). As of 2006 (May), the average rate for electricity in the US is approximately $0.106 per kWh.

Magnetic Energy Storage Efforts in Japan

Integrating energy storage into an electric power system has long been recognized as a way to maximize a utility's generation and/or transmission capacity. Electric power can be stored during off-peak periods and then recovered during high-peak conditions to offset the need for larger generation or expanded transmission capacity. In Japan, load-leveling or diurnal usage of stored energy is currently accomplished at pumped-storage installations, which are considered to be fairly inefficient, having only ~70% turn-around efficiency. Also, Japan's river system, highly dense settlement, and expensive real estate do not favor construction of additional pumped-storage facilities. The Japanese have thus been exploring alternative storage technologies, including SMES and flywheels. As discussed earlier, SMES research in Japan has been underway since the early 1970s, with many prototypes built and tested under actual "utility" conditions.

Japan's International Superconductivity Technology Center (ISTEC) conducted a three-year feasibility study starting in 1988 on electric power apparatus, including SMES, under a programme sponsored by MITI's Agency of Natural Resources and Energy. As part of this study, the Subcommittee on Energy Storage recommended in 1989 the detailed design using available technology of a small-scale SMES, with either a solenoid or toroidal field configuration, that would be consistent with future R&D efforts (Katsuya 1990). This SMES programme, much like Super-GM, did not mandate using HTS materials; thus, LTS conductors such as Nb-Ti and Nb_3Sn were the primary choice. Demonstration of a small-scale SMES, whose size is closely related to that needed for power system stabilization, would also address many major technical issues facing the large-scale diurnal storage SMES, such as ac losses, power conditioning, and refrigeration. ISTEC's initial SMES effort was followed by a six-year programme started in 1991 to implement the construction and test of a small-scale SMES pilot plant at a 100 kWh/20 MW rating. This ISTEC programme, also sponsored by MITI's Agency of Natural Resources and Energy, involves Toshiba as the primary magnet manufacturer, the Electric Power Development Corporation providing power conditioning, and various utilities led by Chubu Electric with additional support from Tohoko and Kyushu Electric Power Companies (Kamiyama 1994).

The design approach for the 100 kWh/20 MW system, is a toroidal magnet with an outside diameter for the cryostat of ~12 m (Kamiyama 1994). A half-size prototype coil was constructed by Toshiba and had been recently tested at the time of the WTEC trip to Japan. The test coil used a forced-flow Nb-Ti cable-in-conduit conductor and demonstrated 20 kA at 2.8 T, which is the rated current for the basic design. The initial testing was conducted at the Japan Atomic Energy Research Institute (JAERI), with further tests planned at Lawerence Livermore National Labs (LLNL) in the United States. Additional information on this SMES programme is presented in the ISTEC, Chubu, and Toshiba site reports

Another programme looking at the use of SMES for power system stabilization has been led by Kansai Electric Power Company (KEPCO). A small three-coil torus (400 kJ per coil), was assembled and tested on the KEPCO power system through a transformer and chopper arrangement. Two of the coils were built separately by Sumitomo Electric and Mitsubishi Heavy Industries using Nb-Ti conductors (SuperCom 1996). The third coil, built by Mitsubishi Electric, used Nb_3Sn. This programme also involved Osaka University, which with KEPCO looked at how SMES could be used for power system control and stabilization (Tada, Mitani, and Tsuji 1995, 250).

REFERENCES

American Wind Energy Association. (1992). *Wind Energy for Sustainable Development.* Washington, DC: American Wind Energy Association; 22 pp.

American Wind Energy Association. (1993). *Small Wind Energy Systems Applications Guide.* Washington, DC: American Wind Energy Association; 8 pp.

American Wind Energy Association. (1994). *American Wind Energy Association's 1994 Membership Directory.* Washington, DC: American Wind Energy Association; 42 pp.

Araj, K.J.; Fisher, T.A.; Kronenburg, J.C. (National Research Council, Washington, DC). (1991). *Assessment of Research Needs for Wind Turbine Rotor Materials Technology.* DOE/AL/58181-T1. Washington, DC: U.S. Department of Energy; 108 pp. (NTIS no. DE92018563).

Ashwill, T.D.; Leonard, T.M. (September 1986). *Developments in Blade Shape Design for a Darrieus Vertical Axis Wind Turbine.* SAND-86-1263. Albuquerque, NM: Sandia National Laboratories; 21 pp.

Babcock, B.A.; Conover, K.E. (R. Lynette & Associates, Inc.). (1994). "Design of Cost-Effective Towers for an Advanced Wind Turbine." *Wind Energy 1994.* SED-Vol. 15. New York, NY: American Society of Mechanical Engineers; pp. 261-268.

Baring-Gould, E.I.; Manwell, J.F.; Jeffries, W.Q.; Stein, W.M. (University of Massachusetts). (1994). "Experimental Validation of the University of Massachusetts Wind/Diesel System Simulation Code, HYBRID1." *Wind Energy 1994.* SED-Vol. 15. New York, NY: American Society of Mechanical Engineers; pp. 77–86.

Barnard, J.C. (June 1990). *An Evaluation of Three Models Designed for Siting Wind Turbines in Areas of Complex Terrain.* PNL-7357. Richland, WA: Pacific Northwest Laboratory; 50 pp. (NTIS no. DE90013625).

Barnard, J.C.; Wegley, H.L. (January 1987). *A Preliminary Evaluation of the Performance of Wind Tunnel and Numerical Modeling Simulations of the Wind.* PNL-6105. Richland, WA: Pacific Northwest Laboratory; 20 pp. (NTIS no. DE87004634).

Barnard, J.C.; Wegley, H.L.; Hiester, T.R. (September 1985). *Improving the Performance of Mass- Consistent Numerical Models Using Optimization Techniques.* PNL-5566. Richland, WA: Pacific Northwest Laboratory; 33 pp. (NTIS no. DE86001224). An article based on this report appeared in the Journal of Climate and Applied Meteorology (26:6); pp. 675-686.

Barnwell, R.W.; Hussaini, M.Y. (1992). *Natural Laminar Flow and Laminar Flow Control.* New York, NY: Springer-Verlag; 411 pp.

Buck, J.W.; Renne, D.S. (August 1985). *Observations of Wake Characteristics at the Goodnoe Hills MOD-2 Array.* PNL-5565. Richland, WA: Pacific Northwest Laboratory; 77 pp. (NTIS no. DE85018282).

Butterfield, C.P.; Musial, W.P.; Scott, G.N.; Simms, D.A. (1992). *NREL Combined Experiment Final Report-Phase II (draft).* NREL/TP-442-4807. Golden, CO: National Renewable Energy Laboratory; 32 pp. (plus appendices).

Butterfield, C.P.; Musial, W.P.; Simms, D.A. (1992). *Combined Experiment Phase I: Final Report.* NREL/TP-257-4655. Golden, CO: National Renewable Energy Laboratory; 220 pp. (NTIS no. DE93000012).

Childs, S.E.; Hughes, P.S.; Saeed, A. (Atlantic Orient Corporation). (1993). "Development of a Dynamic Brake Model." *Windpower '93 Conference; July 12-16, 1993; San Francisco, California.* Washington, DC: American Wind Energy Association; pp. 366-373.

Clark, R.N. (1985). "Wind-diesel Hybrid System for Water Pumping." *Proceedings of the Wind Energy Exposition and National Conference of the American Wind Energy Association.* Washington, DC: American Wind Energy Association; pp. 221–225.

Clark, R.N. (1989). "Stand-alone Wind Electric Water Pumping." *Proceedings of the European Wind Energy Conference, EWEC '89.* Amsterdam: Elsevier; pp. 650–654.

Clark, R.N.; Davis, R.G. (1991). "Performance Changes Caused by Rotor Blade Surface

Debris." *Windpower '91 Conference; September 24-27, 1991; Palm Springs, California.* Washington, DC: American Wind Energy Association; pp. 470-474.

Coleman, C. (Northern Power Systems). (1990). *Wind Hybrid System Research and Testing: Telecommunications Systems.* SERI/STR-257-3686. Golden, CO: Solar Energy Research Institute.

Coleman, C; McNiff, B.P. (Nothern Power Systems). (1990). *Dynamic Response Testing of the North Wind 100 Wind Turbine.* SERI/STR-257-3686. Golden, CO: Solar Energy Research Institute.

Connell, J.R. (1988). "A Primer of Turbulence at the Wind Turbine Rotor." *Solar Energy* (41:3); pp. 281–293. New York, NY: Pergamon Press.

Connell, J.R. (Colorado State University); Morris, V.R. (1991). *Calibration Data for Improved Correction of UVW Propeller Anemometers.* PNL-7824. Richland, WA: Pacific Northwest Laboratory.

Connell, J.R.; George, R.L. (November 1987). "Accurate Correlation of Wind Turbine Response With Wind Speed Using a New Characterization of Turbulent Wind." *Journal of Solar Energy Engineering;* Vol. 109. New York, NY: American Society of Mechanical Engineers.

Connell, J.R.; Morris, V.R. (1990). "Turbulent Wind at the Equatorial Segment of an Operating Darrieus Wind Turbine Blade." *Ninth ASME Wind Energy Symposium.* SED-Vol. 9. New York, NY: American Society of Mechanical Engineers.

Connell, J.R.; Powell, D.C. (August 1989). *User's Guide for Personal Computer Model of Turbulence at a Wind Turbine Rotor.* PNL-6959. Richland, WA: Pacific Northwest Laboratory.

Cramer, G.; Kleinkauf, W.; Schott, T. (1994). "Wind/Diesel/Battery Systems-Applications, Experience, Economy of Different System Configurations and Sizes." *Wind Engineering* (17:5); p. 228. Brentwood, England: Multi-Science Publishing Company, LTD.

Design Criteria Recommended Practices: Wind Energy Conversion Systems. (1988). AWEA Standard: AWEA 3.1-1988. Arlington, VA: American Wind Energy Association; 25 pp.

Doran, J.C. (August 1981). *Inherent Uncertainties in Meteorological Parameters for Wind Turbine Design.* PNL-SA-736. Richland, WA: Pacific Northwest Laboratory; 18 pp.

Doran, J.C.; Packard, K.R. (October 1982). *Comparison of Model and Observations of the Wake of a MOD-0A Wind Turbine.* PNL-4433. Richland, WA: Pacific Northwest Laboratory; 43 pp.

Eggleston, D.M.; Starcher, K.L. (November 1990). "A Comparative Study of the Aerodynamics of Several Wind Turbines Using Flow Visualization." *Journal of Solar Energy Engineering.* Vol. 112. New York, NY: American Society of Mechanical Engineers; pp. 301-309.

Eggleston, D.M.; Stoddard, F.S. (1987). *Wind Turbine Engineering Design.* New York, NY: Van Nostrand Reinhold Company, Inc.; 352 pp.

Electrical Behavior of Wind Power Stations. (Zaininger Engineering Company, Inc.). (1992). Windpower '92 Conference; October 19-23, 1992; Seattle, Washington. Washington DC: American Wind Energy Association.

Elliott, D.L. (September 1991). "Status of Wake and Array Loss Research." *Windpower '91*

Conference; September 24-27, 1991; Palm Springs, California. Washington, DC: American Wind Energy Association.

Elliott, D.L.; Barnard, J.C. (November 1990). "Effects of Trees on Wind Flow Variability and Turbulence." *Journal of Solar Energy Engineering;* Vol. 112. New York, NY: American Society of Mechanical Engineers.

Elliott, D.L.; Buck, J.W.; Barnard, J.C. (April 1988). *An Examination of Wake Effects and Power Production for a Group of Large Wind Turbines.* PNL-6528. Richland, WA: Pacific Northwest Laboratory; 58 pp. (NTIS no. DE88009906).

Elliott, D.L.; Cadogan, J.B. (1990). "Effects of Wind Shear and Turbulence on Wind Turbine Power Curves." *European Community Wind Energy Conference and Exhibition; September 10-14, 1990; Madrid, Spain.* Bedford, England: H.S. Stephens & Associates.

Eppler, R. (1990). *Airfoil Design and Data.* New York, NY: Springer-Verlag; 562 pp.

Fardoun, A.A.; Fuchs, E.F. (University of Colorado); Carlin, P.W. (1993). "A Variable-Speed, Direct- Drive Transmission Wind Power Plant." *Windpower '93 Conference; July 12-16, 1993; San Francisco, California.* Washington, DC: American Wind Energy Association; pp. 134-141.

Frost, W.; Long, B.H.; Turner, R.E. (December 1978). *Engineering Handbook on the Atmospheric Environmental Guidelines for Use in Wind Turbine Generator Development.* NASA-TP-1359. Huntsville, AL: National Aeronautics and Space Administration, Marshall Space Flight Center; 372 pp.

Gipe, P. (1993). *Wind Power for Home & Business.* Post Mills, VT: Chelsea Green Publishing Company; 413 pp.

Hansen, A.C. (1992). *Yaw Dynamics of Horizontal Axis Wind Turbines: Final Report.* NREL/TP-442- 4822. Golden, CO: National Renewable Energy Laboratory; 175 pp. (NTIS no. DE92001245).

Hansen, A.C. (University of Utah). (January 1987). *The TEETER Program-Analysis and User's Guide.* SERI/STR-217-3050. Golden, CO: Solar Energy Research Institute.

Hansen, A.C.; Butterfield, C.P. (1993). "Aerodynamics of Horizontal-Axis Wind Turbines." *Annual Review of Fluid Mechanics.* Vol. 25, pp. 115-149.

Herrera, J.I.; Reddoch, T.W. (Electrotek Concepts, Inc.). (February 1988). *Analysis of the Electrical Harmonic Characteristics of a Slip Recovery Variable Speed Generating System for Wind Turbine Applications.* SERI/STR-217-3268. Golden, CO: Solar Energy Research Institute.

Herrera, J.I.; Reddoch, T.W. (Electrotek Concepts, Inc.). (February 1988). *Analysis of the Electrical Harmonic Characteristics of a Westinghouse Variable Speed Generating System for Wind Turbine Applications.* SERI/STR-217-3133. Golden, CO: Solar Energy Research Institute.

Horizontal-Axis Wind System Rotor Performance Model Comparison-A Compendium. (February 1983). RFP-3508. Golden, CO: Rocky Flats Wind Energy Research Center; 381 pp. (NTIS no. DE83012265).

Hughes, P.S.; Johnson, B.; Sherwin, R.; Stern, W. (Atlantic Orient Corporation). (1990).

System Stability and Penetration Study for Wind Diesel Hybrid Systems. SERI/STR-257-3982. Golden, CO: Solar Energy Research Institute.

Interstate Renewable Energy Council; Solar Energy Industries Association; Sandia National Laboratories. (1993). *Procurement Guide for Renewable Energy Systems;* 140 pp. Available from American Solar Energy Society, 2400 Central Avenue, G-1, Boulder, CO 80301.

James, G.H., III; Carne, T.G.; Lauffer, J.P. (1993). *The Natural Excitation Technique (NExT) for Modal Parameter Extraction from Operating Wind Turbines.* SAND-92-1666. Albuquerque, NM: Sandia National Laboratories; 46 pp. (NTIS no. DE93010611).

Kelley, D.N. (November 1992). *Full Vector (3-D) Inflow Simulation in Natural and Wind Farm Environments Using an Expanded Version of the SNLWIND (Veers) Turbulence Code.* NREL/TP-442- 5225. Golden, CO: National Renewable Energy Laboratory; 8 pp.

Kelley, N.D. (September 1989). *An Initial Look at the Dynamics of the Microscale Flow Field Within a Large Windfarm in Response to Variations in the Natural Inflow.* SERI/TP-257-3591. Golden, CO: National Renewable Energy Laboratory; 5 pp.

Kelley, N.D.; Wright, A.D. (October 1991). *A Comparison of Predicted and Observed Turbulent Wind Fields Present in Natural and Internal Wind Park Environments.* NREL/TP-257-4508. Golden, CO: National Renewable Energy Laboratory; 9 pp.

King, W.R.; Johnson, B.L., III. (Science Applications International Corporation) (1991). *Worldwide Wind / Diesel Hybrid Power System Study: Potential Applications and Technical Issues.* SERI/TP-257-3757. Golden, CO: Solar Energy Research Institute; 104 pp. (NTIS no. DE91002160).

Klimas, P.C. (November 1984). *Tailored Airfoils for Vertical Axis Wind Turbines.* SAND-84-1062. Albuquerque, NM: Sandia National Laboratories; 12 pp.

Kocurek, D. (Computational Methodology Associates). (June 1987). *Lifting Surface Performance Analysis for Horizontal Axis Wind Turbines.* SERI/STR-217-3163. Golden, CO: Solar Energy Research Institute; 319 pp. (NTIS no. DE87001176).

Lauw, H.K.; Weigand, C.H.; Marckx, D.A. (Electronic Power Conditioning, Inc.). (October 1993). *Variable-Speed Wind System Design: Final Report.* DOE/BP/99893-TI. Washington, DC: U.S. Department of Energy; 81 pp. (NTIS no. DE94006210).

Lissaman, P.B.S.; Gyatt, G.W.; Zalay, A.D. (AeroVironment). (June 1982). *Numeric Modeling Sensitivity Analysis of the Performance of Wind Turbine Arrays.* PNL-4183. Richland, WA: Pacific Northwest Laboratory; 97 pp.

Liu, H.-T. (FloWind Corporation). (1988). Field Investigation of a Wake Structure Downwind of a VAWT in a Wind Farm Array. SERI/STR-217-3370. Golden, CO: Solar Energy Research Institute.

Lundsager; Norgaard. (August 1988). *The 55/30 kW Experimental Wind/Diesel System at Riso National Laboratory.* RISO-M-2717. Roskilde, Denmark: Riso National Laboratory.

Manwell, J.F.; McGowan, J.G. (January 1992). "Advances in Quasi-Steady Wind Diesel System Simulation." *11th ASME Wind Energy Symposium.* New York, NY: American Society of Mechanical Engineers.

McNerney, G.M. (U.S. Windpower, Inc.). (1994). "U.S. Windpower 56-100 Stall Operation."

Wind Energy 1994. SED-Vol. 15. New York, NY: American Society of Mechanical Engineers; pp. 219-226.

McNiff, B.P. (Second Wind Inc.); Musial, W.P. (National Renewable Energy Laboratory); Errichello, R. (GEARTECH). (September 1990). "Gear Fatigue Life Variations for Different Wind Turbine Braking Strategies." *Windpower '90 Conference; September 24-28, 1990; Washington, DC*. Washington, DC: American Wind Energy Association; 10 pp. (NTIS no. DE91002161).

Miley, S.J. (Texas A&M University). (December 1984). *Addendum to a Catalog of Low Reynolds Number Airfoil Data for Wind Turbine Applications*. SERI/STR-217-2584. Golden, CO: Solar Energy Research Institute.

Miley, S.J. (Texas A&M University). (February 1982). *A Catalog of Low Reynolds Number Airfoil Data for Wind Turbine Applications*. RFP-3387. Golden, CO: Rocky Flats Wind Energy Research Center; 615 pp. (NTIS no. DE82021712).

Miller, A.H.; Dugundji, J.; Martinez-Sanchez, M.; Gohard, J.; Chung, S.; Wendell, J. (September 1978). *Wind Energy Conversion, Volume 1: Methods for Design Analysis of Horizontal Axis Wind Turbines*. COO-4131-Tq (Vol. I). Washington, DC: U.S. Department of Energy; 253 pp.

Morris, V.R.; Barnard, J.C.; Wendell, L.L.; Tomich, S.D. (October 1992). "Comparison of Anemometers for Turbulence Characterization." *Windpower '92 Conference; October 19-23, 1992; Seattle, Washington*. Washington, DC: American Wind Energy Association; 9 pp. (NTIS no. DE93004336).

Nierenberg, R. (Altamont Energy Corporation). (1989). *Free-Flow Variability on the Jess and Souza Ranches, Altamont Pass*. SERI/STR-217-3404. Golden, CO: Solar Energy Research Institute.

Nierenberg, R. (Altamont Energy Corporation). (1990). *Wake Deficit Measurements on the Jess and Souza Ranches, Altamont Pass*. SERI/STR-257-3455. Golden, CO: Solar Energy Research Institute.

Noll, R.B.; Ham, N.D. (Aerospace Systems, Inc.). (February 1983). *Dynamic Stall of Small Wind Systems*. RFP-3523. Golden, CO: National Renewable Energy Laboratory. (NTIS no. DE84006827).

NREL/TP-442-5225. Golden, CO: National Renewable Energy Laboratory; 8 pp.

Ostowari, C.; Niak, D. (Texas A&M University). (January 1985). *Post-Stall Wind Tunnel Data for NACA 44xx Series Airfoil Sections*. SERI/STR-217-2559. Golden, CO: Solar Energy Research Institute.

Qi, G. "Mathematic Programming Models for Economic Design and Assessment of Wind/Diesel Systems." *Windpower '93 Conference; July 12-16, 1993; San Francisco, California*. Washington, DC: American Wind Energy Association; pp. 501-507.

Reactive Power Management Device Assessment. (University of Washington). (August 1987). EPRI-AP-5210. Palo Alto, CA: Electric Power Research Institute; 96 pp.

Rogers, A; Manwell, J.F.; Stein, W.M.; McGowan, J.G. (1993). "Experimental Testing of Variable Speed Diesels for Wind/Diesel System Applications." *Wind Energy-1993*;

January 31-February 4, 1993; Houston, Texas. SED-Vol. 14. New York, NY: American Society of Mechanical Engineers.

Rohrbach, C.; Wainauski, H.; Worabel, R. (December 1977). *Experimental and Analytical Research on the Aerodynamics of Wind Driven Turbines*. COO-2615-T2. Windsor Locks, CT: Hamilton Standard; 250 pp.

Schluter, L.L. (Sandia National Laboratories); Vachon, W.A. (W.A. Vachon & Associates). (1993) "Optimizing Wind Turbine Control System Parameters." *Windpower '93 Conference; July 12-16, 1993; San Francisco, California*. Washington, DC: American Wind Energy Association; pp. 175-182.

Selig, M.S.; Donovan, J.F.; Fraser, D.B. (1989). *Airfoils at Low Speeds*. Soartech 8. Virginia Beach, VA: H.A. Stokely; 398 pp.

Simms, D.A. (National Renewable Energy Laboratory); Cousineau, K.L. (Zond Systems, Inc.). (June 1993). *A Rugged, Low-Cost, Advanced Data Acquisition System for Field Test Projects*. NREL/TP-442-5617. Golden, CO: National Renewable Energy Laboratory; 11 pp. (NTIS no. DE93010041).

Simms, D.A.; Butterfield, C.P. (April 1990). *A Low-Cost PC-Based Telemetry Data-Reduction System*. SERI/TP-257-3737. Golden, CO: Solar Energy Research Institute.

Simon, R.L. (Fayette Manufacturing Corporation). (April 1988). *Wake Effects in a Fayette 85-IIS Wind Turbine Array*. SERI/STR-217-3186A. Golden, CO: Solar Energy Research Institute; 75 pp.

Simon, R.L.; Matson, D.F.; Fuchs, J.M. (Fayette Manufacturing Corporation). (1987). *Wake Effects in a Fayette 95-IIS Wind Turbine Array*. SERI/STR-217-3186. Golden, CO: Solar Energy Research Institute.

Smith, J.C. (1991). *Studies on the Modeling of Wind-Diesel Systems and the Integration of Wind Energy Systems with Electric Utilities*. Prepared for DOE/EC Wind Energy Assessment and Review, Albuquerque, New Mexico, June 4, 1991, by Electrotek Concepts, Inc.

Snow, A.L.; Heberling, C.F., III; Van Bibber, L.E. (Westinghouse Electric Corporation). (March 1989). *The Dynamic Response of a Westinghouse 600-kW Wind Turbine*. SERI/STR-217-3405. Golden, CO: Solar Energy Research Institute; 185 pp.

Specification of Test Equipment for Power Electronic Drive Research. (OMNION Power Engineering Corporation). (1987). AP-5219. Palo Alto, CA: Electric Power Research Institute; 112 pp.

Spera, D.A. (May 1994). *Wind Turbine Technology: Fundamental Concepts of Wind Turbine Engineering*. No. 100368. Fairfield, NJ: American Society of Mechanical Engineers; 700 pp.

Spera, D.A. (May 1994). *Wind Turbine Technology: Fundamental Concepts of Wind Turbine Engineering*. 100368. Fairfield, NJ: American Society of Mechanical Engineers; 700 pp.

Stoddard, F.S. (1990). "Wind Turbine Blade Technology: A Decade of Lessons Learned." *Energy and the Environment, Into the 1990's, 1st World Renewable Energy Congress*. Edited by A.A.M. Sayigh. Vol. 3. Tarrytown, NY: Pergamon Press; 1570 pp.

Stoddard, F.S.; Porter, B.K. (January 1986). *Wind Turbine Aerodynamics Research Needs Assessment*. DOE/ER/30075-H1. Washington, DC: U.S. Department of Energy; 126 pp. (NTIS no. DE86009189).

Tangler, J.L. (1987). *Status of the Special-Purpose Airfoil Families*. SERI/TP-217-3264. Golden, CO: Solar Energy Research Institute; 9 pp. (NTIS no. DE88001124/XSP).

Tangler, J.L.; Smith, B.; Jager, D. (1992). *SERI Advanced Wind Turbine Blades*. NREL/TP-257-4492. Golden, CO: National Renewable Energy Laboratory. (NTIS no. DE92001216).

Tangler, J.L.; Smith, B.; Kelley, N.D.; Jager, D. (1992). *Measured and Predicted Rotor Performance for the SERI Advanced Wind Turbine Blades*. NREL/TP-257-4594. Golden, CO: National Renewable Energy Laboratory. (NTIS no. DE92001215).

Thresher, R.W.; Holley, W.E.; Wright, A.D. (January 1989). *Prediction of Stochastic Blade Responses Using a Filtered Noise Turbulence Model in the FLAP Code*. SERI/TP-217-3413. Golden, CO: Solar Energy Research Institute.

Thresher, R.W.; Wright, A.D. (January 1989). *Prediction of Stochastic Blade Responses Using Measured Wind-Speed Data as Input to the FLAP Code*. SERI/TP-217-3394. Golden, CO: Solar Energy Research Institute.

Torrey, D.A. (Rensselaer Polytechnic Institute); Childs, S.E. (Atlantic Orient Corporation). (1993). "Development of Variable-Reluctance Wind Generators." *Windpower '93 Conference; July 12-16, 1993; San Francisco, California*. Washington, DC: American Wind Energy Association; pp. 258-265.

Vachon, W.A. (January 1990). "Smart Control Algorithms for Operation of Variable-Speed Wind Turbines." *Proceedings of the Ninth ASME Wind Energy Symposium*. Edited by D.E. Berg. SED- Vol. 9. New York, NY: American Society of Mechanical Engineers.

Veers, P.S. (1988). *Three-Dimensional Wind Simulation*. SAND-88-0152. Albuquerque, NM: Sandia National Laboratories; 36 pp. (NTIS no. DE88009793/XSP).

Viterna, L.A.; Janetzke, D.C. (September 1982). *Theoretical and Experimental Power from Large Horizontal-Axis Wind Turbines*. DOE/NASA/20320-41. Cleveland, OH: NASA Lewis Research Center; 18 pp. (NTIS no. DE83001099).

Walker, S.N.; Wade, J.E. (Pacific Wind Energy). (1988). *Effects of Precipitation on Wind Turbine Performance*. SERI/STR-217-3287. Golden, CO: Solar Energy Research Institute.

Weber, T.L.; Wilson, R.E.; Walker, S.N. (Oregon State University). (June 1991). *Analysis of a Teetered, Variable-Speed Rotor*. SERI/TP-267-4366. Golden, CO: Solar Energy Research Institute; 59 pp. (NTIS no. DE91002190).

Wehrey, M.; Redmond, I.; Anderson, C.; Jamieson, R. (Southern California Edison, James Howden and Company Ltd.). (February 1988). *Dynamic Response of a 330-kW Horizontal Axis Wind Turbine Generator*. SERI/STR-217-3203. Golden, CO: Solar Energy Research Institute; 103 pp.

Wendell, L.L; Barnard, J.C.; Morris, V.R.; Gower, G.L. (1993). "Turbulence Characteristics in Complex Terrain." *Wind Energy 1993*. SED-Vol. 14, Wind Energy. New York, NY: The American Society of Mechanical Engineers.

Wilreker, V.F.; Stiller, P.H.; Scott, G.W.; Kruse, V.J.; Smith, R.F. (February 1984). *Wind*

Turbine Generator Interaction with Conventional Diesel Generators on Block Island, Rhode Island. Vol. 1, Executive Summary. DOE/NASA-0354-1. Cleveland, OH: NASA Lewis Research Center; 45 pp. (NTIS no. DE84015874).

Wilson, R.E.; Walker, S.N. (1985). "State-of-the-Art in Wind Turbine Aerodynamics." *Windpower '85 Conference; August 27, 1985; San Francisco, California.* Alexandria, VA: American Wind Energy Association; pp. 30-38.

Wilson, R.E.; Walker, S.N.; Smith, C.E.; Freeman, L.N. (1993). "Advanced Dynamics Code." *Wind Energy 1993;* January 31-February 4, 1993; Houston, Texas. SED-Vol. 14. New York, NY: American Society of Mechanical Engineers; pp. 119-124.

Wind Energy Applications and Training Symposium. Final Report. (1990). Conference proceedings. Arlington, VA: American Wind Energy Association; 120 pp. (NTIS no. TI91006703).

Wind/Diesel Systems Architecture Guidebook. (1991). AWEA Standard: AWEA 10.1-1991. Washington, DC: American Wind Energy Association; 41 pp.

Wright, A.D.; Buhl, M.L., Jr.; Elliott, A.S. (1992). "Development of a Wind Turbine System's Dynamics Model Using the Automatic Dynamic Analysis of Mechanical Systems (ADAMS) Software." *Windpower '92 Conference; October 19-23, 1992; Seattle, Washington.* Washington, DC: American Wind Energy Association; pp. 329-338.

Wright, A.D.; Buhl, M.L., Jr.; Thresher, R.W. (January 1988). *FLAP Code Development and Validation.* SERI/TR-217-3125. Golden, CO: Solar Energy Research Institute.

Wright, A.D.; Butterfield, C.P. (1991). *The NREL Teetering Hub Rotor Code: Final Results and Conclusions.* NREL/TP-257-4517. Golden, CO: National Renewable Energy Laboratory; 10 pp. (NTIS no. DE92001187).

Wyatt, A. (Research Triangle Institute). (1992). *Wind Electric Pumping Systems: Sizing and Cost Estimation.* Washington, DC: American Wind Energy Association; 39 pp.

8

Research Methodology in Nano & Micro Sciences

INTRODUCTION

The two constructional technologies of microengineering are microelectronics and micromachining. Microelectronics, producing electronic circuitry on silicon chips, is a very well developed technology. Micromachining is the name for the techniques used to produce the structures and moving parts of microengineered devices. Microengineering refers to the technologies and practice of making three dimensional structures and devices with dimensions in the order of micrometers. One of the main goals of Microengineering is to be able to integrate microelectronic circuitry into micromachined structures, to produce completely integrated systems (microsystems). Such systems could have have the same advantages of low cost, reliability and small size as silicon chips produced in the microelectronics industry. When considering such small devices, a number of physical effects have different significance on the micrometer scale compared to macroscopic scales. Interest in microengineering has spawned or renewed interest in a number of areas dealing with the study of these effects on microscopic scales. This includes such topics as micromechanics, which deals with the moving parts of microengineered devices, and microfluidics, etc.

The remainder of this document introduces three of the micromachining techniques that are in use / under development. Silicon micromachining is given most prominence, since this is one of the better developed micromachining techniques. Silicon is the primary substrate material used in the production microelectronic circuitry (ie, bettersilicon chips), and so is the most suitable candidate for the eventual production of microsystems. The Excimer laser is an ultraviolet laser which can be used to micromachine a number of materials without heating them, unlike many other lasers which remove material by burning or vaporising it. The Excimer laser lends itself particularly to the machining of organic materials (polymers, etc). LIGA is a technique that can be used to produce moulds for the fabrication of micromachined components. Microengineered components can be made from a variety of materials using this technique, however it does suffer the disadvantage that currently the technique requires X-rays from a synchrotron source. A quick introduction to mask design is provided following discussion of techniques and structures, rather than directly following the photolithography section. This is so that the reader is able to become acquainted with the concept of creating structures by sequential photolithography and machining steps first, which hopefully makes it easier to understand what mask design software is trying to achieve.

Microsystems & Microsystem Components

A microsystem can be considered to be any device or unit made up of a number of microengineered components. A convenient model of a microsystem is that of a control system; many proposed microsystems take this form. Microsensors detect changes in the parameter to be controlled, electronic control logic then operates microactuators based on information from the sensors, to bring the parameter to be controlled within the desired limits. An example of such a system would be to refresh the medium in a small cell culture dish. Sensors could detect changes in pH, pO2, or pCO2, and a micropump could deliver new culture medium from a reservoir as required. Not all devices need follow this control-system scheme. For instance an accelerometer designed to inflate an air-bag in the event of a car crash may not only incorporate a micromachined acceleration sensor, but also electronics to condition the signal and detect a rapid deceleration, and microactuators that put a force on the sensor allowing the device to be tested before the driver moves off. Microsystems may be constructed from parts produced using different technologies on different substrates, connected together; i.e. a hybrid system. For example, a silicon chip would be used to implement control circuitry, whereas the actuators it controlled could be micromoulded in plastic, or electroplated metal (using the LIGA technique, perhaps). Alternatively, all components of a system could be constructed on a single substrate using one technology (a monolithic system). Hybrid systems have the advantages that the most appropriate technology for each component can be selected to optimise system performance. This will often lead to a shorter development time since microfabrication techniques for each component may already exist, and compromises will not have to be made to ensure that each component can be fabricated without damaging components already existing on the substrate. Monolithic devices will typically be more compact than hybrid devices, and more reliable (fewer interconnections that can go wrong, for example).

Further, once the fabrication process has been developed, they can be manufactured more cheaply since less assembly is required. A significant problem facing microengineers is that of assembling many microscopic components. Potential solutions include self assembling systems, and desk-top factories staffed by microrobots! At some point many microsystems will have to interact with macroscopic systems. Often it may be that only a critical component of a system has to be microengineered, and supported by a complex system produced using more conventional engineering techniques. It is easy to underestimate the problems involved in mounting and packaging microdevices, and integrating them with macroscopic supporting systems. The remainder of this document goes on to introduce two components of microsystems; microsensors and microactuators. Microsystem "intelligence", in the form of microelectronics is a well developed subject, and covered in many textbooks. The document is rounded out with some comments on passivation, packaging, and assembly of microsystems. Micromachining techniques by which various components may be manufactured are introduced in the companion document: "Introduction to Microengineering". A microsystem will typically comprise of components from one or more of three classes: microsensors to detect changes in the system's environment; an intelligent component that makes decisions based on changes detected by the sensors; and microactuators, by which the system changes its environment. Obviously a system need not be a closed loop.

Even if it is, some (most ?) elements may be implemented in macroscopic technology; for instance, microsensors may be used as cheap disposable components to control large actuators. Currently, the intelligent component of the system would be implemented using microelectronic components; this is a well established technology, and is not discussed in this document. The subject of "integrated optics" — microsystems constructed of optical microcomponents — is also neglected in this document; this is quite a broad topic which needs to be introduced separately. Some of the microsensors described below are micromechanical devices that have to be driven into resonance. This is an additional role for the microactuators described in the section following the sensor section.

What is Microprocessor?

A microprocessor (sometimes abbreviated μP) is a programmable digital electronic component that incorporates the functions of a central processing unit (CPU) on a single semiconducting integrated circuit (IC). The microprocessor was born by reducing the word size of the CPU from 32 bits to 4 bits, so that the transistors of its logic circuits would fit onto a single part. One or more microprocessors typically serve as the CPU in a computer system, embedded system, or handheld device. Microprocessors made possible the advent of the microcomputer in the mid-1970s. Before this period, electronic CPUs were typically made from bulky discrete switching devices (and later small-scale integrated circuits) containing the equivalent of only a few transistors. By integrating the processor onto one or a very few large-scale integrated circuit packages (containing the equivalent of thousands or millions of discrete transistors), the cost of processor power was greatly reduced. Since the advent of the IC in the mid-1970s, the microprocessor has become the most prevalent implementation of the CPU, nearly completely replacing all other forms. See History of computing hardware for pre-electronic and early electronic computers. The evolution of microprocessors has been known to follow Moore's Law when it comes to steadily increasing performance over the years. This law suggests that *the complexity of an integrated circuit, with respect to minimum component cost, doubles every 24 months*. This dictum has generally proven true since the early 1970s. From their humble beginnings as the drivers for calculators, the continued increase in power has led to the dominance of microprocessors over every other form of computer; every system from the largest mainframes to the smallest handheld computers now uses a microprocessor at its core.

First Microprocessors

As with many advances in technology, the microprocessor was an idea whose time had come. Three projects arguably delivered a complete microprocessor at about the same time, Intel's 4004, Texas Instruments' TMS 1000, and Garrett AiResearch's Central Air Data Computer. In 1968, Garrett was invited to produce a digital computer to compete with electromechanical systems then under development for the main flight control computer in the US Navy's new F-14 Tomcat fighter. The design was complete by 1970, and used a MOS-based chipset as the core CPU. The design was smaller and much more reliable than the mechanical systems it competed against, and was used in all of the early Tomcat models. However, the system was considered so advanced that the Navy refused to allow publication of the design until 1997. For this reason the CADC, and the MP944 chipset it used, are fairly unknown even today. TI developed the 4-bit TMS 1000 and stressed pre-programmed embedded applications, introducing a version called the TMS1802NC on September 17, 1971, which implemented a calculator on a chip. The Intel chip was the 4-bit 4004, released on November 15, 1971, developed by Federico Faggin. TI filed for the patent on the microprocessor. Gary Boone was awarded U.S. Patent 3,757,306 for the single-chip microprocessor architecture on September 4, 1973. It may never be known which company actually had the first working microprocessor running on the lab bench. In both 1971 and 1976, Intel and TI entered into broad patent cross-licensing agreements, with Intel paying royalties to TI for the microprocessor patent. A nice history of these events is contained in court documentation from a legal dispute between Cyrix and Intel, with TI as intervenor and owner of the microprocessor patent. Interestingly, a third party claims to have been awarded a patent which might cover the "microprocessor". See a webpage claiming an inventor pre-dating both TI and Intel, describing a "microcontroller", which may or may not count as a "microprocessor". A computer-on-a-chip is a variation of a microprocessor which combines the microprocessor core (CPU), some memory, and I/O (input/output) lines, all on one chip. The computer-on-a-chip patent, called the "microcomputer patent" at the time, U.S. Patent 4,074,351, was awarded to Gary Boone and Michael J. Cochran of TI. Aside from this patent, the standard meaning of microcomputer is a computer using one or more microprocessors as its CPU(s), while the concept defined in the patent is perhaps more akin to a microcontroller. According to *A History of Modern Computing*, (MIT Press), pp. 220–21, Intel entered into a contract with Computer Terminals Corporation, later called Datapoint, of San Antonio TX, for a chip for a terminal they were designing. Datapoint later decided not to use the chip, and Intel marketed it as the 8008 in April, 1972. This was the world's first 8-bit microprocessor. It was the basis for the famous "Mark-8" computer kit advertised in the magazine Radio-Electronics in 1974. The 8008 and its successor, the world-famous 8080, opened up the microprocessor component marketplace.

Notable 8-bit Designs

The 4004 was later followed in 1972 by the 8008, the world's first 8-bit microprocessor. These processors are the precursors to the very successful Intel 8080 (1974), Zilog Z80 (1976), and derivative Intel 8-bit processors. The competing Motorola 6800 was released August 1974. Its architecture was cloned and improved in the MOS Technology 6502 in 1975, rivaling the Z80 in popularity during the 1980s.

Both the Z80 and 6502 concentrated on low overall cost, through a combination of small packaging, simple computer bus requirements, and the inclusion of circuitry that would normally have to be provided in a separate chip (for instance, the Z80 included a memory controller). It was these features that allowed the home computer "revolution" to take off in the early 1980s, eventually delivering semi-usable machines that sold for US$99. The Western Design Center, Inc. (WDC) introduced the CMOS 65C02 in 1982 and licensed the design to several companies which became the core of the Apple IIc and IIe personal computers, medical implantable grade pacemakers and defibrilators, automotive, industrial and consumer devices. WDC pioneered the licensing of microprocessor technology which was later followed by ARM and other microprocessor Intellectual Property (IP) providers in the 1990's. Motorola trumped the entire 8-bit world by introducing the MC6809 in 1978, arguably one of the most powerful, orthogonal, and clean 8-bit microprocessor designs ever fielded – and also one of the most complex hardwired logic designs that ever made it into production for any microprocessor. Microcoding replaced hardwired logic at about this point in time for all designs more powerful than the MC6809 – specifically because the design requirements were getting too complex for hardwired logic. Another early 8-

bit microprocessor was the Signetics 2650, which enjoyed a brief flurry of interest due to its innovative and powerful instruction set architecture. A seminal microprocessor in the world of spaceflight was RCA's RCA 1802 (aka CDP1802, RCA COSMAC) (introduced in 1976) which was used in NASA's Voyager and Viking spaceprobes of the 1970s, and onboard the Galileo probe to Jupiter (launched 1989, arrived 1995). RCA COSMAC was the first to implement C-MOS technology.The CDP1802 was used because it could be run at very low power,* and because its production process (Silicon on Sapphire) ensured much better protection against cosmic radiation and electrostatic discharges than that of any other processor of the era. Thus, the 1802 is said to be the first radiation-hardened microprocessor.

16-bit Designs

The first multi-chip 16-bit microprocessor was the National Semiconductor IMP-16, introduced in early 1973. An 8-bit version of the chipset was introduced in 1974 as the IMP-8. In 1975, National introduced the first 16-bit single-chip microprocessor, the PACE, which was later followed by an NMOS version, the INS8900. Other early multi-chip 16-bit microprocessors include one used by Digital Equipment Corporation (DEC) in the LSI-11 OEM board set and the packaged PDP 11/03 minicomputer, and the Fairchild Semiconductor MicroFlame 9440, both of which were introduced in the 1975 to 1976 timeframe. The first single-chip 16-bit microprocessor was TI's TMS 9900, which was also compatible with their TI-990 line of minicomputers. The 9900 was used in the TI 990/4 minicomputer, the TI-99/4A home computer, and the TM990 line of OEM microcomputer boards. The chip was packaged in a large ceramic 64-pin DIP package package, while most 8-bit microprocessors such as the Intel 8080 used the more common, smaller, and less expensive plastic 40-pin DIP. A follow-on chip, the TMS 9980, was designed to compete with the Intel 8080, had the full TI 990 16-bit instruction set, used a plastic 40-pin package, moved data 8 bits at a time, but could only address 16KB. A third chip, the TMS 9995, was a new design. The family later expanded to include the 99105 and 99110. The Western Design Center, Inc. (WDC) introduced the CMOS 65816 16-bit upgrade of the WDC CMOS 65C02 in 1984. The 65816 16-bit microprocessor was the core of the Apple IIgs and later the Super Nintendo Entertainment System, making it one of the most popular 16-bit designs of all time. Intel followed a different path, having no minicomputers to emulate, and instead "upsized" their 8080 design into the 16-bit Intel 8086, the first member of the x86 family which powers most modern PC type computers. Intel introduced the 8086 as a cost effective way of porting software from the 8080 lines, and succeeded in winning much business on that premise. The 8088, a version of the 8086 that used an external 8-bit data bus, was the microprocessor in the first IBM PC, the model 5150. Following up their 8086 and 8088, Intel released the 80186, 80286 and, in 1985, the 32-bit 80386, cementing their PC market dominance with the processor family's backwards compatibility. The integrated microprocessor memory management unit (MMU) was developed by Childs et al. of Intel, and awarded US patent number 4,442,484.

32-bit Designs

16-bit designs were in the market only briefly when full 32-bit implementations started to appear. The most famous of the 32-bit designs is the MC68000, introduced in 1979. The 68K, as it was widely known, had 32-bit registers but used 16-bit internal data paths, and a 16-bit external data bus to reduce pin count, and supported only 24-bit addresses. Motorola generally described it as a 16-bit processor, though it clearly has 32-bit architecture. The combination of high speed, large (16 megabyte) memory space and fairly low costs made it the most popular CPU design of its class. The Apple Lisa and Macintosh designs made use of the 68000, as did a host of other designs in the mid-1980s, including the Atari ST and Commodore Amiga. The world's first single-chip fully-32-bit microprocessor, with 32-bit data paths, 32-bit buses, and 32-bit addresses, was the AT&T Bell Labs BELLMAC-32A, with first samples in 1980, and general production in 1982. After the divestiture of AT&T in 1984, it was renamed the WE 32000 (WE for Western Electric), and had two follow-on generations, the WE 32100 and WE 32200. These microprocessors were used in the AT&T 3B5 and 3B15 minicomputers; in the 3B2, the world's first desktop supermicrocomputer; in the "Companion", the world's first 32-bit laptop computer; and in "Alexander", the world's first book-sized supermicrocomputer, featuring ROM-pack memory cartridges similar to today's gaming consoles. All these systems ran the UNIX System V operating system.

Intel's first 32-bit microprocessor was the iAPX 432, which was introduced in 1981 but was not a commercial success. It had an advanced capability-based object-oriented architecture, but poor performance compared to other competing architectures such as the Motorola 68000. Motorola's success with the 68000 led to the MC68010, which added virtual memory support. The MC68020, introduced in 1985 added full 32-bit data and address busses. The 68020 became hugely popular in the Unix supermicrocomputer market, and many small companies (e.g., Altos, Charles River Data Systems) produced desktop-size systems. Following this with the MC68030, which added the MMU into the chip, the 68K family became *the* processor for everything that wasn't running DOS. The continued success led to the MC68040, which included an FPU for better math performance. A 68050 failed to achieve its performance goals and was not released, and the follow-up MC68060 was released into a market saturated by much faster RISC designs. The 68K family faded from the desktop in the early 1990s.

Other large companies designed the 68020 and follow-ons into embedded equipment. At one point, there were more 68020s in embedded equipment than there were Intel

Pentiums in PCs (See this webpage for this embedded usage information). The ColdFire processor cores are derivatives of the venerable 68020. During this time (early to mid 1980s), National Semiconductor introduced a very similar 16-bit pinout, 32-bit internal microprocessor called the NS 16032 (later renamed 32016), the full 32-bit version named the NS 32032, and a line of 32-bit industrial OEM microcomputers. By the mid-1980s, Sequent introduced the first symmetric multiprocessor (SMP) server-class computer using the NS 32032. This was one of the design's few wins, and it disappeared in the late 1980s. The MIPS R2000 (1984) and R3000 (1989) were highly successful 32-bit RISC microprocessors. They were used in high-end workstations and servers by SGI, among others. Other designs included the interesting Zilog Z8000, which arrived too late to market to stand a chance and disappeared quickly. In the late 1980s, "microprocessor wars" started killing off some of the microprocessors. Apparently, with only one major design win, Sequent, the NS 32032 just faded out of existence, and Sequent switched to Intel microprocessors. From 1985 to 2003, the 32-bit x86 architectures became increasingly dominant in desktop, laptop, and server markets, and these microprocessors became faster and more capable. Intel had licensed early versions of the architecture to other companies, but declined to license the Pentium, so AMD and Cyrix built later versions of the architecture based on their own designs.During this span, these processors increased in complexity (transistor count) and capability (instructions/second) by at least a factor of 1000.

64-bit Microchips on the Desktop

While 64-bit microprocessor designs have been in use in several markets since the early 1990s, the early 2000s have seen the introduction of 64-bit microchips targeted at the PC market. With AMD's introduction of the first 64-bit IA-32 backwards-compatible architecture, AMD64, in September 2003, followed by Intel's own x86-64 chips, the 64-bit desktop era began. Both processors can run 32-bit legacy apps as well as the new 64-bit software. With 64-bit Windows XP, Linux and Mac OS X (to a certain extent) that run 64-bit native, the software too is geared to utilise the full power of such processors. The move to 64 bits is more than just an increase in register size from the IA-32 as it also doubles the number of general-purpose registers for the aging CISC designs. The move to 64 bits by PowerPC processors had been intended since the processors' design in the early 90s and was not a major cause of incompatibility. Existing integer registers are extended as are all related data pathways, but, as was the case with IA-32, both floating point and vector units had been operating at or above 64 bits for several years. Unlike what happened with IA-32 was extended to x86-64, no new general purpose registers were added in 64-bit PowerPC, so any performance gained when using the 64-bit mode for applications making no use of the larger address space is minimal.

Multi-core Processors

A different approach to improving a computer's performance is to add extra processors, as in symmetric multiprocessing designs which have been popular in servers and workstations since the early 1990's. Keeping up with Moore's Law is becoming increasingly challenging as chip-making technologies approach the physical limits of the technology. In response, the microprocessor manufacturers look for other ways to improve performance, in order to hold on to the momentum of constant upgrades in the market. A multi-core processor is simply a single chip containing more than one microprocessor core, effectively multiplying the potential performance with the number of cores (as long as the operating system and software is designed to take advantage of more than one processor). Some components, such as bus interface and second level cache, may be shared between cores. Because the cores are physically very close they interface at much faster clock speeds compared to discrete multiprocessor systems, improving overall system performance. In 2005, the first mass-market dual-core processors were announced and as of 2006 dual-core processors are widely used in high-end servers and workstations while quad-core processors for servers are beginning to become available.

RISC

In the mid-1980s to early-1990s, a crop of new high-performance RISC (reduced instruction set computer) microprocessors appeared, which were initially used in special purpose machines and Unix workstations, but have since become almost universal in all roles except the Intel-standard desktop.

The first commercial design was released by MIPS Technologies, the 32-bit R2000 (the R1000 was not released). The R3000 made the design truly practical, and the R4000 introduced the world's first 64-bit design. Competing projects would result in the IBM POWER and Sun SPARC systems, respectively. Soon every major vendor was releasing a RISC design, including the AT&T CRISP, AMD 29000, Intel i860 and Intel i960, Motorola 88000, DEC Alpha and the HP-PA. Market forces have "weeded out" many of these designs, leaving the PowerPC as the main desktop RISC processor, with the SPARC being used in Sun designs only. MIPS continues to supply some SGI systems, but is primarily used as an embedded design, notably in Cisco routers. The rest of the original crop of designs have either disappeared, or are about to. Other companies have attacked niches in the market, notably ARM, originally intended for home computer use but since focussed at the embedded processor market. Today RISC designs based on the MIPS, ARM or PowerPC core power the vast majority of computing devices. As of 2006, several 64-bit architectures are still produced. These include x86-64, MIPS, SPARC, Power Architecture, and IA-64.

Special-purpose Microprocessors

Though the term "microprocessor" has traditionally referred to a single- or multi-chip CPU or System-on-a-chip (SoC), several types of specialized processing devices have followed from the technology. The most common examples are microcontrollers, Digital Signal Processors (DSP) and Graphics processing units (GPU). Many examples of these are either not programmable, or have limited programming facilities. For example, in general GPUs through the 1990s were mostly non-programmable and have only recently gained limited facilities like programmable vertex shaders. There is no universal consensus on what defines a "microprocessor", but it is usually safe to assume that the term refers to a general-purpose CPU of some sort and not a special-purpose processor unless specifically noted. The RCA 1802 had what is called a *static design*, meaning that the clock frequency could be made arbitrarily low, even to 0 Hz, a total stop condition. This let the Voyager/Viking/Galileo spacecraft use minimum electric power for long uneventful stretches of a voyage. Timers and/or sensors would awaken/speed up the processor in time for important tasks, such as navigation updates, attitude control, data acquisition, and radio communication.

Market Statistics

In 2003, about $44 billion (USD) worth of microprocessors were manufactured and sold. Although about half of that money was spent on CPUs used in desktop or laptop personal computers, those count for only about 0.2% of all CPUs sold. About 55% of all CPUs sold in the world are 8-bit microcontrollers. Over 2 billion 8-bit microcontrollers were sold in 1997. Less than 10% of all the CPUs sold in the world are 32-bit or more. Of all the 32-bit CPUs sold, about 2% are used in desktop or laptop personal computers, the rest are sold in household appliances such as toasters, microwaves, vacuum cleaners and televisions. "Taken as a whole, the average price for a microprocessor, microcontroller, or DSP is just over $6."

Common Microprocessors and Architectures

- AMD K5, K6, K6-2, K6-III, Duron, Athlon, Athlon XP, Athlon MP, Athlon XP-M (IA-32 32-bit Intel x86 architecture)
- AMD Athlon 64, Athlon 64 FX, Athlon 64 X2, Opteron, Sempron, Turion 64 (AMD64 architecture)
- ARM family, StrongARM, Intel PXA2xx
- Altera Nios, Nios II
- Atmel AVR architecture (purely microcontrollers)
- EISC

- RCA 1802 (aka RCA COSMAC, CDP1802)
- Cyrix M1, M2 (IA-32 32-bit Intel x86 architecture)
- DEC Alpha
- Intel 4004, 4040
- Intel 8080, 8085, Zilog Z80
- Intel 8086, 8088, 80186, 80188 (16-bit real mode-only Intel x86 architecture)
- Intel 80286 (16-bit real mode and protected mode Intel x86 architecture)
- Intel 80386, 80486, Pentium, Pentium Pro, Celeron, Intel Core Duo, Pentium II, Pentium III, Xeon, Pentium 4, Pentium M, Celeron M (IA-32 32-bit Intel x86 architecture; parents of IA-64, with HP PA-RISC)
- Later Xeon, later Pentium 4, Pentium D, Celeron D, Intel Core 2 (EM64T architecture)
- Intel Itanium (IA-64 architecture)
- Intel i860
- Intel i960
- M32R architecture
- MIPS architecture
- Motorola 6800, MOS Technology 6502, Motorola 6809
- Motorola 68000 family, ColdFire
- Motorola 88000 (parents of the PowerPC family, with POWER)
- NexGen Nx586 (IA-32 32-bit Intel x86 architecture)
- IBM POWER (parents of the PowerPC family, with 88000)
- NSC 320xx
- OpenCores OpenRISC architecture
- PA-RISC family (HP, parents to the IA-64 architecture, with IA-32)
- PowerPC family, G3, G4, G5
- National Semiconductor SC/MP ("scamp")
- Signetics 2650
- SPARC, UltraSPARC, UltraSPARC II–IV
- SuperH family
- Transmeta Crusoe, Efficeon (VLIW architectures, IA-32 32-bit Intel x86 emulator)

- INMOS Transputer
- VIA's C3,C7,Eden Series (IA-32 32-bit Intel x86 architecture)
- Western Design Center 65xx
- XAP processor from Cambridge Consultants

Microcomponent Assembly and Packaging

Microengineered devices have the potential to be as inexpensive as silicon chips are today. This, however, will only be true when two conditions are met: 1) the fabrication process has a high yield (most of the devices on a wafer function properly, and will continue to do so after packaging); and 2) batch processing techniques are used for as much of the process as possible (i.e. large numbers of devices per silicon wafer, and a large number of wafers are processed at the same time at each fabrication step). When developing microengineered devices, and with complicated microsystems, it is difficult to achieve high yields. However these must be achieved before putting the device into production, with few exceptions. If the device does something that is very important, and cannot be done any other way, then perhaps a low yield and expensive devices can be justified. Assembling complex devices from many microscopic parts and, in particular, packaging these devices so they can be handled and connected to other components / systems, will generally involve handling the devices individually. This can add significantly to the cost of the finished part (tens to hundreds of times the cost of the actual active part of the device — depending on the complexity and requirements of packaging). Consequently the assembly and packaging of devices for commercial manufacture has to be carefully considered.

Digital Signal Processor

A digital signal processor (DSP) is a specialized microprocessor designed specially for digital signal processing, generally in real-time.

Architectural Features of Digital Signal Processors

Digital signal processing can be done on general-purpose microprocessors. However, a digital signal processor contains architectural optimizations to speed up processing. These optimizations are also important to lower costs, heat-emission and power-consumption.

Memory Architecture

- DSPs often use special memory architectures that are able to fetch multiple data and/or instructions at the same time:
- Harvard architecture

- modified von Neumann architecture
- Use of direct memory access
- Memory-address calculation unit

Characteristics of Digital Signal Processors

- Designed for real-time processing
- Optimum performance with streaming data
- Separate programme and data memories (Harvard architecture)
- Special Instructions for SIMD (Single Instruction, Multiple Data) operations
- No hardware support for multitasking
- The ability to act as a direct memory access device if in a host environment
- Processes digital signals converted (using an Analog-to-digital converter (ADC)) from analog signals. Output is then converted back to analog form using a Digital-to-analog converter (DAC)

Programme Flow

- Floating-point unit integrated directly into the data-path.
- Pipelined architecture
- Highly parallel accumulator and multiplier
- Special looping hardware. Low-overhead or Zero-overhead looping capability

Data Operations

- Saturation arithmetic, in which operations that produce overflows will accumulate at the maximum (or minimum) values that the register can hold rather than wrapping around (maximum+1 doesn't overflow to minimum as in many general-purpose CPUs, instead it stays at maximum). Sometimes various sticky bits operation modes are available.
- Fixed-point arithmetic is often used to speed up arithmetic processing.
- Single-cycle operations to increase the benefits of pipelining.

Instruction Sets

- Multiply-accumulate (MAC) operations, which are good for all kinds of matrix operations, such as convolution for filtering, dot product, or even polynomial evaluation (see Horner scheme, also fused multiply-add).
- Instructions to increase parallelism: SIMD, VLIW, superscalar architecture.

- Specialized instructions for modulo addressing in ring buffers and bit-reversed addressing mode for FFT cross-referencing.
- Digital signal processors sometimes use time-stationary encoding to simplify hardware and increase coding efficiency

In 1978, Intel released the 2920 as an "analog signal processor". It had an on-chip ADC/DAC with an internal signal processor, but it didn't have a hardware multiplier and was not successful in the market. In 1979, AMI released the S2811. It was designed as a microprocessor peripheral, and it had to be initialized by the host. The S2811 was likewise not successful in the market. In 1979, Bell Labs introduced the first single chip DSP, the Mac 4 Microprocessor. Then, in 1980 the first stand-alone, complete DSPs — the NEC μPD7720 and AT&T DSP1 — were presented at the IEEE International Solid-State Circuits Conference '80. Both processors were inspired by the research in PSTN telecommunications. The Altamira DX-1 was another early DSP, utilizing a quad integer pipelines with delayed branches and branch prediction. The first DSP produced by Texas Instruments (TI), the TMS32010 presented in 1983, proved to be an even bigger success. It was based on the Harvard architecture, and so had separate instruction and data memory. It already had a special instruction set, with instructions like load-and-accumulate or multiply-and-accumulate. It could work on 16-bit numbers and needed 390ns for a multiply-add operation. TI is now the market leader in general purpose DSPs. Another successful design was the Motorola 56000. About five years later, the second generation of DSPs began to spread. They had 3 memories for storing two operands simultaneously and included hardware to accelerate tight loops, they also had an addressing unit capable of loop-addressing. Some of them operated on 24-bit variables and a typical model only required about 21ns for a MAC (multiply-accumulate). Members of this generation were for example the AT&T DSP16A or the Motorola DSP56001. The main improvement in the third generation was the appearance of application-specific units and instructions in the data path, or sometimes as coprocessors. These units allowed direct hardware acceleration of very specific but complex mathematical problems, like the fourier-transform or matrix operations. Some chips, like the Motorola MC68356, even included more than one processor cores to work in parallel. Other DSPs from 1995 are the TI TMS320C541 or the TMS 320C80. The fourth generation is best characterized by the changes in the instruction set and the instruction encoding/decoding. SIMD and MMX extensions were added, VLIW and the superscalar architecture appeared. As always, the clock-speeds have increased, a 3ns MAC became now possible.

DSPs in 2007

Today's signal processors provide a very higher performance thanks to both technological and architectural advancements like lower design rules, fast-access two-level cache, (E)DMA circuit and a wider bussystem. Of course, not all DSPs provide the same speed and many-many kind of signal processors exist, each one of them being better suited for a specific task, ranging in price from about 1,50 to 300 dollars. A Texas Instruments C6000 series DSP clocks at 1Ghz and implements separate instruction and data caches as well as a 8Mbyte 2nd level cache, and its I/O speed is rapid thanks to its 64 EDMA channels. The top models are capable of even 8000 MIPS (million instructions per second), use VLIW encoding, perform eight operations per clock-cycle and are compatible with a broad range of external peripherals and various buses (PCI/serial/etc). Another big signal processor manufacturer today is Analog Devices. The company provides a broad range of DSPs, but its main portfolio is multimedia processors, such as codecs, filters and digital-analog converters. Its SHARC-based processors range in performance from 66Mhz/198MFLOPS (million floating-point operations per second) to 400Mhz/2400MFLOPS. Some models even support multiple multipliers and ALUs, SIMD instructions and audioprocessing-specific components and peripherials.Another product of the company is the Blackfin family of embedded digital signal processors, with models like the ADSP-BF531 to ADSP-BF536. These processors combine the features of a DSP with those of a general use processor. As a result, these processors can run simple operating systems like ìCLinux, velOSity and Nucleus while operating relatively efficiently on real-time data. Most DSPs use fixed-point arithmetic, because in real world signal processing, the additional range provided by floating point is not needed, and there is a large speed benefit; however, floating point DSPs are common for scientific and other applications where additional range or precision may be required. General purpose CPU's have ideas and influences from digital signal processors with extensions such as the MMX extensions in the Intel IA-32 architecture instruction set (ISA). Generally, DSPs are dedicated integrated circuits, however DSP functionality can also be realised using Field Programmable Gate Array chips.

Microcontroller

A microcontroller (or MCU) is a computer-on-a-chip. It is a type of microprocessor emphasizing self-sufficiency and cost-effectiveness, in contrast to a general-purpose microprocessor (the kind used in a PC). A typical microcontroller contains all the memory and interfaces needed for a simple application, whereas a general purpose microprocessor requires additional chips to provide these functions. A microcontroller is a single integrated circuit, commonly with the following features:

- central processing unit — ranging from small and simple 4-bit processors to sophisticated 32- or 64-bit processors
- input/output interfaces such as serial ports
- peripherals such as timers and watchdog circuits and signal conversion circuits.

- RAM for data storage
- ROM, EPROM, EEPROM or Flash memory for programme storage
- clock generator — often an oscillator for a quartz timing crystal, resonator or RC circuit

This integration drastically reduces the number of chips and the amount of wiring and PCB space that would be needed to produce equivalent systems using separate chips and have proved to be highly popular in embedded systems since their introduction in the 1970s.

Microcontrollers take the largest share of sales in the wider microprocessor market. Over 50% are "simple" controllers, and another 20% are more specialized digital signal processors (DSPs). A typical home in a developed country is likely to have only one or two general-purpose microprocessors but somewhere between one and two dozen microcontrollers. A typical mid range vehicle has as many as 50 or more microcontrollers. They can also be found in almost any electrical device: washing machines, microwave ovens, telephones etc.

Manufacturers have often produced special versions of their microcontrollers in order to help the hardware and software development of the target system. These have included EPROM versions that have a "window" on the top of the device through which programme memory can be erased by ultra violet light, ready for reprogramming after a programming ("burn") and test cycle. Other versions may be available where the ROM is accessed as an external device rather than as internal memory. A simple EPROM programr, rather than a more complex and expensive microcontroller programr, may then be used, however there is a potential loss of functionality through pin outs being tied up with external memory addressing rather than for general input/output. These kind of devices usually carry a cost up in part prices but if the target production quantities are small, certainly in the case of a hobbyist, they can be the most economical option compared with the set up charges involved in mask programmed devices. A more rarely encountered development microcontroller is the "piggy back" version. This device has no internal ROM memory; instead pin outs on the top of the microcontroller form a socket into which a standard EPROM programme memory device may be installed. The benefit of this approach is the release of microcontroller pins for input and output use rather than programme memory. These kinds of device are normally expensive and are impractical for anything but the development phase of a project.

Development Platforms for Hobbyists

For almost every bare microcontroller manufacturer, there are a dozen little companies repacking its products into a more hobbyist friendly package. Their product is often an MCU preloaded with a BASIC or similar interpreter, soldered onto a Dual Inline Pin board along with a power regulator and other goodies. PICmicros seem to be very popular here, possibly due to good static protection. More powerful examples (e.g. faster execution, more RAM and code space) seem to be based on Atmel AVR or Hitachi chips and now ARM.

Description

In addition to the key features listed above, most microcontrollers today take further advantage of not needing external pins for memory buses. They can afford to use a Harvard architecture: separate memory buses for instructions and data, allowing accesses to take place concurrently. Microcontrollers also usually have a variety of input/output interfaces. Serial I/O (UARTs) are very common, and many include analog-to-digital converters, timers, or specialized serial communications interfaces like I²C, Serial Peripheral Interface and Controller Area Network. Originally, microcontrollers were only programmed in assembly language, or later in C code. Recent microcontrollers integrated with on-chip debug circuitry accessed by In-circuit emulator via JTAG enables a programr to debug the software of an embedded system with a debugger. Some microcontrollers have begun to include a built-in high-level programming language interpreter for greater ease of use. The Intel 8052 and Zilog Z8 were available with BASIC very early on, and BASIC is more recently used in the popular BASIC Stamp MCUs. Microcontrollers trade speed and flexibility against ease of equipment design and low cost. Manufacturers have to balance the need to minimize the chip size against additional functionality. Finally, it must be mentioned that microcontroller architectures are available from many different vendors in so many varieties that they could rightly belong to a category of their own. Chief among these are the 8051, Z80 and ARM derivatives.

Arduino

Arduino is an open-source physical computing platform based on a simple i/o board and a development environment that implements the Processing/Wiring language. Arduino can be used to develop stand-alone interactive objects or can be connected to software on your computer (e.g. Flash, Processing, MaxMSP). The boards can be assembled by hand or purchased preassembled; the open-source IDE can be downloaded for free. Arduino uses an Atmega-8 microcontroller from Atmel's Atmel_AVR series.

PICAXE

This PICAXE range of controllers from Revolution Education Limited are based upon Microchip PICmicro's programmed with a BASIC interpreter. Using internal EEPROM or Flash to store the user's programme they deliver a single-chip solution and are quite inexpensive. A PICAXE programr is simply a serial plug plus two resistors. Complete development software, comprehensive documentation and application notes are all available free of charge. The BASIC-like programming language is almost identical to that used by Parallax's Basic Stamp 1 (BS1) but has been enhanced to support on-chip hardware and additional functionality. In common with the BS1 programming language, the PICAXE has support only for a limited number of variables, but allows access to internal RAM for storage which helps overcome that limitation. The 5.0.X versions of the Visual IDE (the Programming Editor) introduced 'enhanced compilers' which support block-structured programming constructs plus conditional compilation and other directives. Initially targeted at the UK educational sector, use of the PICAXE has spread to hobbyists, semi-professionals and it can also be found inside commercial products. With its user base in many countries, the PICAXE has steadily gained a good international reputation.

A-WIT Technologies, Inc.

A-WIT Technologies, Inc. has a microcontroller module named the C STAMP, along with support boards, kits, and software tools and infrastructure. The C STAMP is designed around a PIC microcontroller, and is programmed in a very user friendly subset of the standard C language called WC that is easy and powerful, because it relies on A-WIT's supplied software infrastructure. This microcontroller module is very affordable, and it has 48 pins, 35 KB of memory, and runs at 40 MHz. The C STAMP also has a vast array of accessories and components, which are supported by A-WIT's software interfaces that enables seamless connectivity. This, in turn, enhances the ease of complete system development.

Platforms from Parallax, Inc.

BASIC Stamp by Parallax, is the 'big name' in BASIC microcontrollers. They are Microchip PIC micros programmed with an interpreter that processes the programme stored in an external EEPROM. Several different modules are available of varying processing speeds, RAM, and EEPROM sizes. Most popular is the original BASIC Stamp 2 module. The BASIC Stamp is used by Parallax as a platform for introductory programming and robotic kits. SX-Key, Parallax's development tool for the SX line of microcontrollers, supporting every SX chip commercially available. Using free SX-Key software (Assembly language), or the SX/B Compiler (BASIC-style language) from Parallax, the SX-Key programming tool can programme SX chips in-system and perform in-circuit source-level debugging. Propeller, A multi-core microcontroller developed by Parallax, Inc. It features eight 32bit cores and 32 I/O pins in the currently released version. Each core operates independently at 80Mhz, it is programmed in a language named SPIN(tm) which was developed by Parallax to support this unique micro.

Coridium ARMexpress

ARMexpress is the first of a new family of DIP-24 (stamp-sized) controllers that combine a 60 MHz ARM CPU with a builtin BASIC compiler to achieve new levels of performance in this form factor. This combination makes this simple to use but very fast controller a good choice for the prototype builder or system integrator. 40K of code and 40K of data are available to the user, and code speed rivals that of programmes written in C. The dialect of BASIC conforms more to Visual BASIC, but has hardware extensions like PBASIC.

Comfile Technology Inc.

Comfile Technology Inc. produces a series of microcontrollers branded as CUBLOC and CuTOUCH, using the Atmel ATmega128 processor. They are very price competitive, being aimed at industrial applications, and include some nice features such as Ladder Logic in addition to BASIC, a huge 80Kbyte programme memory, and hardware pulse width modulation. Their focus is on developing industrial controllers which are fast, easy-to-use, and versatile. Comfile Technology's CuTOUCH is a visual Touch-screen controller that can be programmed in BASIC and Ladder Logic. This product is the first of its kind in the world yet.

ZX-24, ZX-40, ZX-44

The ZX series MCUs are based on the Atmel ATmega32 and ATmega644 processors. The devices run a field-upgradable Virtual Machine that features built-in multi-tasking, 32-bit floating point math and 1.5K to 3.5K of RAM for user's programmes. Multi-tasking facilitates a more structured approach to coding for interface devices that require prompt service, e.g. serial devices, infrared remotes, etc. The programming language for the ZX series is ZBasic, a modern dialect of Basic modeled after Microsoft's Visual Basic. The biggest improvement over the typical MCU Basic dialect is the availability of parameterized subroutines/functions that support local variables. Strong type checking is another improvement that aids in writing correct programmes more quickly. User-defined types (structures) are also supported along with aliases, based variables, sub-byte data types (Bit and Nibble) and other advanced capabilities.

Microtechnology

Microtechnology is technology with features near one micrometre (one millionth of a metre, or 10^{-6} metre, or 1ìm). In the 1960s, scientists learned that by arraying large numbers of microscopic transistors on a single chip, microelectronic circuits could be built that dramatically improved performance, functionality, and reliability, all while reducing cost and decreasing volume. This development led to the Information Revolution. More recently, scientists have learned that not only electrical devices, but also mechanical devices, may be miniaturized and batch-fabricated, promising the same benefits to the mechanical world as integrated circuit technology has given to the electrical world. While electronics now provide the 'brains' for today's advanced systems and products, micromechanical devices can provide the sensors and actuators — the eyes and ears, hands and feet — which interface to the outside world. Today, micromechanical devices are the key components in a wide range of products such as automobile airbags, ink-jet printers, blood pressure monitors, and projection display systems. It seems clear that in the not-too-distant future these devices will be as pervasive as electronics.

Micro Electro Mechanical Systems

The term MEMS, for Micro Electro Mechanical Systems, was coined in the 1980's to describe new, sophisticated mechanical systems on a chip, such as micro electric motors, resonators, gears, and so on. Today, the term MEMS in practice is used to refer to any microscopic device with a mechanical function, which can be fabricated in a batch process (for example, an array of microscopic gears fabricated on a microchip would be considered a MEMS device but a tiny laser-machined stent or watch component would not). In Europe, the term MST for Micro System Technology is preferred, and in Japan MEMS are simply referred to as Micromachines. The distinctions in these terms are relatively minor and are often used interchangeably. Though MEMS processes are generally classified into a number of categories – such as surface machining, bulk machining, LIGA, and EFAB – there are indeed thousands of different MEMS processes. Some produce fairly simple geometries, while others offer more complex 3-D geometries and more versatility. A company making accelerometers for airbags would need a completely different design and process to produce an accelerometer for inertial navigation. Changing from an accelerometer to another inertial device such as a gyroscope requires an even greater change in design and process, and most likely a completely different fabrication facility and engineering team. MEMS technology has generated a tremendous amount of excitement, due to the vast range of important applications where MEMS can offer previously unattainable performance and reliability standards. In an age where everything must be smaller, faster, and cheaper, MEMS offers a compelling solution. MEMS have already had a profound impact on certain applications such as automotive sensors and inkjet printers. The emerging MEMS industry is already a multi-billion dollar market. It is expected to grow rapidly and become one of the major industries of the 21st century. Cahners In-Stat Group has projected sales of MEMS to reach $12B by 2005. The European NEXUS group projects even larger revenues, using a more inclusive definition of MEMS. Microtechnology is often constructed using photolithography. Lightwaves are focused through a mask onto a surface. They solidify a chemical film. The soft, unexposed parts of the film are washed away. Then acid etches away the material not protected. Microtechnology's most famous success is the integrated circuit. It has also been used to construct micromachinery.

Micro-Technology Devices

Technology on the micro-scale has its origins in the development of microelectronic circuits in the 1960s. More recently, engineers have realised that not only electronic devices, but also mechanical devices can be miniaturized, promising the same benefits to the mechanical world as integrated circuit technology has given to the electrical world. Today, micromechanical devices are key components in a wide range of products such as vehicle airbags, ink-jet printers, blood pressure monitors and projection display systems. Tecan draws upon its 30-year experience of manufacturing precision parts to offer a solution to the most demanding micromechanical requirements – from technology development through to rapid prototyping and ultimately volume production. Using advanced precision photolithography and Photo Electro Forming techniques, Tecan produces micro parts and components with feature sizes and tolerance measured in microns, nanometers and even angstroms.

What is Nanotechnology?

Nanotechnology is the design, characterization, production and application of structures, devices and systems by controlling the shape and size at the nanometer scale. A nanometer (nm) is one billionth of a meter. For comparison, a single human hair is about 80,000 nm wide, a red blood cell is approximately 7,000 nm wide and a water molecule is almost 0.3nm across. The nanoscale is defined from 100 nm down to the size of atoms (approximately 0.2 nm). The understanding and control of matter at the nanoscale is of interest because a material's properties at the nanoscale can be very different from those at a larger scale. The unique physical, chemical, and biological properties of materials at the nanoscale enable novel applications and functions with the potential to promote enormous societal and economic benefits. Some current innovative applications of nanotechnology include the use of nanomaterials in liquid filtration and water purification, as catalysts in petroleum refining and catalytic converters, and in nanoscale biological imaging. In the near future of two to five years, nanotechnology will be integrated into advanced drug delivery systems, real-time medical diagnostic tools, sensors for airborne chemicals or other toxins, and photovoltaics (solar cells), fuel cells and portable power to provide inexpensive, clean energy. The properties of materials at the nanoscale are attributed to two main factors. First, nanomaterials have a relatively larger surface area when compared to the same mass of material produced in a larger form. This can make materials more chemically reactive (in some cases materials that are inert in their larger form are reactive when produced in their nanoscale form), and affect their strength or electrical properties. Second, quantum effects can begin to dominate the behavior of matter at the nanoscale — particularly at the lower end — affecting the optical, electrical and magnetic behavior of materials. The photo on the left shows an example of quantum dots. Particles of inorganic semiconducting crystals with nanometer scale dimensions ("quantum dots") exhibit size-dependent optical properties. Companies are investigating quantum dots to develop composite materials such as those used as light-emitting diodes (LEDs) where the color is determined by the size of the quantum dots. The superior optical properties of quantum dots are being studied for use in diode lasers, optical amplifiers and switches, biological sensors, and solid-state quantum computing.

Current Applications in Indstry

Although nanotechnology is considered by most science and technology experts as an oncoming industrial revolution that will have enormous societal and economic impact, the applied use of nanotechnology and nanomaterials is limited currently. Companies have introduced nanotechnology in several consumer products such as cosmetics, food and clothing. The Woodrow Wilson International Center for Scholars developed searchable databases listing the uses of nanotechnology in consumer products and the research being conducted on agrifood (e.g., food, agriculture, forestry, and agroecosystems). Other applications include new detection systems, filtration media, and electronic chip design.

Why is DTSC is Interest in Nanotechnology

The Technology Development staff sees nanotechnology as the new "plastic", i.e., materials that will show up in many industrial application and consumer products. Materials and devices designed at the nanoscale level are being used or considered for use in applications as diverse as cancer treatment to scratch-resistance automotive coatings. Due to the unique properties of nanomaterials, staff began to research whether the current hazardous waste classification, handling and disposal procedures were appropriate for wastestreams associated with nanomaterials. In December 2004, a Blue Ribbon Task Force on Nanotechnology (BRTFN) was formed to address the major areas of concern in promoting nanotechnology research and development, commercialization, facilities and infrastructure, policies and ethics, and education. The members of the Task Force came from many backgrounds including academia, government, corporate California, start-up firms and consulting firms, non-profit groups and industry associations. Several recommendations called for CalEPA involvement in understanding and communicating the ethical, environmental and societal implications of nanotechnology.

(i) The Royal Society & the Royal Academy of Engineering, Nanoscience and nanotechnologies: opportunities and uncertainties, July 2004

(ii) Blue Ribbon Task Force on Nanotechnology, Thinking Big About Thinking Small An Action Agenda for California, December 2005

(iii) U.S. EPA, Draft Nanotechnology White Paper, December 2005

Remediation

(i) U.S. Navy, Nanoscale Zerovalent Iron Tool, January 2005

(ii) U.S. EPA, U.S. EPA Workshop on Nanotechnology for Site Remediation, October 2005

(iii) Nanoforum, Nano and the Environment, December 2005

(iv) U.S. EPA, Nanotechnology and OSWER: New Opportunities and Challenges, July 2006 NEW!

Research

(i) U.S. EPA, Nanotechnology: Research Projects

(ii) Environmental Health and Safety/Ecotoxicology

(iii) International Council on Nanotechnology, EHS Database web page

(iv) National Institute of Occupational Health, Nanoparticle Information Library (NIL)

(v) ETC Group, From Genomes to Atoms the Big Down Atomtech: Technologies Converging at the Nano-Scale, January 2003

(vi) Bell, Trudy E., Reporting Risk Assessment of Nanotechnology: A reporter's guide to sources and research issues, Prepublication Issue, June 2006

(vii) Oberdorster, Eva, Manufactured Nanomaterials (Fullerenes, C60) Induce Oxidative Stress In Brain Of Juvenile Largemouth Bass, Environ. Health Perspect. 112:1058 — 1062 (2004) NEW!

Nanotechnology today is an emerging set of tools, techniques and unique applications involving the structure and composition of materials on a nanoscale. IBM is one of the true pioneers in nanotechnology. Among IBM's many nanotechnology milestones, its scientists have invented the scanning tunneling microscope (STM) capable of imaging individual atoms, they have positioned atoms one-by-one for the first time, and incorporated sub-nanometer material layers into commercially mass-produced hard disk drive recording heads and magnetic disk coatings. IBM's current nanotechnology research aims to devise new atomic- and moleculear-scale structures and methods for enhancing information technologies, as well as discover and understand their scientific foundations. In particular, carbon nanotubes and scanning probes derived from the atomic force microscope and the STM show particular promise in enabling dramatically improved circuits and data storage devices.

When it comes to matter, size matters. The properties of materials that we notice — color, hardness, conductivity, and so on — all depend on the nature and structure of the constituent atoms and molecules. With increasing ability to design and build on an atomic and molecular scale — a reasonable definition of "nanotechnology" — we are becoming better and better at developing materials with entirely new properties. Those materials, in turn, become the building blocks for more complex systems and entirely new products. But when an emerging technology is the subject of as much hype as nanotech, it's easy to tune out and stop listening. That would be a big mistake. There are solid reasons to expect significant long-term developments in what the National Science Foundation estimates to be a trillion-dollar plus industry over the next 10 to 15 years. Why nanotech? With it, today's supercomputer could become tomorrow's wristwatch personal assistant. Buildings and machines could signal when they need maintenance, and perhaps repair themselves. Our clothing could monitor our health and alert us to environmental hazards. All of these wonders, and many more, are scientifically possible. The difficulty comes in figuring out how and when these things will happen.

The term "nanotechnology" has evolved over the years via terminology drift to mean "anything smaller than microtechnology," such as nano powders, and other things that are nanoscale in size, but not referring to *mechanisms* that have been purposefully built from nanoscale components. See our "Current Uses" page for examples. This evolved version of the term is more properly labeled "nanoscale bulk technology," while the original meaning is now more properly labeled "molecular nanotechnology" (MNT), or "nanoscale engineering," or "molecular mechanics," or "molecular machine systems," or "molecular manufacturing." Recently, the Foresight Institute has suggested an alternate term to represent the original meaning of nanotechnology: zettatechnology. At the most basic *technical* level, MNT is building, with intent and design, and molecule by molecule, these two things: 1) incredibly advanced and extremely capable nano-scale and micro-scale machines and computers, and 2) ordinary size objects, using other incredibly small machines called assemblers or fabricators (found inside nanofactories). In a nutshell, by taking advantage of quantum-level properties, MNT allows for unprecedented control of the material world, at the nanoscale, providing the means by which systems and materials can be built with exacting specifications and characteristics. Or, as Dr. K. Eric Drexler puts it "large-scale mechanosynthesis based on positional control of chemically reactive molecules." MNT represents the state of the art in advances in biology, chemistry, physics, engineering, computer science and mathematics. The major research objectives in MNT are the design, modeling, and fabrication of molecular machines and molecular devices. The emergence of MNT — both infant and mature — has numerous social, legal, cultural, ethical, religious, philosophical and political implications. At the most basic social level, MNT is going to be responsible for massive changes in the way we live, the way we interact with one another and our environment, and the things we are capable of doing. For more information, read What is Nanotechnology? by the Center for Responsible Nanotechnology, this explanation from CORDIS for a "child's eye view" of nanotechnology, and What is Nanotechnology? by Tim Harper. See also Introduction to Nanoscience by Prof. Vicki Colvin, Rice University Department of Chemistry and Center for Nanoscale Science and Technology. See also What is Nanotechnology? from LANL, and Dr. Mihail Roco's presentation titled National Nanotechnology Initiative Overview from the 3rd Integrated Nanosystems Conference in Pasadena, California, held on September 22nd, 2004. At the end of the day, it is not the meaning behind the terms that is important, it is the fact that all the many definitions suggest that we have been and are on a rapidly accelerating technological rollercoaster, and rapid change is the track it rides. Once MNT develops to the stage where we've built the two most essential machines — called the Universal Assembler and the Nanocomputer — *everything has a near-term possibility of significant change.* "A key ingredient in understanding nanotechnology is realising precisely what it is and what it isn't.... we are talking about research and development in the length scale of.1 nanometers to 100 nanometers to create unique structures, devices, and systems. In many instances the actual structures, devices, and systems will be much larger, but they will be classified as nanotechnology due to the fact that they will either be created at the nanoscale or nanotechnology will enable them to perform new and/or improved functions. Many materials, once they are individually reduced below 100 nanometers, begin displaying a set of unique characteristics based on quantum mechanical forces that are exhibited at the level. Due to these quantum mechanical effects, materials may become more conducting, be able to transfer heat better, or have modified mechanical properties." From "The Next Big Thing Is Really Small: How Nanotechnology Will Change the Future of Your Business." By Jack Uldrich & Deb Newberry. March 2003 Read our review

"Imagine a world in which microscopic procreating robots are sent into the human body with the mission of detecting cancer cells, disassembling them, and sending them out into the bloodstream as waste products. Then imagine similar robots in the hands of a sinister force that decides to turn an entire continent into gray dust. Science fiction or reality?" From Souls, Slavery, and Survival in the Molenotech Age: An Alien's Version by Lin Sten. Comprised of three independent and interdependent movements: Dry, Wet, and Computational Nanotechnology; MNT represents the state of the art in advances in biology, chemistry, physics, engineering, computer science and mathematics. The major *research objectives* in MNT are the design, modeling, and fabrication of molecular machines and molecular devices. The emergence of MNT—both infant and mature—has numerous social, legal, cultural, ethical, religious, philosophical and political implications. Much as the invention of electricity and transistors were enabling technologies, so too is Nanotechnology (more precisely, nanoscale technologies) enabling — it will enable us to do radical new things in virtually every technological and scientific arena. It will also change things in unpredictable and unanticipated ways. Having learned lessons from their experiences with other revolutionary technologies, scientists (technologists and social scientists) are collaborating in examinating the implications of the developments that are beginning to take place, in an effort to both smooth the transitions, and to head off potential negative consequences (such as Gray Goo or government intervention in research and development).

- Nearly free consumer products
- PC's billions of times faster then today
- Safe and affordable space travel
- Virtual end to illness, aging, death
- No more pollution and automatic cleanup of existing pollution
- End of famine and starvation
- Superior education for every child on Earth
- Reintroduction of many extinct plants and animals
- Terraforming Earth and the Solar System

The following devices and capabilities appear to be both physically possible and practically realisable:

- Programmable positioning of reactive molecules with ~0.1 nm precisio
- Mechanosynthesis at $>10^6$operations/device · second
- Mechanosynthetic assembly of 1 kg objects in $<10^4$
- Nanomechanical systems operating at $\sim 10^9$ Hz
- Logic gates that occupy $\sim 10^{-26}$ m^3 ($\sim 10^{-8}$ m^3)
- Logic gates that switch in ~0.1 ns and dissipate $<10^{-21}$ J
- Computers that perform 10^{16} instructions per second per watt
- Cooling of cubic-centimeter, $\sim 10^5$ W systems at 300 K
- Compact 10^{15} MIPS parallel computing systems
- Mechanochemical power conversion at $>10^9$ W/m^3
- Electromechanical power conversion at $>10^{15}$ W/m^3
- Macroscopic components with tensile strengths $>5\times10^{10}$ Pa
- Production systems that can double capital stocks in $<10^4$ s

Of these capabilities, several are qualitatively novel and others improve on present engineering practice by one or more orders of magnitude. Each is an aspect or a consequence of molecular manufacturing. While there is great debate as to when MNT will start to seriously impact us, best guesses range from around 2015 to as late as 2025. Certainly within the lifetime of most everyone currently under 60 years of age. Along with the development of Nanotechnology comes the necessity to develop reasonable guidelines, procedures, and laws in order to protect humanity from misuse of the technologies. With that in mind, we bring you: Assumptions, principles, and some specific recommendations intended to provide a basis for responsible development of molecular nanotechnology.

Development Principles

1. Artificial replicators must not be capable of replication in a natural, uncontrolled environment.
2. Evolution within the context of a self-replicating manufacturing system is discouraged.
3. Any replicated information should be error free.
4. MNT device designs should specifically limit proliferation and provide traceability of any replicating systems.
5. Developers should attempt to consider systematically the environmental consequences of the technology, and to limit these consequences to intended effects. This requires significant research on environmental models, risk management, as well as the theory, mechanisms, and experimental designs for built-in safeguard systems.
6. Industry self-regulation should be designed in whenever possible. Economic incentives could be provided through discounts on insurance policies for MNT development organisations that certify Guidelines compliance. Willingness to provide self-regulation should be one condition for access to advanced forms of the technology.
7. Distribution of molecular manufacturing *development* capability should be restricted, whenever possible, to responsible actors that have agreed to use the Guidelines. No such restriction need apply to end products of the development process that satisfy the Guidelines.

Specific Design Guidelines

1. Any self-replicating device which has sufficient onboard information to describe its own manufacture should encrypt it such that any replication error will randomize its blueprint.
2. Encrypted MNT device instruction sets should be utilized to discourage irresponsible proliferation and piracy.

3. Mutation (autonomous and otherwise) outside of sealed laboratory conditions, should be discouraged.
4. Replication systems should generate audit trails.
5. MNT device designs should incorporate provisions for built-in safety mechanisms, such as: 1) absolute dependence on a single artificial fuel source or artificial "vitamins" that don't exist in any natural environment; 2) making devices that are dependent on broadcast transmissions for replication or in some cases operation; 3) routing control signal paths throughout a device, so that subassemblies do not function independently; 4) programming termination dates into devices, and 5) other innovations in laboratory or device safety technology developed specifically to address the potential dangers of MNT.
6. MNT developers should adopt systematic security measures to avoid unplanned distribution of their designs and technical capabilities.

There is one very important and simple method to insure the safe development of MNT: become involved in the debate. Regardles of your opionion, you need to both express it and stand behind it, in forums both public and virtual. Educate yourself — we will provide the tools, right here. And if there is ever anything you need regarding information, just ask.

Another great resource is The Center for Responsible Nanotechnology (CRN) A non-profit organisation, formed to advance the safe use of molecular nanotechnology. CRN was founded by Chris Phoenix and Mike Treder in December 2002. The vision of CRN is a world in which nanotechnology is widely used for productive and beneficial purposes, and where malicious uses are limited by effective administration of the technology. Or go to the Google sci.nanotech Group and ask your questions.

What is the Use of Nanotechnology?

Nanotechnology is set to change our lives. Less than 10 years ago most people had never heard of the Internet and had difficulty understanding the concept. Nanotechnology is at that same stage today, but in the very near future nanotechnology will be part of our everyday vocabulary and everyday lives and we will all enjoy the incredible benefits. Nanotechnology has enabled the development of coatings that can make normal materials ultrastrong. The steel screws holding this fractured bone together are coated with a layer of nano-sized diamond crystals a 1000th of a millimetre thick. Bodies are less likely to reject the screws because diamond is a pure form of carbon and so are our bodies. One of the biggest areas for nanotechnology is in clothing and fabrics. Treated with a nano-scale coating, liquids turn into beads which roll off the fabric. This kind of coating is especially useful for the military and medical professions. The electronic Digicharm Tag uses a new display technology called PLEDs. A layer of plastic only a few nanometres thick is between two thin layers of metal. When electricity passes through the plastic it glows. PLED displays have a higher contrast than LCD screens and can be printed onto anything. Ordinary tennis rackets are weakened by tiny gaps between individual carbon fibres. Now rackets can be reinforced with nano-sized silicon-dioxide crystals that plug the holes. Nanotech has also made balls bouncier for longer. Inside, a coating of nano-thin platelets forms a barrier that keeps air in for longer. Hydrogen Solar's Tandem Cell uses sunlight to split water into hydrogen and oxygen. The cells have nanocrystalline metal particles on their surface, giving them a vast surface area with which to collect the Sun's energy. Currently, it converts just over 8% of the Sun's energy into pure hydrogen. Hydrogen is potentially a clean source of fuel.

Concept & Appreals

Nanotechnology is a field of applied science and technology covering a broad range of topics. The main unifying theme is the control of matter on a scale smaller than one micrometre, as well as the fabrication of devices on this same length scale. It is a highly multidisciplinary field, drawing from fields such as colloidal science, device physics, and supramolecular chemistry. Much speculation exists as to what new science and technology might result from these lines of research. Some view nanotechnology as a marketing term that describes pre-existing lines of research. Despite the apparent simplicity of this definition, nanotechnology actually encompasses diverse lines of inquiry. Nanotechnology cuts across many disciplines, including colloidal science, chemistry, applied physics, biology. It could variously be seen as an extension of existing sciences into the nanoscale, or as a recasting of existing sciences using a newer, more modern term. Two main approaches are used in nanotechnology: one is a "bottom-up" approach where materials and devices are built from molecular components which assemble themselves chemically using principles of molecular recognition; the other being a "top-down" approach where nano-objects are constructed from larger entities without atomic-level control. The impetus for nanotechnology has stemmed from a renewed interest in colloidal science, coupled with a new generation of analytical tools such as the atomic force microscope (AFM) and the scanning tunneling microscope (STM). Combined with refined processes such as electron beam lithography, these instruments allow the deliberate manipulation of nanostructures, and in turn led to the observation of novel phenomena. Nanotechnology is also an umbrella description of emerging technological developments associated with sub-microscopic dimensions. Despite the great promise of numerous nanotechnologies such as quantum dots and nanotubes, real applications that have moved out of the lab and into the marketplace have mainly utilized the advantages of colloidal nanoparticles in bulk form, such as suntan lotion, cosmetics, protective coatings, and stain resistant clothing.

Fundamental Concepts

Nanotechnology is an umbrella term that is used to describe a variety of techniques to fabricate materials and devices on the nanoscale. The genesis for nanotechnology has its roots in the colloidal science of the late 19th century. These early innovations have been combined with more recent developments in device manufacture. The term has served in some regards as a means to generate new lines of funding from government agencies. One nanometer (nm) is one billionth, or 10^{-9} of a meter. For comparison, typical carbon-carbon bond lengths, or the spacing between these atoms in a molecule, are in the range.12-.15 nm, and a DNA double-helix has a diameter around 2 nm. On the other hand, the smallest cellular lifeforms, the bacteria of the genus Mycoplasma, are around 200 nm in length. Nanotechnological techniques include those used for fabrication of nanowires, those used in semiconductor fabrication such as deep ultraviolet lithography, electron beam lithography, focused ion beam machining, nanoimprint lithography, atomic layer deposition, and molecular vapor deposition, and further including molecular self-assembly techniques such as those employing di-block copolymers. However, all of these techniques preceded the nanotech era, and are extensions in the development of scientific advancements rather than techniques which were devised with the sole purpose of creating nanotechnology or which were results of nanotechnology research. General fields involved with proper characterization of these systems include physics, chemistry, and biology, as well as mechanical and electrical engineering. However, due to the inter- and multidisciplinary nature of nanotechnology, subdisciplines such as physical chemistry, materials science, or biomedical engineering are considered significant or essential components of nanotechnology. The design, synthesis, characterization, and application of materials are dominant concerns of nanotechnologists. The manufacture of polymers based on molecular structure, or the design of computer chip layouts based on surface science are examples of nanotechnology in modern use. Colloidal suspensions also play an essential role in nanotechnology.

Technologies currently branded with the term 'nano' are little related to and fall far short of the most ambitious and transformative technological goals of the sort in molecular manufacturing proposals, but the term still connotes such ideas. Thus there may be a danger that a "nano bubble" will form from the use of the term by scientists and entrepreneurs to garner funding, regardless of (and perhaps despite a lack of) interest in the transformative possibilities of more ambitious and far-sighted work. The National Science Foundation (a major source of funding for nanotechnology in the United States) funded researcher David Berube to study the field of nanotechnology. His findings are published in the monograph "Nano-Hype: The Truth Behind the Nanotechnology Buzz". This published study (with a foreword by Mihail Roco, head of the NNI) concludes that much of what is sold as "nanotechnology" is in fact a recasting of straightforward materials science, which is leading to a "nanotech industry built solely on selling nanotubes, nanowires, and the like" which will "end up with a few suppliers selling low margin products in huge volumes."

Larger to Smaller: A Materials Perspective

A unique aspect of nanotechnology is the vastly increased ratio of surface area to volume present in many nanoscale materials which opens new possibilities in surface-based science, such as catalysis. A number of physical phenomena become noticeably pronounced as the size of the system decreases. These include statistical mechanical effects, as well as quantum mechanical effects, for example the "quantum size effect" where the electronic properties of solids are altered with great reductions in particle size. This effect does not come into play by going from macro to micro dimensions. However, it becomes dominant when the nanometer size range is reached. Additionally, a number of physical properties change when compared to macroscopic systems. One example is the increase in surface area to volume of materials. This catalytic activity also opens potential risks in their interaction with biomaterials. Nanotechnology can be thought of as extensions of traditional disciplines towards the explicit consideration of these properties. Additionally, traditional disciplines can be re-interpreted as specific applications of nanotechnology. This dynamic reciprocation of ideas and concepts contributes to the modern understanding of the field. Broadly speaking, nanotechnology is the synthesis and application of ideas from science and engineering towards the understanding and production of novel materials and devices. These products generally make copious use of physical properties associated with small scales. Materials reduced to the nanoscale can suddenly show very different properties compared to what they exhibit on a macroscale, enabling unique applications. For instance, opaque substances become transparent (copper); inert materials become catalysts (platinum); stable materials turn combustible (aluminum); solids turn into liquids at room temperature (gold); insulators become conductors (silicon). Materials such as gold, which is chemically inert at normal scales, can serve as a potent chemical catalyst at nanoscales. Much of the fascination with nanotechnology stems from these unique quantum and surface phenomena that matter exhibits at the nanoscale.

Nanosize powder particles (a few nanometres in diameter, also called nanoparticles) are potentially important in ceramics, powder metallurgy, the achievement of uniform nanoporosity and similar applications. The strong tendency of small particles to form clumps ("agglomerates") is a serious technological problem that impedes such applications. However, a few dispersants such as ammonium citrate (aqueous) and imidazoline or oleyl alcohol (nonaqueous) are promising additives for deagglomeration. (Dispersants are discussed in "Organic Additives And Ceramic Processing," by Daniel J. Shanefield, Kluwer Academic Publ., Boston.)

Another concern is that the volume of an object decreases as the third power of its linear dimensions, but the surface area only decreases as its second power. This somewhat subtle and unavoidable principle has huge ramifications. For example the power of a drill (or any other machine) is proportional to the volume, while the friction of the drill's bearings and gears is proportional to their surface area. For a normal-sized drill, the power of the device is enough to handily overcome any friction. However, scaling its length down by a factor of 1000, for example, decreases its power by 1000^3 (a factor of a billion) while reducing the friction by only 1000^2 (a factor of "only" a million). Proportionally it has 1000 times less power per unit friction than the original drill. If the original friction-to-power ratio was, say, 1%, that implies the smaller drill will have 10 times as much friction as power. The drill is useless. This is why, while super-miniature electronic integrated circuits can be made to function, the same technology cannot be used to make functional mechanical devices in miniature: the friction overtakes the available power at such small scales. So while you may see microphotographs of delicately etched silicon gears, such devices are curiosities only, not actually usable parts. Surface tension increases in the same way, causing very small objects tend to stick together. This could possibly make any kind of "micro factory" impractical: even if robotic arms and hands could be scaled down, anything they pick up will tend to be impossible to put down. All these scaling issues have to be kept in mind while evaluating any kind of nanotechnology.

Simple to complex: a molecular perspective

Modern synthetic chemistry has reached the point where it is possible to prepare small molecules to almost any structure. These methods are used today to produce a wide variety of useful chemicals such as pharmaceuticals or commercial polymers. The ability of this is to extend the control to the next, seeking methods to assemble these single molecules into supramolecular assemblies consisting of many molecules arranged in a well defined manner. These approaches utilize the concepts of molecular self-assembly and/or supramolecular chemistry to automatically arrange themselves into some useful conformation through a bottom-up approach. The concept of molecular recognition is especially important: molecules can be designed so that a specific conformation or arrangement is favored due to non-covalent intermolecular forces. The Watson-Crick basepairing rules are a direct result of this, as is the specificity of an enzyme being targeted to a single substrate, or the specific folding of the protein itself. Thus, two or more components can be designed to be complementary and mutually attractive so that they make a more complex and useful whole. Such bottom-up approaches should, broadly speaking, be able to produce devices in parallel and much cheaper than top-down methods, but could potentially be overwhelmed as the size and complexity of the desired assembly increases.

Molecular Nanotechnology: A Long-term View

Advanced nanotechnology, sometimes called molecular manufacturing, is a term given to the concept of engineered nanosystems (nanoscale machines) operating on the molecular scale. By the countless examples found in biology it is currently known that billions of years of evolutionary feedback can produce sophisticated, stochastically optimized biological machines, and it is hoped that developments in nanotechnology will make possible their construction by some shorter means, perhaps using biomimetic principles. However, K Eric Drexler and other researchers have proposed that advanced nanotechnology, although perhaps initially implemented by biomimetic means, ultimately could be based on mechanical engineering principles. When the term "nanotechnology" was independently coined and popularized by Eric Drexler, who at the time was unaware of an earlier usage by Norio Taniguchi, it referred to a future manufacturing technology based on molecular machine systems. The premise was that molecular-scale biological analogies of traditional machine components demonstrated that molecular machines were possible, and that a manufacturing technology based on the mechanical functionality of these components (such as gears, bearings, motors, and structural members) would enable programmable, positional assembly to atomic specification (see the original reference PNAS-1981). The physics and engineering performance of exemplar designs were analyzed in the textbook Nanosystems. Another view, put forth by Carlo Montemagno, is that future nanosystems will be hybrids of silicon technology and biological molecular machines, and his group's research is directed toward this end. The seminal experiment proving that positional molecular assembly is possible was performed by Ho and Lee at Cornell University in 1999. They used a scanning tunneling microscope to move an individual carbon monoxide molecule (CO) to an individual iron atom (Fe) sitting on a flat silver crystal, and chemically bound the CO to the Fe by applying a voltage. Though biology clearly demonstrates that molecular machine systems are possible, non-biological molecular machines are today only in their infancy. Leaders in research on non-biological molecular machines are Dr. Alex Zettl and his colleagues at Lawrence Berkeley Laboratories and UC Berkeley. They have constructed at least three distinct molecular devices whose motion is controlled from the desktop with changing voltage: a nanotube nanomotor, a molecular actuator, and a nanoelectromechanical relaxation oscillator. Manufacturing in the context of productive nanosystems is not related to, and should be clearly distinguished from, the conventional technologies used to manufacture nanomaterials such as carbon nanotubes and nanoparticles. There exists the potential to design and fabricate artificial structures analogous to natural cells and even organisms.

Current Research

As nanotechnology is a very broad term, there are many disparate but sometimes overlapping subfields that could fall under its umbrella. The following avenues of research could be considered subfields of nanotechnology. Note that these categories are fairly nebulous and a single subfield may overlap many of them, especially as the field of nanotechnology continues to mature.

Nanomaterials

This includes subfields which develop or study materials having unique properties arising from their nanoscale dimensions.

- Colloid science has given rise to many materials which may be useful in nanotechnology, such as carbon nanotubes and other fullerenes, and various nanoparticles and nanorods.
- Nanoscale materials can also be used for bulk applications; most present commercial applications of nanotechnology are of this flavor.
- Headway has been made in using these materials for medical applications; see Nanomedicine.

Bottom-up Approaches

These seek to arrange smaller components into more complex assemblies.

- DNA Nanotechnology utilizes the specificity of Watson-Crick basepairing to construct well-defined structures out of DNA and other nucleic acids.
- More generally, molecular self-assembly seeks to use concepts of supramolecular chemistry, and molecular recognition in particular, to cause single-molecule components to automatically arrange themselves into some useful conformation.

Top-down Approaches

These seek to create smaller devices by using larger ones to direct their assembly.

- Many technologies descended from conventional solid-state silicon methods for fabricating microprocessors are now capable of creating features smaller than 100 nm, falling under the definition of nanotechnology. Giant magnetoresistance-based hard drives already on the market fit this description, as do atomic layer deposition (ALD) techniques.
- Solid-state techniques can also be used to create devices known as nanoelectromechanical systems or NEMS, which are related to microelectromechanical systems or MEMS.

- Atomic force microscope tips can be used as a nanoscale "write head" to deposit a chemical on a surface in a desired pattern in a process called dip pen nanolithography. This fits into the larger subfield of nanolithography.

Functional Approaches

These seek to develop components of a desired functionality without regard to how they might be assembled.

- Molecular electronics seeks to develop molecules with useful electronic properties. These could then be used as single-molecule components in a nanoelectronic device. For an example see rotaxane.
- Synthetic chemical methods can also be used to create synthetic molecular motors, such as in a so-called nanocar.

Speculative

These subfields seek to anticipate what inventions nanotechnology might yield, or attempt to propose an agenda along which inquiry might progress. These often take a big-picture view of nanotechnology, with more emphasis on its societal implications than the details of how such inventions could actually be created.

- Molecular nanotechnology is a proposed approach which involves manipulating single molecules in finely controlled, deterministic ways. This is more theoretical than the other subfields and is beyond current capabilities.
- Nanorobotics centers on self-sufficient machines of some functionality operating at the nanoscale.
- Programmable matter based on artificial atoms seeks to design materials whose properties can be easily and reversibly be externally controlled.

Tools and Techniques

Nanoscience and nanotechnology only became possible in the 1910's with the development of the first tools to measure and make nanostructures. But the actual development started with the discovery of electrons and neutrons which showed scientists that matter can really exist on a much smaller scale than what we normally think of as small, and/or what they thought was possible at the time. It was at this time when curiosity for nanostructures had originated.

The atomic force microscope (AFM) and the Scanning Tunneling Microscope (STM) are two early versions of scanning probes that launched nanotechnology. There are other types of scanning probe microscopy, all based on the idea of the

STM, that make it possible to see structures at the nanoscale. The tip of scanning probes can also be used to manipulate nanostructures (a process called positional assembly). However, this is a very slow process. This led to the development of various techniques of nanolithography such as dip pen nanolithography, electron beam lithography or nanoimprint lithography. Lithography is a top-down fabrication technique where a bulk material is reduced in size to nanoscale pattern.

The top-down approach anticipates nanodevices that must be built piece by piece in stages, much as manufactured items are currently made. Scanning probe microscopy is an important technique both for characterization and synthesis of nanomaterials. Atomic force microscopes and scanning tunneling microscopes can be used to look at surfaces and to move atoms around. By designing different tips for these microscopes, they can be used for carving out structures on surfaces and to help guide self-assembling structures. Atoms can be moved around on a surface with scanning probe microscopy techniques, but it is cumbersome, expensive and very time-consuming. For these reasons, it is not feasible to construct nanoscaled devices atom by atom. Assembling a billion transistor microchip at the rate of about one transistor an hour is inefficient. However, these techniques may eventually be used to make primitive nanomachines, which in turn can be used to make more sophisticated nanomachines. In contrast, bottom-up techniques build or grow larger structures atom by atom or molecule by molecule. These techniques include chemical synthesis, self-assembly and positional assembly. Another variation of the bottom-up approach is molecular beam epitaxy or MBE. Researchers at Bell Telephone Laboratories like John R. Arthur. Alfred Y. Cho, and Art C. Gossard developed and implemented MBE as a research tool in the late 1960s and 1970s. Samples made by MBE were key to to the discovery of the fractional quantum Hall effect for which the 1998 Nobel Prize in Physics was awarded. MBE allows scientists to lay down atomically-precise layers of atoms and, in the process, build up complex structures. Important for research on semiconductors, MBE is also widely used to make samples and devices for the newly emerging field of spintronics. Newer techniques such as Dual Polarisation Interferometry are enabling scientists to measure quantitatively the molecular interactions that take place at the nano-scale.

Societal Implications

Potential risks of nanotechnology can broadly be grouped into three areas:

- the risk to health and environment from nanoparticles and nanomaterials;
- the risk posed by molecular manufacturing (or advanced nanotechnology);
- societal risks.

In August 2005, a task force consisting of 50+ international experts from various fields was organised by the Center for Responsible Nanotechnology to study the societal implications of molecular nanotechnology. In October 2005, the National Science Foundation announced that it would fund two national centers to research the potential societal implications of nanotechnology. Located at the University of California, Santa Barbara and Arizona State University, researchers at these two centers are exploring a wide range of issues including nanotechnology's historical context, technology assessment, innovation and globalization issues, and societal perceptions of risk. A book by Geoffrey Hunt and Michael Mehta (2006) entitled *Nanotechnology: Risk, Ethics and Law* (London: Earthscan Book) provides a global overview of the state of nanotechology and society in Europe, the USA, Japan and Canada, and examines the ethics, the environmental and public health risks, and the governance and regulation of this most promising, and potentially most dangerous, of all technologies. Determining a set of pathways for the development of molecular nanotechnology is now an objective of a broadly based technology roadmap project led by Battelle (the manager of several U.S. National Laboratories) and the Foresight Institute. That roadmap should be completed by early 2007.

Risks from Nanoparticles

The mere presence of nanomaterials (materials that contain nanoparticles) is not in itself a threat. It is only certain aspects that can make them risky, in particular their mobility and their increased reactivity. Only if certain properties of certain nanoparticles were harmful to living beings or the environment would we be faced with a genuine hazard. In this case it can be called Nanopollution.

In addressing the health and environmental impact of nanomaterials we need to differentiate two types of nanostructures: (1) Nanocomposites, nanostructured surfaces and nanocomponents (electronic, optical, sensors etc.), where nanoscale particles are incorporated into a substance, material or device ("fixed" nano-particles); and (2) "free" nanoparticles, where at some stage in production or use individual nanoparticles of a substance are present. These free nanoparticles could be nanoscale species of elements, or simple compounds, but also complex compounds where for instance a nanoparticle of a particular element is coated with another substance ("coated" nanoparticle or "core-shell" nanoparticle). There seems to be consensus that, although one should be aware of materials containing fixed nanoparticles, the immediate concern is with free nanoparticles. Because nanoparticles are very different from their everyday counterparts, their adverse effects cannot be derived from the known toxicity of the macro-sized material. This poses significant issues for addressing the health and environmental impact of free nanoparticles. To complicate things further, in talking about nanoparticles it is important that a powder or liquid containing nanoparticles is almost never monodisperse, but will contain a range of particle sizes. This complicates the experimental analysis as larger nanoparticles might have different properties than smaller ones. Also, nanoparticles show a tendency to aggregate and such aggregates often behave differently from individual nanoparticles.

The lethal dose over six months for lab rats, of different kinds of nanoparticles are often characterized by a Skov Kjaer index, named after the scientist Kasper Skov Kjaer.

Health Issues

There are several potential entry routes for nanoparticles into the body. They can be inhaled, swallowed, absorbed through skin or be deliberately injected during medical procedures (or released from implants). Once within the body they are highly mobile and in some instances can even cross the blood-brain barrier.

How these nanoparticles behave inside the organism is one of the big issues that needs to be resolved. The behavior of nanoparticles is a function of their size, shape and surface reactivity with the surrounding tissue. They could cause overload on phagocytes, cells that ingest and destroy foreign matter, thereby triggering stress reactions that lead to inflammation and weaken the body's defense against other pathogens. Apart from what happens if non-degradable or slowly degradable nanoparticles accumulate in organs, another concern is their potential interaction with biological processes inside the body: because of their large surface, nanoparticles on exposure to tissue and fluids will immediately adsorb onto their surface some of the macromolecules they encounter. This may, for instance, affect the regulatory mechanisms of enzymes and other proteins.

Environmental Issues

Not enough data exists to know for sure if nanoparticles could have undesirable effects on the environment. Two areas are relevant here: (1) In free form nanoparticles can be released in the air or water during production (or production accidents) or as waste byproduct of production, and ultimately accumulate in the soil, water or plant life. (2) In fixed form, where they are part of a manufactured substance or product, they will ultimately have to be recycled or disposed of as waste. It is not known yet whether certain nanoparticles will constitute a completely new class of non-biodegradable pollutant. In case they do, it is not known how such pollutants could be removed from air or water because most traditional filters are not suitable for such tasks (their pores are too big to catch nanoparticles).

Health and environmental issues combine in the workplace of companies engaged in producing or using nanomaterials and in the laboratories engaged in nanoscience and nanotechnology research. It is safe to say that current workplace exposure standards for dusts cannot be applied directly to nanoparticle dusts.To properly assess the health hazards of engineered nanoparticles the whole life cycle of these particles needs to be evaluated, including their fabrication, storage and distribution, application and potential abuse, and disposal. The impact on humans or the environment may vary at different stages of the life cycle.

Regarding the risks from molecular manufacturing, an often cited worst-case scenario is "grey goo", a hypothetical substance into which the surface of the earth might be transformed by self-replicating nanobots running amok. This concept has been analyzed by Freitas in "Some Limits to Global Ecophagy by Biovorous Nanoreplicators, with Public Policy Recommendations" With the advent of nanbiotech, a different scenario called green goo has been forwarded. Here, the malignant substance is not nanobots but rather self-replicating organisms engineered through nanotechnology.

A Need for Regulation?

Regulatory bodies such as the Environmental Protection Agency and the Food and Drug Administration in the U.S. or the Health & Consumer Protection Directorate of the European Commission have started dealing with the potential risks posed by nanoparticles. So far, neither engineered nanoparticles nor the products and materials that contain them are subject to any special regulation regarding production, handling or labeling. The Material Safety Data Sheet that must be issued for certain materials often does not differentiate between bulk and nanoscale size of the material in question and even when it does these MSDS are advisory only.

Studies of the health impact of airborne particles are the closest thing we have to a tool for assessing potential health risks from free nanoparticles. These studies have generally shown that the smaller the particles get, the more toxic they become. This is due in part to the fact that, given the same mass per volume, the dose in terms of particle numbers increases as particle size decreases.

Looking at all available data, it must be concluded that current risk assessment methodologies are not suited to the hazards associated with nanoparticles; in particular, existing toxicological and eco-toxicological methods are not up to the task; exposure evaluation (dose) needs to be expressed as quantity of nanoparticles and/or surface area rather than simply mass; equipment for routine detecting and measuring nanoparticles in air, water, or soil is inadequate; and very little is known about the physiological responses to nanoparticles.

Regulatory bodies in the U.S. as well as in the EU have concluded that nanoparticles form the potential for an entirely new risk and that it is necessary to carry out an extensive analysis of the risk. The challenge for regulators is whether a matrix can be developed which would identify nanoparticles and more complex nanoformulations which are likely to have special toxicological properties or whether it is more reasonable for each particle or formulation to be tested separately. Since null hypotheses are unfalsifiable in a true sense, nanoparticles and nanoformulations can never be proven safe. A truly precautionary approach to regulation would severely impede development in the field of nanotechnology if we require safety studies for each and every nanoscience application. Consequently, the rush seems to be on to establish a research needs assessment in the nanocommunity to preclude universal safety studies. While the outcome of these studies can form the basis for government and international regulations, a more reasonable approach might be development of a risk matrix that indentifes likely culprits.

Possible Military Applications

Societal risks from the use of nanotechnology have also been raised. On the instrumental level, these include the possibility of military applications of nanotechnology (for instance, as in implants and other means for soldier enhancement like those being developed at the Institute for Soldier Nanotechnologies at MIT) as well as enhanced surveillance capabilities through nano-sensors.

Potential benefits and risks for developing countries

Nanotechnologies may provide new solutions for the millions of people in developing countries who lack access to basic services, such as safe water, reliable energy, health care, and education. The United Nations has set Millennium Development Goals for meeting these needs. The 2004 UN Task Force on Science, Technology and Innovation noted that some of the advantages of nanotechnology include production using little labor, land, or maintenance, high productivity, low cost, and modest requirements for materials and energy.

Many developing countries, for example Costa Rica, Chile, Bangladesh, Thailand, and Malaysia, are investing considerable resources in research and development of nanotechnologies. Emerging economies such as Brazil, China, India and South Africa are spending millions of US dollars annually on R&D, and are rapidly increasing their scientific output as demonstrated by their increasing numbers of publications in peer-reviewed scientific publications.

Potential opportunities of nanotechnologies to help address critical international development priorities include improved water purification systems, energy systems, medicine and pharmaceuticals, food production and nutrition, and information and communications technologies. Nanotechnologies are already incorporated in products that are on the market. Other nanotechnologies are still in the research phase, while others are concepts that are years or decades away from development. Applying nanotechnologies in developing countries raises similar questions about the environmental, health, and societal risks described in the previous section. Additional challenges have been raised regarding the linkages between nanotechnology and development. Protection of the environment, human health and worker safety in developing countries often suffers from a combination of factors that can include but are not limited to lack of robust environmental, human health, and worker safety regulations; poorly or unenforced regulation which is linked to a lack of physical (e.g., equipment) and human capacity (i.e., properly trained regulatory staff). Often, these nations require assistance, particularly financial assistance, to develop the scientific and institutional capacity to adequately assess and manage risks, including the necessary infrastructure such as laboratories and technology for detection.

Very little is known about the risks and broader impacts of nanotechnology. At a time of great uncertainty over the impacts of nanotechnology it will be challenging for governments, companies, civil society organisations, and the general public in developing countries, as in developed countries, to make decisions about the governance of nanotechnology.

Companies, and to a lesser extent governments and universities, are receiving patents on nanotechnology. The rapid increase in patenting of nanotechnology is illustrated by the fact that in the US, there were 500 nanotechnology patent applications in 1998 and 1,300 in 2000. Some patents are very broadly defined, which has raised concern among some groups that the rush to patent could slow innovation and drive up costs of products, thus reducing the potential for innovations that could benefit low income populations in developing countries.

There is a clear link between commodities and poverty. Many least developed countries are dependent on a few commodities for employment, government revenue, and export earnings. Many applications of nanotechnology are being developed that could impact global demand for specific commodities. For instance, certain nanoscale materials could enhance the strength and durability of rubber, which might eventually lead to a decrease in demand for natural rubber. Other nanotechnology applications may result in increases in demand for certain commodities. For example, demand for titanium may increase as a result of new uses for nanoscale titanium oxides, such as titanium dioxide nanotubes that can be used to produce and store hydrogen for use as fuel. Various organisations have called for international dialogue on mechanisms that will allow developing countries to anticipate and proactively adjust to these changes. In 2003, Meridian Institute began the Global Dialogue on Nanotechnology and the Poor: Opportunities and Risks (GDNP) to raise awareness of the opportunities and risks of nanotechnology for developing countries, close the gaps within and between sectors of society to catalyze actions that address specific opportunities and risks of nanotechnology for developing countries, and identify ways that science and technology can play an appropriate role in the development process. The GDNP has released several publicly accessible papers on nanotechnology and development, including "Nanotechnology and the Poor: Opportunities and Risks — Closing the Gaps Within and Between Sectors of Society"; "Nanotechnology, Water, and Development"; and "Overview and Comparison of Conventional and Nano-Based Water Treatment Technologies".

Intellectual Property Issues

On the structural level, critics of nanotechnology point to a new world of ownership and corporate control opened up by nanotechnology. The claim is that, just as biotechnology's ability to manipulate genes went hand in hand with the patenting of life, so too nanotechnology's ability to manipulate molecules has led to the patenting of matter. The last few years has seen a gold rush to claim patents at the nanoscale. Over 800 nano-related patents were granted in 2003, and the numbers are increasing year to year. Corporations are already taking out broad-ranging patents on nanoscale discoveries and inventions. For example, two corporations, NEC and IBM, hold the basic patents on carbon nanotubes, one of the current cornerstones of nanotechnology. Carbon nanotubes have a wide range of uses, and look set to become crucial to several industries from electronics and computers, to strengthened materials to drug delivery and diagnostics. Carbon nanotubes are poised to become a major traded commodity with the potential to replace major conventional raw materials. However, as their use expands, anyone seeking to manufacture or sell carbon nanotubes, no matter what the application, must first buy a license from NEC or IBM.

Opportunities and Challenges

This section of the review describes current limits on the biological aspects of nanotechnology. It outlines selected broad areas of research opportunity, and it sets forth challenges for the intermediate and long-term future.

Enabling Technologies

We owe much of the recent progress in the biological aspects of nanotechnology and nanoparticles to important enabling technologies. Many of these enabling technologies are described in other parts of this report. Those that have had a significant impact on biological measurements include techniques and instrumentation such as optical traps, laser tweezers, and "nano-pokers" to measure femtoNewton forces (Svoboda and Block 1994), and AFM and STM (Wildoer et al. 1998).

There is also a need to move detection away from the ensemble, to the single molecule scale. This includes driving instrumentation toward singlemolecule spectroscopy; single-spin detection (Rugar et al. 1994); chemical analysis of nanoliter volumes (Hietpas et al. 1996, 139-158); nuclear magnetic resonance microcoils;

nanoscale electrode arrays and chemical sensing and detecting technology (McConnell 1996, 97-102); separations technology; chemical analysis of single cells (Hietpas et al. 1996); new biological transformations; and chemical probes of nanostructures. Enhanced computational infrastructure, both hardware and algorithm development, will also be required to investigate supramolecular assemblies and to understand interactions that occur over a wide variety of time and length scales. At present, we are fairly well equipped to handle quantum mechanical calculations. At the next level of computational complexity, the molecular dynamics force fields and atomic charges can presently handle up to about a million particles and still retain their accuracy. The next scale of complexity, mesoscale modeling, currently requires the use of pseudo atoms. Much more research is required to improve the accuracy of finite element calculations and the modeling of materials applications (Goddard 1998).

Surface Interactions and the Interface Between Biomolecules and Substrates

Much of the progress in the biological aspects of nanotechnology has come from research on surface interactions. It will be important to continue to develop a better understanding of the interface between biomolecules and surfaces. Of particular interest are questions about whether surfaces cause biomolecules to denature, the optimum length of linker groups, and ways to communicate from the biomolecule, through the linker, to the substrate. Another challenge is to produce nanometer ultrathin films that have stable order (Jaworek et al. 1998).

Robustness of Biomolecules and Their Interactions in Aqueous Solution

Although much progress has been made on ways to confer additional robustness on biomolecules, it will be necessary to continue to improve the stability and reliability of biomolecular assemblies. Such research may be along the lines of understanding the thermal stability of biomolecules, perhaps by examining extremophiles and molecules produced by directed evolution. It will also be necessary to produce molecules that can work in the absence of water, or to devise ways in which aqueous solutions can be used reliably.

Assemblers or Templating

The hierarchical self-assembly of biomolecules is often touted as one of their key attributes for nanotechnology applications. However, there remains much to be learned about how to make periodic arrays of biomolecular assemblies, how to use biological templating in an efficient manner, how to mimic biological self-assembly with nonbiological molecules, and how to exploit differences between biological and nonbiological self-assembly.

Combinatorial Approaches to Nanotechnology

Combinatorial chemistry has had an enormous impact on drug discovery and development in the past several years. Ability to develop molecularly sensitive screening techniques is one of the limiting factors of combinatorial chemistry. It would be intriguing to discover screening methods that are particularly sensitive for picking up new phenomena exhibited by biological nanoparticles.

Biomimetics and Polymer-Biopolymer Chimeras

The largest impact of biological sciences on nanotechnology may prove to be through the lessons one can learn from biology, rather than from directly employing the biological molecules themselves. To this end, it is important to support research on various aspects of biomimetics, including activities in which synthetic and biologically-based molecules are combined in chimeric fashion.

Nanword

So, Aanotechnology is defined as fabrication of devices with atomic or molecular scale precision. Devices with minimum feature sizes less than 100 nanometers (nm) are considered to be products of nanotechnology. A nanometer is one billionth of a meter (10-9 m) and is the unit of length that is generally most appropriate for describing the size of single molecules. The nanoscale marks the nebulous boundary between the classical and quantum mechanical worlds; thus, realisation of nanotechnology promises to bring revolutionary capabilities. Fabrication of nanomachines, nanoelectronics and other nanodevices will undoubtedly solve an enormous amount of the problems faced by mankind today. Nanotechnology is currently in a very infantile stage. However, we now have the ability to organise matter on the atomic scale and there are already numerous products available as a direct result of our rapidly increasing ability to fabricate and characterize feature sizes less than 100 nm. Mirrors that don't fog, biomimetic paint with a contact angle near 180°, gene chips and fat soluble vitamins in aqueous beverages are some of the first manifestations of nanotechnology. However, immenant breakthroughs in computer science and medicine will be where the real potential of nanotechnology will first be achieved. Nanoscience is an interdisciplinary field that seeks to bring about mature nanotechnology. Focusing on the nanoscale intersection of fields such as physics, biology, engineering, chemistry, computer science and more, nanoscience is rapidly expanding. Nanotechnology centers are popping up around the world as more funding is provided and nanotechnology market share increases. The rapid progress is apparent by the increasing appearance of the prefix "nano" in scientific journals and the news. Thus, as we increase our ability to fabricate computer chips with smaller features and improve our ability to cure disease at the molecular level, nanotechnology is here.

Nanotechnology for Bio Materials and Tissues

Technology Focus

Historically, synthetic materials have not served as sufficient implants.For example, the current average lifetime of an orthopedic implant, such as hip, knee, ankle, etc., is only 15 years. Similarly, for the vascular community, small diameter vascular grafts are only functional 25% of time past 5 years of use. The same limited lifetimes for a number of failing organs could be stated for cartilage, bladder, and the peripheral and central nervous systems. Clearly, conventional materials, or those materials with constituent dimensions greater than 1 micron have not invoked proper cellular responses to regenerate tissue that allows for these devices to be successful for long periods of times. In contrast, due to their ability to mimic the dimensions of constituent components of natural tissues, like proteins, nanophase materials may be an exciting successful alternative. Nanophase materials are defined as materials with constituent dimension less than 100 nm in at least one direction. Materials investigated to date include nanophase ceramics, metals, polymers, and composites. Nanophase materials have revolutionized many traditional engineering fields such as those involving catalysis, electronics, optics, magnetics, etc. For example, in the catalytic field, higher surface reactivity of nanophase materials allows for faster reaction times. To date,there has also been the emergence of data that nanophase materials may be optimal materials for tissue engineering applications. This is not only due to their ability to simulate dimensions of proteins that comprise tissues, but also because of their higher reactivity for interactions of proteins that control cell adhesion and, thus, the ability to regenerate tissues. There are numerous reasons to consider using nanophase materials as better tissue engineering materials.While most of the promise is conjecture, limited scientific evidence does exist and is generating wide support.

Course Objectives

The first objective of this course is to define current problems with implants. You will be familiar with issues related to implant failure particular for orthopedic,dental, cartilage,vascular,bladder,and the nervous systems. The second objective is to understand how cells interact with implanted materials to generate a biological response.We will describe how proteins interact with materials to control cell function leading to either a desirable or undesirable biological response to an implanted material. The third objective is to understand how nanophase materials could be used as better implants. What are their promising properties and what experimental evidence exists? We will sort through the experimental data and separate promise from “hype”. The discussion will make you be able to indicate what are the promising properties of nanophase materials and how to make them to control initial protein interactions leading to successful regeneration of tissue. The last objective is to understand what further information will be needed to elucidate if nanophase materials will make better biomaterials for these organ systems. While there is no golden standard chemistry for various implant applications, this course will discuss the possibility of a standard topography that may be useful to regenerate tissue for a wide range of materials including ceramics, metals, and polymers. You will be able to identify future experiments needed for the further consideration of nanophase materials to regenerate tissue

This course will begin with a general overview of current problems with orthopedic, dental, cartilage, vascular, bladder, and central as well as peripheral nervous system devices. These organs will be chosen since they are the ones in which the most implantable devices have been used to date. One day will be spent describing situations in which these devices are needed and why we have not been able to develop sufficient materials to successfully regenerate these tissues. To do this, we will cover the basic anatomy, physiology, and cellular structures of these organs. This will help us understand the complexity of our organs and consequently to design materials that match the most important properties of tissue to promote tissue regeneration. Next, we will cover how the body responds to implanted materials by first adsorbing a thin layer of proteins and then through select cellular responses. Topics such as how proteins adsorb to surfaces,how cells recognize proteins, how cells adhere and function on materials, and how tissues may or may not regrow will be the focus of this day.The biological response section will be complemented with results from experiments that have delineated fundamental material properties that control such responses. By understanding how the body originally responds to implanted materials, we can begin to develop materials to obtain a desirable response, such a tissue regrowth. We will then interject a new classification of materials that are beginning to be considered to serve as better biomaterials: nanophase materials. One day will be spent defining these materials and highlighting their special properties pertinent for implants. This includes their mechanical, optical, electrical, and surface properties. We will also discuss various routes to synthesize nanophase materials. Although, to date, the special properties of nanophase materials have been used in non-biological applications, we will emphasize their role in biological applications. We will constantly compare theory will experimental evidence in this section. The day will end with the most promising results obtained thus far in terms of increased tissue regeneration on nanophase materials. Lastly, we will discuss the most promising future directions for the use of nanophase materials in biological applications. This day will focus on key unanswered questions if nanophase materials are to be used in the body. Central to this question is the immunological response to nanoparticles, which will comprise a majority of the day. To enhance learning as well as stimulate discussion, the course will be interactive with participation from the learners throughout.

What is Nanomotor?

A nanomotor is a molecular device capable of converting energy into movement and forces on the order of the piconewtons. A proposed branch of research is the integration of molecular motor proteins found in living cells into molecular motors implanted in artificial devices. Such a motor protein would be able to move a "cargo" within that device, similarly to how kinesin moves various molecules along tracks of microtubules inside cells. Starting and stopping the movement of such motor proteins would involve caging the ATP in molecular structures sensitive to UV light, pulses of UV illumination would thus provide pulses of movement. Nanomotors have also be made using synthetic materials and chemical methods, as described in the following section. See also Nanocar and Synthetic molecular motors.

Nanotube Nanomotor

Researchers at University of California, Berkeley, have developed rotational bearings based upon multiwall carbon nanotubes. By attaching a gold plate (with dimensions of order 100nm) to the outer shell of a suspended multiwall carbon nanotube (like nested carbon cylinders), they are able to electrostatically rotate the outer shell relative to the inner core. These bearings are very robust; Devices have been oscillated thousands of times with no indication of wear. The work was done in situ in an SEM. These nanoelectromechanical systems (NEMS) are the next step in miniaturization that may find their way into commercial aspects in the future. The thin vertical string seen in the middle, is the nanotube to which the rotor is attached. When the outer tube is sheared, the rotor is able to spin freely on the nanotube bearing. The process and technology can be seen in this render.

- Physicists build world's smallest motor using nanotubes and etched silicon
- Research project
- Carbon nanotube
- Electrostatic motor

Understanding Nanosensor

Nanosensors are a technology that may exist in the future. The closest thing in existence to these currently are MEMS (Micro Electro Mechanical Systems) devices. A MEMS biosensor may be based upon a thin film bulk electric wave resonator, with some components of it less than 100 nm. They work in a similar way to quartz crystal microbalances, but at much higher frequencies, 3 GHz and beyond. They have been fabricated by the famous biological nanotechnologist Dr. Gregory Ashley at Cranfield University as part of his doctoral thesis. They can be used in nanorobots which when injected inside a human being could be used to detect inimical human tissues.

Nanocrystalline Silicon

Nanocrystalline silicon (nc-Si) — an allotropic form of silicon — is similar to amorphous silicon (a-Si), in that it has an amorphous phase. Where they differ, however, is that nc-Si has small grains of crystalline silicon within the amorphous phase. This is in contrast to polycrystalline silicon (poly-Si) which consists solely of crystalline silicon grains, separated by grain boundaries. nc-Si is sometimes also known as microcrystalline silicon (μc-Si). The difference comes solely from the grain size of the crystalline grains. Most materials with grains in the micrometre range are actually fine-grained polysilicon, so nanocrystalline silicon is a better term. The term Nanocrystalline silicon refers to a range of materials around the transition region from amorphous to microcrystalline phase in the silicon thin film. The crystalline volume fraction (as measured from Raman spectroscopy) is another criteria to describe the materials in this transition zone. It also shows increased absorption in the red and infrared wavelengths, which make it an important material for use in a-Si solar cells. One of the most important advantages of nanocrystalline silicon, however, is that it has increased stability over a-Si, one of the reasons being because of its lower hydrogen concentration. Although it currently cannot attain the mobility that poly-Si can, it has the advantage over poly-Si that it is easier to fabricate, as it can be deposited using conventional low temperature a-Si deposition techniques, such as PECVD, as opposed to laser annealing or high temperature CVD processes, in the case of poly-Si. The main application of this novel material is in the field of silion thin film solar cells.

What is Nanopore?

A nanopore is a small pore in an electrically insulating membrane, that can be used as a single-molecule detector. It can be a biological protein channel in a lipid bilayer or a pore in a solid-state membrane. The detection principle is based on monitoring the ionic current of an electrolyte solution passing through the nanopore as a voltage is applied across the membrane. When the nanopore is of molecular dimensions, passage of molecules (e.g., DNA) cause interruptions of the "open" current level, leading to a "translocation event" pulse. The passage of single-stranded DNA molecules through the membrane-embedded alpha-hemolysin channel (1.5 nm diameter), for example, causes a ~90% blockage of the current (measured at 1 M KCl solution). The observation that a passing strand of DNA containing different bases results in different blocking levels has led to the nanopore sequencing hypothesis. Such sequencing, if successful, could revolutionize the field of genomics, as sequencing could be carried out in a matter of seconds.Apart from rapid DNA sequencing, other applications include separation of single stranded and double stranded DNA in solution, and the determination of length of polymers. At this stage, nanopores are making contributions to the understanding of polymer biophysics, as well as to single-molecule analysis of DNA-protein interactions.

Solid-state nanopores are generally made in silicon compound membranes, one of the most common being Si_3N_4. Solid-state nanopores can be manufactured with several techniques including ion-beam sculpting and electron beams.

Nanomedicine Appliences

Nanomedicine is the medical application of nanotechnology and related research. It covers areas such as nanoparticle drug delivery and possible future applications of molecular nanotechnology (MNT) and nanovaccinology.

Current problems for nanomedicine involve understanding the issues related to toxicity and environmental impact of nanoscale materials.

Direct funding for nanomedicine projects has begun, and the US National Institute of Health received funding in 2005 to set up four nanomedicine centres. In April 2006, the journal Nature Materials estimated that 130 nanotech-based drugs and delivery systems were being developed worldwide.

The first thorough analysis of possible applications of MNT to medicine can be read in *Nanomedicine*, a book series by Robert Freitas; it analyzes a wide range of possible nanotechnology-based medical devices, and explains the relevant science behind their design.

Cancer

Nanoparticles of cadmium selenide (quantum dots) glow when exposed to ultraviolet light. When injected, they seep into cancer tumors. The surgeon can see the glowing tumor, and use it as a guide for more accurate tumor removal.

Jim Heath, a Caltech chemist, is developing nano-sized sensors that can detect and diagnose cancer in the early stages, when there are only a few thousand cancer cells in the body. A few drops of the patient's blood are placed on the sensor test chip. The chip contains tens of thousands of nanowires that can detect proteins and other biomarkers left behind by cancer cells. Cancer is curable in the early stages, so this test could save lives once perfected.

Jennifer West, a bioengineer, used nanoshells coated with gold to kill cancer tumors in mice. The nanoshells are 120 nanometers in diameter, 170 times smaller than a cancer cell. The nanoshells are injected into the mouse. The nanoshells become lodged in the cracks of the tumors. Then the mouse is shot with an infrared laser. The ray passes through the flesh harmlessly, but heats up the gold. The gold burns the cancer cells to death, without harming the healthy cells. No mice have died, even when injected with large doses of nanoshells as per Food and Drug Administration requirement. This Nanomedicine method is more accurate, cheaper, faster, free of side effects, and less dangerous than surgery, chemotherapy, and radiation treatment (4/19/2004).

One scientist, University of Michigan's James Baker, believes he has discovered a highly efficient and successful way of delivering cancer-treatment drugs that is less harmful to the surrounding body. Baker has developed a nanotechnology that can locate and then eliminate cancerous cells. He looks at a molecule called a dendrimer. This molecule has over a hundred hooks on it that allow it to attach to cells in the body for a variety of purposes. Baker then attaches folic-acid to a few of the hooks (folic-acid, being a vitamin, is recepted by cells in the body). Cancer cells have more vitamin receptors than normal cells, so Baker's vitamin-laden dendrimer will be absorbed by the cancer cell. To the rest of the hooks on the dendrimer, Baker places anti-cancer drugs that will be absorbed with the dendrimer into the cancer cell, thereby delivering the cancer drug to the cancer cell and nowhere else (Bullis 2006).

Surgery

At Rice University, a flesh welder is used to fuse two pieces of chicken meat into a single piece. The two pieces of chicken are placed together touching. A greenish liquid containing gold-coated nanoshells is dribbled along the seam. An infrared laser is traced along the seam, causing the two side to weld together. This could solve the difficulties and blood leaks caused when the surgeon tries to restitch the arteries he/she has cut during a kidney or heart transplant. The flesh welder could meld the artery into a perfect seal. Nanomedicine has varied applications in Cardiac Surgery also.

Nanorobots

The somewhat speculative claims about the possibility of using nanorobots in medicine, advocates say, would totally change the world of medicine once it is realised. Nanomedicine would make use of these nanorobots, introduced into the body, to repair or detect damages and infections. A typical blood borne medical nanorobot would be between 0.5-3 micrometres in size, because that is the maximum size possible due to capillary passage requirement. Carbon would be the primary element used to build these nanorobots due to the inherent strength and other characteristics of some forms of carbon (diamond/fullerene composites). Cancer can be treated very effectively, according to nanomedicine advocates. Nanorobots could counter the problem of identifying and isolating cancer cells as they could be introduced into the blood stream. These nanorobots would search out cancer affected cells using certain molecular markers. Medical nanorobots would then destroy these cells, and only these cells. Nanomedicines could be a very helpful and hopeful theraphy for patients, since current treatments like radiation therapy and chemotherapy often end up destroying more healthy cells than cancerous ones. From this point of view, it provides a non-depressed therapy for cancer patients. Nanorobots could also be useful in treating vascular disease, physical trauma, and even biological aging.

Nanocomposites, Nanocrystals & Nanoparticles

Researchers at Pacific Northwest National Laboratory have developed a coating process to make sponge-like silica latch onto toxic metals in water. Self-Assembled Monolayers on Mesoporous Supports easily captures such metals as lead and mercury, which are then recovered for reuse or contained in-place forever. © PNNL One example of a SAMMS nanocomposite (Self-Assembled Monolayers on Mesoporous Supports). An hexagonally close-packed cluster of tubular pores (end view) is shown in the foreground. A single pore, in this case coated with a mercaptopropylsiloxy monolayer, is shown in the background. A model of one surfactant molcule is also shown. See Will copper sop up radioactive pollution? for details. A plastic nanocomposite is being used for "step assists" in the GM Safari and Astro Vans. It is scratch-resistant, light-weight, and rust-proof, and generates improvements in strength and reductions in weight, which lead to fuel savings and increased longevity. And in 2001, Toyota started using nanocomposites in a bumper that makes it 60% lighter and twice as resistant to denting and scratching. Impact: Will likely be used on other GM and Toyota models soon, and in other areas of their vehicles, as well as the other auto manufactures, lowering weight, increasing milage, and creating longer-lasting autos. Likely to impact repair shops (fewer repairs needed) and auto insurance companies (fewer claims). Will also likely soon be seen everywhere weight, weather-proofing, durability, and strength are important factors. Expect NASA, the ESA, and other space-faring organisations to take a serious look, soon, which will eventually result in lower lift costs, which will result in more material being lifted into space. Electron micrograph of typical silicon nanocomposite cross section showing uniform distribution of conductive carbon nanotube network. Photo courtesy of U.S. Air Force. See Electrically Conductive Polymer Nanocomposite Materials for details. Strong and light sounds like the perfect recipe for a golf club, of course, which is why so many golf club manufacturers are now devoting big bucks to nano R&D. Thus far, the only company among the big boys to convert research into tangible products is Wilson, which offers three drivers, a fairway wood, four balls and even a golf bag made using nano-materials.

The Chicago-based firm began looking into the potential of nanotechnology six years ago, but didn't come out with a product — the NCODE series of tennis rackets — until 2004. Later in the year, it started shipping drivers and fairway woods, the Pd5, Dd5, Td5 (MSRP — $300) and FwC (MSRP — $200), whose crowns are constructed with nano carbon which, Wilson claims, creates a low-density, high-strength clubhead. Angus Moir, global business director, says use of nano materials sets Wilson apart from the competition. "They make our products more user friendly," he says.

"Metal nanocrystals might be incorporated into car bumpers, making the parts stronger, or into aluminum, making it more wear resistant. Metal nanocrystals might be used to produce bearings that last longer than their conventional counterparts, new types of sensors and components for computers and electronic hardware. Nanocrystals of various metals have been shown to be 100 percent, 200 percent and even as much as 300 percent harder than the same materials in bulk form. Because wear resistance often is dictated by the hardness of a metal, parts made from nanocrystals might last significantly longer than conventional parts." http://news.uns.purdue.edu/UNS/html4ever/020816.Chandrasekar.nano.html Nanocrystals absorb then re-emit the light in a different color — the size of the nanocrystal (in the Angstrom scale) determines the color. Six different quantum dot solutions are shown, excited with a long-wave UV lamp. Quantum dots are molecular-scale optical beacons. "Nanocrystals are an ideal light harvester in photovoltaic devices. (They) absorb sunlight more strongly than dye molecules or bulk semiconductor material, therefore high optical densities can be achieved while maintaining the requirement of thin films. Perfectly crystalline CdSe nanocrystals are also an artificial reaction center, separating the electron hole pair on a femtosecond timescale. Fluorescent nanocrystals have several advantages over organic dye molecules as fluorescent markers in biology. They are incredibly bright and do not photodegrade. Drug-conjugated nanocrystals attach to the protein in an extracellular fashion, enabling movies of protein trafficking. (They) also form the basis of a high-throughput fluorescence assay for drug discovery."

Stain-repellent Eddie Bauer Nano-Care™ khakis, with surface fibers of 10 to 100 nanometers, uses a process that coats each fiber of fabric with "nano-whiskers." Developed by Nano-Tex, a Burlington Industries subsidiary. Dockers also makes khakis, a dress shirt and even a tie treated with what they call "Stain Defender", another example of the same nanoscale cloth treatment. *Impact:* Dry cleaners, detergent and stain-removal makers, carpet and furniture makers, window covering makers.... See Nano-Tex products BASF's annual sales of aqueous polymer dispersion products amount to around $1.65 billion. All of them contain polymer particles ranging from ten to several hundred nanometers in size. Polymer dispersions are found in exterior paints, coatings and adhesives, or are used in the finishing of paper, textiles and leather. Nanotechnology also has applications in the food sector. Many vitamins and their precursors, such as carotinoids, are insoluble in water. However, when skillfully produced and formulated as nanoparticles, these substances can easily be mixed with cold water, and their bioavailability in the human body also increases. Many lemonades and fruit juices contain these specially formulated additives, which often also provide an attractive color. In the cosmetics sector, BASF has for several years been among the leading suppliers of UV absorbers based on nanoparticulate zinc oxide. Incorporated in sun creams, the small particles filter the high-energy radiation out of sunlight. Because of their tiny size, they remain invisible to the naked eye and so the cream is transparent on the skin. From Nanotechnology at BASF Sunscreens are utilizing nanoparticles that are extremely effective at absorbing light, especially in the ultra-violet (UV) range. Due to the particle size, they spread more easily, cover better, and save money since you use less. And they are transparent, unlike traditional screens which are white. These sunscreens are so successful that by 2001 they had captured 60% of the Australian sunscreen market. Makers of sunscreen have to convert to using nanoparticles. And other product manufactures, like packaging makers, will find ways to incorporate them into packages to reduces UV exposure and subsequent spoilage. The $480B packaging and $300B plastics industries will be directly effected. See Big Opportunities for Small Particles Using aluminum nanoparticles, Argonide has created rocket propellants that burn at double the rate. They also produce copper nanoparticles that are incorporated into automotive lubricant to reduce engine wear. AngstroMedica has produced a nanoparticulate-based synthetic bone. "Human bone is made of a calcium and phosphate composite called Hydroxyapatite. By manipulation calcium and phosphate at the molecular level, we have created a patented material that is identical in structure and composition to natural bone. This novel synthetic bone can be used in areas where natural bone is damaged or removed, such as in the in the treatment of fractures and soft tissue injuries."

Nanostructured Materials, Nanoclays and Nanocomposites

Nanodyne makes a tungsten-carbide-cobalt composite powder (grain size less than 15nm) that is used to make a sintered alloy as hard as diamond, which is in turn used to make cutting tools, drill bits, armor plate, and jet engine parts. Every industry that makes parts or components whose properties must include hardness and durability. See Nanostructured Materials Get Tough A PDF document Kodak is producing OLED color screens (made of nanostructured polymer films) for use in car stereos and cell phones. OLEDs (organic light emitting diodes) may enable thinner, lighter, more flexible, less power consuming displays, and other consumer products such as cameras, PDAs, laptops, televisions, and other as yet undreamt of applications. All current makers of CRTs, liquid crystal displays (LCDs), and other display types. See OLEDs get ready to light up the market for flexible screens and KODAK OLED technical details.

Used in packaging, like beer bottles, as a barrier, allowing for thinner material, with a subsequently lighter weight, and greater shelf-life. $480B packaging and $300B plastics industries. Reduced weight means transportation costs decline.

Changing from glass and aluminum — think beer and soda bottles — to plastic reduces production costs. Nanoclays help to hold the pressure and carbonation inside the bottle, increasing shelf life. It is estimated that beer in these containers will gain an extra 60 days (from 120 to 180) of shelf life, reducing spoilage, and decreasing overall costs to the end user. Nanocor is one company producing nanoclays and nanocomposites, for a variety of uses, including flame retardants, barrier film (as in juice containers), and bottle barrier (as shown above). "They are not only used to improve existing products, but also are extending their reach into areas formerly dominated by metal, glass and wood." See Nanocor

Nanotubes

Nanoledge makes carbon nanotubes for commercial uses, of which one mundane (marketing tactic) use is in a tennis racket, made by Babolat. The yoke of the racket bends less during ball impact, improving the player's performance. Once companies like Nanoledge can scale-up their production from grams, to pounds, to tons, and can do so while controlling the type of nanotube they produce, the world becomes their oyster: everywhere strength and weight are a factor — such as in the aerospace, automobile, and airplane industries — they will make a major (disruptive) impact. See French firm hopes to get PR bounce out of nanotubes in tennis rackets Applied Nanotech recently demonstrated a 14" monochrome display based on electron emission from carbon nanotubes. Once the process is perfected, costs will go down, and the high-end market will start being filled. Shortly thereafter, and hand-in-hand with the predictable drop in price of CNTs, production economies-of-scale will enable the costs to drop further still, at which time we will see nanotube-based screens in use everywhere CRTs and view screens are used today. See Applied Nanotech demonstrates carbon nanotube TV

Nanocatalysts & Nanofilters

China's largest coal company (Shenhua Group) has licensed technology from Hydrocarbon Technologies that will enable it to liquify coal and turn it into gas. The process uses a gel-based nanoscale catalyst, which improves the efficiency and reduces the cost. "If the technology lives up to its promise and can economically transform coal into diesel fuel and gasoline, coal-rich countries such as the U.S., China and Germany could depend far less on imported oil. At the same time, acid-rain pollution would be reduced because the liquefaction strips coal of harmful sulfur." See Very Small Business One of the characteristic properties of all nanoparticles has been used from the outset in the manufacture of automotive catalytic converters: The surface area of the particles increases dramatically as the particle size decreases and the weight remains the same. A variety of chemical reactions take place on the surface of the catalyst, and the larger the surface area, the more active the catalyst. Nanoscale catalysts thus open the way for numerous process innovations to make many chemical processes more efficient and resource-saving – in other words more competitive. From Nanotechnology at BASF Argonide Nanomaterials, an Orlando based manufacturer of nanoparticles and nanofiltration products, makes a filter that is capable of filtering the smallest of particles. The performance is due to it's nano size alumina fiber, which attracts and retains sub-micron and nanosize particles. This disposable filter retains 99.9999+% of viruses at water flow rates several hundred times greater than virus-rated ultra porous membranes. It is useful for sterilization of biological, pharmaceutical and medical serums, protein separation, collector/concentrator for biological warfare detectors, and several other applications. In the future, for one application, sterilizing drinking water, this product may have an impact on so-called Third World peoples, who only have access to dubious sources of water. These are just a few of the many ways in which nanotechnology is working itself into our everyday lives. At present, there are no nanobots, no molecular-scale machines, and no assemblers — these are still in the basic research stages, and may not be seen for decades (although many would argue that a concerted effort would bring them to fruition in just a few years). "What we are seeing is the beginning of a revolution, caused by our ability to work on the same scale as nature. Nanotechnology will affect every aspect of our lives, from the medicines we use, to the power of our computers, the energy supplies we require, the food we eat, the cars we drive, the buildings we live in, and the clothes we wear. And it will happen sooner than most people think. By 2010 you won't be able to count the number of businesses affected by nanotechnology." iTim Harper, Founder and Chief Executive Director of the European NanoBusiness Association, and CEO CMP Cientifica

Nano, Bio and Quantum Technologies: An Overview

The size of a single transistor has been reducing in an exponential manner for several decades, leading to integrated circuits containing tens of millions of transistors. But as the size of the transistor decreases a physical limit is encountered where the transistor becomes too small and quantum effects become significant. When this limit is reached the exponential growth in computing power that has been characteristic of the 1980s and 1990s will come to an end. This event is expected to occur somewhere between 2010 and 2020. This will be the end of the road for pure silicon technology. At this point completely new technologies will be needed. So the question is, what is coming next? The field of Nano, Bio and Quantum is concerned with the future, with technologies that do not yet exist. The field at present is therefore one of science, but interestingly, it is technology that is enabling this science of the very small. The technological tools now exist that provide a means of working at the level beyond micro. These tools are allowing scientists to interact with and to manipulate matter at a level that has not been previously possible. At the technological level however, there is an increasing realisation that there is not much to be gained in a practical sense by manipulating single atoms or molecules. Therefore novel concepts are needed such as self fabrication and self assembly, which raise problems both at the physics level but also at the level of architecture, computer science and systems engineering.

What are the challenges of miniaturisation beyond the level of micro technologies? One issue is when to stop investing in the current micro technologies, which can be taken down to the nano scale. Miniaturisation of microelectronics will probably be able to take the size of a transistor down to 30 nanometres. The next issue would be how to take the transistor size down to 5 nanometres, for which entirely new concepts would be needed. The success of microelectronics has been in its scalability. As things became smaller the principles and underlying physics did not change. New steps in the technology built on previous knowledge and the financing of development came from the profits from sales of the previous generation of the technology. Technology development was non-disruptive. The problem with nanotechnologies is that they are not completely scalable. Also there is no miniaturisation perspective. In microelectronics the goal has been to make things smaller, faster and cheaper. In the nano regime one is at the level of single electron transistors, single atoms, or single molecules. So adding value becomes important and that will mean making things more intelligent. Interfacing on the nano scale is going to be a big problem. At this level as soon as an interface is made to a nano object, its properties change. The new concepts in the field of nanotechnology will come from nature. The nano scale operates at the transition point between condensed matter and molecular behaviour and quantum effects. Ideas will therefore come from the fields of biology, chemistry and quantum physics.

Some Useful Contribution of Quantum Physics

One of the new concepts that may have a role to play in the post microelectronics world is quantum computation, which is about using the states of atoms as a basis for computation. Unlike classical logical devices, which only exist in two states (0 or 1), atoms can have three states (0 or 1 or 01 where the latter is a superposition of the first two states). There are several different reasons for the interest in quantum computation. Technologists are interested because they want to address the problems that arise when the physical limits of silicon are reached. Physicists on the other hand, want to understand more about quantum mechanics. Computer scientists however are interested in changes to complexity classes, where problems that are today seen as difficult may become easy to solve. Logicians are also interested in this area because it opens up the possibility of proving logical propositions through a physical process. It is very difficult to build a quantum computer. Single quantum bit (qubit) logic gates are possible. Two qubit gates (a controlled NOT logic gate) are currently being investigated. Building single qubit gates using trapped ion technology is feasible, but the big problem and major difficulty is creating a 2 qubit gate. The difficulty lies in controlling the state of a target atom based upon the state of the control atom, and keeping the system stable over time. Most of the ideas that have gone into quantum computing have been focused on entangling two atoms to create a 2 qubit gate. The successful development of a 2 qubit gate is now very near and it is to be expected that within a few years 10 qubit gates will have been developed. A number of competing technologies are being considered including trapped ions, superconductors and quantum dots. Another issue is how to construct quantum memory. Quantum memory exists and is workable in laboratory environments. This involves storing and trapping single atoms in space. Part of the problems in building quantum computers lies in the area of engineering the components. A major difficulty lies in controlling the quantum effects. There is knowledge about how to control quantum optics in a laboratory environment, but in the long term the emerging technology may be solid state physics. At the moment however this field needs a lot of development work. Another element of quantum computation is quantum communications, which involves transmitting quantum states in a reliable way. One way to do this is to use photons, and to write the quantum state of an atom onto photons and then to transmit these. At the receiver these photons are then used to write the state back onto atoms. The technologies for building this kind of quantum communication system, including error correction, will probably be available in a few years time.

Some Useful Contribution of Biology

Moving on from the area of quantum physics, another discipline that may have a role to play in shaping future generations of computer technologies, is biology. Can chemical processes at the bio cell and molecular level be used for information processing? Can knowledge of biological molecules help with the development of new information technologies? These are very much, open questions. Life has a big advantage in terms of technology. There is 3.5 billion years of evolutionary development. DNA provides an example of long term information storage. It is very compact and replicable, however it is not very fast. So its use as a model for information processing seems to be limited. Short-term information storage is handled in biological systems by energy consuming processes. Brain activity is, for example, linked to increased energy consumption, but again the time scales are very slow when compared with microelectronics. However, in photosynthesis, photon inputs result in virtually instantaneous (pico second range) charge separation which then drives the energy making process in the system. This is an example where biological systems are very fast. Many biological systems are very efficient electron or ion carriers. Life provides examples of sophisticated biological information mechanisms that might provide models for technological concepts. But which components of biological systems provide useful models? Looking to nature for models is a different approach from trying to incorporate biological materials into computers. This latter objective may in fact not work because biological systems tend to be too slow for most information processing applications. For this reason the emphasis should be more on using biology to discover how to improve information processing.

Some Useful Contribution of Chemistry

Another disciplinary area that potentially provides useful concepts for computation is chemistry. Here the focus is on molecular electronic components based on single molecules. A few years ago this idea of molecular electronic was seen as fantasy, but the field is now established and a significant amount of industrial funding is being directed at the area. A lot of work in the field of molecular electronics is focused on reinvesting the transistor at the molecular level. This seems to be the wrong thing to do, as the scale is wrong, since transistors are not molecules. Another way forward is to try and reinvent the computer, to build computers from molecules. The basic architecture for such a computer already exists. An interesting feature of the architecture is that it is default tolerant. Defects in the computer can be ignored. This is a key requirement for a molecular computer as manufacturing processes working at the molecular level are always going to produce defects. Eliminating the defects in manufacturing would be too difficult and too expensive. To make this architecture work switches and wires need to be manufactured using molecules. And this is now possible. One current question is how small these switches and wires can be made.

The semiconductor industry needs to decide when to stop investing in silicon technologies and when to switch over to the new concepts. There is always a temptation to go a little further with existing technologies, when the most appropriate response would be to move on to new technologies. One of the implications of quantum computing is a change to complexity classes, where previously difficult problems become easy to solve. This has major implications for security and privacy as current cryptography techniques rely on current complexity classes. As soon as difficult computational problems become easy to solve today's security systems become useless. In the field of biology the most fruitful path to follow is using nature as a model to find better ways of doing information processing. Biology can provide models that lead to new concepts which would be implemented with non-biological materials. Of great interest are the self-organising capabilities of biological system. If such processes can be better understood, then this might be an area where biology can make a significant input to computer technology. Molecular electronics is based on building computers out of molecular switches and wires. A combination of molecular electronics and conventional CMOS seems to have the potential to extend the exponential growth in computer power for another 50 years. Three dimensions systems have fundamental problems. The interconnections needed to pass information through three dimensions are very complex. A three dimensional system is not defect tolerant. Separation of bits is also difficult in three dimensions. Interfacing at the nano scale is difficult. There is a need for protection from fluctuations in the environment. The nano scale is more than just about developing new computer technology.

The nano scale is also about new materials that may serve other purposes, for example in medical treatments. A big issue is whether the computing power that is foreseen will actually be needed. The answer to this question is that no one really knows. The same question could have been asked about electricity or Charles Babbage's mechanical analytical engine. It is an issue of human curiosity. If more powerful computers are not built then the potential will never be explored or discovered. One potential benefit of more powerful computers would be to get rid of the keyboard and to be able to hold intelligent conversations with a computer. Building powerful computers that are seen as being intelligent is an issue that needs to be discussed. Does society want computers that are capable of thinking? This is an open question. The issue of knowing when to stop long term research was raised. This is important as there are examples of research that have been stopped just as the results were about to become highly relevant. There is a need for faith in ones work and results and determination to bring technologies to market. Many things are open-ended. It is difficult to say something is impossible. Problems that are hard should not be dropped because often the solution to hard problems is very valuable. Technology roadmaps for technology can be constructed. But one problem is how to create roadmaps for algorithms. It is very difficult to predict algorithms. At the moment there are not many good algorithms for quantum computation. Computers have improved significantly over the years, but programrs have not improved to the same extent. There is a need to build machines that can decide how to work in parallel with programrs. The notion of self-assembly is important. To be able to manufacture regular but defective structures and then to be able to work around these defects through self-assembly is a power concept. One of the benefits of biology will be to show how to control complex systems and how to handle complexity, not how to do computing. Biology will provide the insights into how to do things differently.

Research that is high risk in its nature and long term and deals with strange ideas needs public funding. The question of return on investment should not be asked or a return expected. Public money should support high quality high-risk work regardless of return. It is an investment in society. It is impossible to state what is and is not valuable. The research should just be supported. In the longer-term quantum computation may replace classical computation. These are challenges that need to be addressed in this particular field. An important question is how to design larger qubit gates. Scalability is a central and difficult issue. What is likely is a convergence between nanotechnology concepts and quantum optics. However whatever technology is used, the ability to scale up to larger and larger numbers of qubit gates will be a fundamental requirement of the technology.

Small-scale qubit gates are very near in time, but large scale is far away and very large scale is a very far way. Biology shows us that nature uses many different systems for storage and processing of information. Generally life does not care about the system as long as it works. This is an important point and its implications need to be understood. The direct incorporation of biological molecules in current microelectronic components is generally difficult if not impossible. The value of biological systems and molecules is that they offer interesting models that might provide new concepts for information technology. The industrial interest in the area of molecular electronics is concerned with developing strategic technologies that might be useful in 10 to 15 year's time. These technologies are high risk, but are ones where the investment is counted in terms of millions of dollars and the potential payoff comes in billions of dollars. The past several decades have seen a massive improvement in computer technology in terms of size shrinkage and reduction in the power cost of information processing. The question arises how much better can things get and how much further can developments move without resorting to quantum computing? The answer is that there is no physical law that says that efficiency improvements of another factor of one billion cannot be achieved! This could mean that there is no reason why in one handheld computer, consuming one watt of power, it should not be possible to achieve computer power equivalent to all the computers that are in existence in the year 2000! The challenge is how to do this — avoiding problems with physics and economics (in terms of the high costs of building fabrication facilities). The challenge for molecular electronics is not the making switches, or wires or circuits. The main problem is interfacing between the micro scale world and the nano scale world.

Nanotechnology Feature

Visions of self-replicating nanomachines that could devour the Earth in a "grey goo" are probably wide of the mark, but "radical nanotechnology" could still deliver great benefits to society. The question is how best to achieve this goal Nanotechnology is slowly creeping into popular culture, but not in a way that most scientists will like. There is a great example in *Dorian* — novelist Will Self's modern reworking of Oscar Wilde's *The Picture of Dorian Gray*. In one scene, set in a dingy industrial building on the outskirts of Los Angeles, we find Dorian Gray and his friends looking across rows of Dewar flasks, in which the heads and bodies of the dead are kept frozen, waiting for the day when medical science has advanced far enough to cure their ailments. Although one of Dorian's friends doubts that technology will ever be able to repair the damage caused when the body parts are thawed out, another friend -- Fergus the Ferret — is more optimistic.

- Course they will, the Ferret yawned; Dorian says they'll do it with nannywhatsit, little robot thingies — isn't that it, Dorian?
- *Nanotechnology, Fergus — you're quite right; they'll have tiny hyperintelligent robots working in concert to repair our damaged bodies.*

This view that nanotechnology will lead to tiny robotic submarines navigating our bloodstream is ubiquitous, and images are frequently used to illustrate stories about nanotechnology in the press. Yet today's products of nanotechnology are much more mundane — stain-resistant trousers, better sun creams and tennis rackets reinforced with carbon nanotubes. There is an almost surreal gap between what the technology is believed to promise and what it actually delivers. The reason for this disparity is that most definitions of nanotechnology are impossibly broad. They assume that any branch of technology that results from our ability to control and manipulate matter on length scales of 1-100 nm can be counted as nanotechnology. However, many successes that are attributed to nanotechnology are merely the result of years of research into conventional fields like materials or colloid science. It is therefore helpful to break up the definition of nanotechnology a little. What we could call "incremental nanotechnology" involves improving the properties of many materials by controlling their nano-scale structure. Plastics, for example, can be reinforced using nano-scale clay particles, making them stronger, stiffer and more chemically resistant. Cosmetics can be formulated such that the oil phase is much more finely dispersed, thereby improving the feel of the product on the skin. These are the sorts of commercially available products that are said to be based on nanotechnology. The science underlying them is sophisticated and the products are often big improvements on what has gone before. However, they do not really represent a decisive break from the past. In "evolutionary nanotechnology" we move beyond simple materials that have been redesigned at the nano-scale to actual nano-scale devices that do something interesting. Such devices can, for example, sense the environment, process information or convert energy from one form to another. They include nano-scale sensors, which exploit the huge surface area of carbon nanotubes and other nano-structured materials to detect environmental contaminants or biochemicals. Other products of evolutionary nanotechnology are semiconductor nanostructures — such as quantum dots and quantum wells — that are being used to build better solid-state lasers. Scientists are also developing ever more sophisticated ways of encapsulating molecules and delivering them on demand for targeted drug delivery. Taken together, incremental and evolutionary nanotechnology are driving the current excitement in industry and academia for all things nano-scale. The biggest steps are currently being made in evolutionary nanotechnology, more and more products of which should appear on the market over the next five years.

Radical Nanotechnology

Even if the most extreme visions of the nanotechnology evangelists do not come to pass, nanotechnology — in the form of machines structured on the nano-scale that do interesting and useful things — will certainly play a growing part in our lives over the next half-century. How revolutionary the impact of these new technologies will be is difficult to say. Scientists almost always greatly overestimate how much can be done over a 10 year period, but underestimate what can be done in 50 years. Sometimes the contrast between the grand visions of nanotechnology — robotic nanosubmarines repairing our bodies — and the reality it delivers — say an improved all-in-one shampoo and conditioner — has a profoundly bathetic quality. But the experience we will gain in manipulating matter on the nano-scale in industrial quantities is going to be invaluable. Similarly, there is no point being dismissive about the fact that lots of early applications of nanotechnology will be essentially toys — whether for children or adults — just as data-storage technology is currently being driven forward by the needs of digital TV recorders and portable music players like Apple's iPod. These apparently frivolous applications will provide the incentive and resources to push the technology further. But which design philosophy of radical nanotechnology will prevail — Drexler's original "diamondoid" visions or something closer to the marvellous contrivances of cell biology? One way of finding the answer would be to simply develop the existing technologies that have driven the relentless miniaturization of microelectronics. This "top-down" approach, which uses techniques like photolithography and etching, has already been used to make so-called microelectromechanical systems (MEMS). Such systems are commercially available and have components on length scales of many microns — the acceleration sensors in airbags being a well known example. All we need to do now is shrink these systems even further to create true nanoelectromechanical systems, or NEMS. The advantage of this top-down approach is that a massive amount of existing technology and understanding is already in place. The investment, both in terms of plant, and research and development, is currently huge, driven as it is by the vast economic power of the electronics and computing industries. But, as we have seen, the disadvantage is that there are both physical and economic bounds to how small this technology can go. Although industry has shown extraordinary ingenuity in overcoming seemingly insurmountable barriers already — new ultraviolet light sources and phase-shifting masks have made feature sizes below 100 nm a commercial reality — maybe its luck will soon run out. A more fundamental problem is the importance in the nanoworld of Brownian motion and surface forces. Strong surface forces may make the moving parts of a NEMS device stick together and seize up.

Grey Goo and Radical Nanotechnology

But where does this leave the original vision of nanotechnology as articulated by Eric Drexler? Back in 1986 Drexler published an influential book called *Engines of Creation: The Coming Era of Nanotechnology*, in which he imagined sophisticated nano-scale machines that could operate with atomic precision. We might call this goal "radical nanotechnology". Drexler envisaged a particular way of achieving radical nanotechnology, which involved using hard materials like diamond to fabricate complex nano-scale structures by moving reactive molecular fragments into position. His approach was essentially mechanical, whereby tiny cogs, gears and bearings are integrated to make tiny robot factories, probes and vehicles Drexler's most compelling argument that radical nanotechnology must be possible is that cell biology gives us endless examples of sophisticated nano-scale machines. These include molecular motors of the kind that make up our muscles, which can convert chemical energy to mechanical energy with astonishingly high efficiencies. There are also ion channels and ion pumps that can control the flow of molecules through membranes. Other examples include ribosomes — molecular structures that can construct protein molecules, amino acid by amino acid, with ultimate precision according to the instructions on DNA. Drexler argued that if biology works as well as it does, researchers ought to be able to do much better. Biology, after all, uses unpromising soft materials — proteins, lipids and polysaccharides — and random design methods that are restricted by the accidents of evolution. Motion is created by changes to the shapes of these molecules, rather than through the cogs and pistons of macroscopic engineering.

Furthermore, molecules are moved around through their continual bombardment by other molecules — what is known as Brownian motion — rather than via pipes and tubes. We researchers, however, have the best materials at our disposal. Surely we can create what are, in effect, synthetic life forms that can reproduce and adapt to the environment and overcome "normal" life in the competition for resources? Drexler's book raised one big spectre. By engineering a synthetic life form that could create runaway self-replicating machines, we might eventually render all normal life extinct. Could we make, by accident or malevolent design, a plague of self-replicating nanorobots that spreads across the biosphere, consuming its resources and rendering life, including ourselves, extinct? This scary possibility was dubbed by Drexler as the "grey goo" scenario. It is what triggered much of the public's doubts about nanotechnology and was the inspiration for Michael Crichton's novel *Prey*, which is shortly to be turned into a film. However, many scientists simply dismissed Drexler's visions of tiny nano-scale robots as science fiction, so self-evidently absurd as not to be worth considering. Indeed, Drexler himself has recently declared that self-replicating machines are not, after all, necessary for molecular nanotechnology.

Flaws in Drexler's Vision

It is nevertheless worth examining the shortcomings of Drexler's original vision because this may give clues as to how we might make radical nanotechnology feasible. Why, for example, do illustrations of nanosubmarines look so absurd to a scientific eye? The reason is that these pictures assume that the engineering that we employ on macroscopic scales can simply be scaled down to the nano-scale. But physics looks very different at such dimensions. Designs that function well in our macroscopic world will work less and less well as they shrink in size. A nanosubmarine would operate in a very different environment to its macroscopic counterpart. Small objects have lower Reynolds numbers — a dimensionless quantity proportional to the ratio of the product of the size and flow speed to viscosity. The dominating force opposing motion therefore arises from viscosity rather than inertia. Fluid molecules, meanwhile, will continually bombard the object because of Brownian motion. The submarine would therefore be perpetually jostled around, while its internal parts and mechanisms would bend and flex in constant random motion. Another difference at the nano-scale is that surface forces are very strong: the nanosubmarine would probably just stick to the first surface it encountered. These three factors — low Reynolds numbers, ubiquitous Brownian motion and strong surface forces — are what makes nano-scale design very challenging, at least at ambient temperatures in the presence of water. So is radical nanotechnology simply impossible? What biology teaches us is that, contrary to Drexler's implicit position, life is highly optimized, by billions of years of evolution, for the particular type of physics that operates at the nano-scale. The principles of self-assembly and molecular shape change that cell biology uses so extensively exploit the special physics of the nanoworld — namely ubiquitous Brownian motion and strong surface forces. In other words, if we want to fulfil the goals of radical nanotechnology, we should use soft materials and biological design paradigms. We should also stop worrying about grey goo, because it is going to be very hard to produce more highly optimized nano-scale organisms than nature has already achieved.

Taking a Lead from Nature

So how could we follow biology's example and work with the "grain" of the nanoworld? The most obvious method is simply to exploit the existing components that nature gives us. One way would be to deliberately remove and isolate from their natural habitats a number of components, such as molecular motors, and then incorporate them into artificial nanostructures. For example, Nadrian Seeman at New York University and others have shown how the self-assembly properties of DNA can be used to create quite complicated nano-scale structures and devices. Another approach would be to start with a whole, living organism — probably a simple bacterium — and then genetically engineer a stripped-down version that contains only the components that we are interested in. One can think of this approach — often called "bionanotechnology" — as the *Mad Max* or *Scrap Heap Challenge* approach to nano-engineering. We are stripping down and then partially reassembling a very complex and only partially understood system to obtain something else that works. This approach exploits the fact that evolution — nature's remarkable optimization tool — has produced very powerful and efficient nanomachines. We now understand enough about biology to be able to separate out a cell's components and to some extent utilize them outside the context of a living cell — as illustrated in the work of Carlo Montemagno at the University of California at Los Angeles and Harold Craighead from Cornell University. This approach is quick and the most likely way to achieve radical nanotechnology soon. As we learn more about how bionanotechnology works, it should be possible to use some of the design methods of biology and apply them to synthetic materials. Like bionanotechnology, such "biomimetic nanotechnology" would work with the grain of the special physics of the nanoworld.

Of course, the task of copying even life's simplest mechanisms is formidably hard. Proteins, for example, function so well as enzymes because the particular sequence of amino acids has been selected by evolution from a myriad of possibilities. So when designing synthetic molecules, we need to take note of how evolution achieved this. But despite the difficulties, biomimetic nanotechnology will let us do some useful — if crude — things.

For example, ALZA, a subsidiary of Johnson and Johnson, has already been able to wrap a drug molecule in a nanoscopic container — in this case a spherical shell made from double layers of phospholipid molecules — and transport it to where it is required in the body. The container can then be made to open and release its bounty. I do not think that Drexler's alternative approach — based on mechanical devices made from rigid materials — fundamentally contradicts any physical laws, but I fear that its proponents underestimate the problems that certain features of the nanoworld will pose for it. The close tolerances that we take for granted in macroscopic engineering will be very difficult to achieve at the nano-scale because the machines will be shaken about so much by Brownian motion. Finding ways for surfaces to slide past each other without sticking together or feeling excessive friction is going to be difficult. Unlike the top-down route using silicon, we have no large base of experience and expertise to draw on, and no big economic pressures driving the research forward. And unlike the bionanotechnological and biomimetic approaches, it is working against the grain, rather than with the grain, of the special physics of the nanoworld. Drexler's approach to radical nanotechnology, in other words, is the least likely to deliver results.

Concerns and Fears

Assuming that some kind of radical nanotechnology is possible and feasible, the question is whether we should even want these developments to take place. Some 50 years ago it was generally taken for granted that scientific progress was good for society, but this is certainly not the case now.

In some quarters, there are calls for a cautious approach to nanotechnology; at the most extreme, there are demands for a complete moratorium on the development of the technology. In the light of these concerns, the UK government last year asked the Royal Society and the Royal Academy of Engineering to carry out a major survey into the benefits and possible problems of nanotechnology. The report, which is based on extensive collaboration with the public, has just been published. There are two key concerns as far as the public is concerned. The first relates to the kind of incremental nanotechnology that is already at or near market — namely that finely divided matter might be intrinsically more toxic than the forms in which we normally encounter it. If the properties of matter are so dramatically affected by size, the argument goes, matter that is harmless in bulk quantities might be more toxic and more effective at getting into our bodies when it is in the form of nano-scale particles. We know that the physical form of a material can drastically affect its toxicity. One sobering example is asbestos, which comes in two chemically identical forms — serpentine and chrysotile asbestos. While the former is a harmless mineral that consists of flat sheets of atoms, the latter contains nano-scale tubes of atoms. Exposure to this tubular form is what has killed so many people from lung cancer and other diseases.

Carbon nanotubes, like chrysotile, are the rolled-up version of a sheet-forming mineral that itself is not toxic — in this case, graphite. Although we have no definitive evidence that carbon nanotubes are dangerously toxic, prudence certainly suggests that we should be careful when handling them. After all, every new material has the potential to be toxic. Regulations controlling the introduction of new materials into the workplace and the environment are, rightly, much stricter now than in the past, and we should appreciate that the properties of materials depend on their physical manifestation as well as their chemical content. But we do not have to assume that all nano-scale materials are inherently dangerous. Imposing a blanket ban would be absurd and unenforceable, simply because we have enough experience of many forms of nanoparticles to know they are safe. If we wanted to avoid nanoparticles completely, we would have to give up drinking milk, full as it is of nano-scale casein particles. Evolutionary nanotechnology is certainly going to lead to far-reaching changes in society, which we should get to grips with now. It will allow computing that is so cheap and powerful that every product or gadget — no matter the price — will be able to process, sense and transmit information.

Radio-frequency identification chips, which are already available, are just the beginning. But the prospect of cheap, powerful, computing — when combined with mass storage and automated image processing -- is a totalitarian's dream and a libertarian's nightmare. The public's second big fear of nanotechnology — beyond these concrete social, environmental and economic factors — concerns the proper relationship between man and nature. Is it right to take living organisms from nature and then reassemble and reconstruct their most basic structures, possibly with additional synthetic components? By replacing living parts of the body with man-made artefacts, are we blurring the line between man and machine? These fears are at the root of the most far-reaching concern about nanotechnology — the grey-goo problem. Of course, fear of loss of control is a primal fear about any technology. The question is whether it is realistic to worry about it. We should be clear about what this proposition implies: that we can out-engineer evolution by making an entirely synthetic form of life that is better adapted to the Earth's environment than life itself is. Such a feat is unrealistic in the next 20 years — and probably for a lot longer. We simply do not have a detailed enough knowledge of how life itself works. We have the "parts list", but very little understanding of how it all fits together and operates as a complex system. Still, our appreciation of how nature engineers at the nano-scale will grow rapidly, and attempts to mimic some of the functions of life will help us to appreciate how biology operates. But is it even possible in principle to develop a different form of life that works better than the one that currently exists? To find out, we need to take a view on how perfectly adapted life is to its environment. We need to know how many times life got started and how many alternative schemes were tried and failed.

CASE STUDY: C-MAC MICROTECHNOLOGY

C-MAC Micro Technology is the world leader high- reliability electronic systems, modules and components for the automotive, medical, communications and aerospace industries. We offer customers unrivalled expertise in the design and manufacture of advanced high-precision microelectronics systems, as well as over 70 years' experience in high-performance oscillators and quartz crystals. C-MAC employs best-in-class customer service, C-MAacturing and engineering techniques through leveraging these combined strengths and capabilities. With manufacturing facilities worldwide, and a technical sales network, C-MAC MicroTechnology provides customers both leading-edge technology and unprecedented levels of engineering support wherever and whenever they are needed. C-MAC MicroTechnology offers world class product solutions in response to customer and market requirements, addressing the complex analog, digital and frequency control requirements of modern systems. Our customers are the world's leading telecommunications, automotive, military,aerospace and medical manufacturers, with whom we have developed long-standing strategic partnerships.

Frequency Products

C-MAC MicroTechnology's design and manufacturing capability stems from its market-leading array of precision frequency control products that address the needs of the communications, commercial, industrial, automotive and military markets. C-MAC MicroTechnology's history with quartz crystal products spans over 65 years and comprises five ISO9001 certified design and manufacturing centers worldwide C-MAC. MicroTechnology has introduced an Online Frequency Products Data Book to ensure that customers have access to C-MAC MicroTechnology's products and information. We have combined the product information, distributor details and product specific solutions, of the traditional printed Data Book, with the speed, flexibility and accessibility of the Internet. Optoelectronics C-MAC MicroTechnology's dedicated Optoelectronics Engineering Group (OEG) has brought together a wide technology portfolio, oriented towards design and manufacture of optical module packaging. C-MAC MicroTechnology offers a range of services, from design of individual analog, digital or mixed-signal modules, to the development, mass production and test of finished produts from customer concept.

Microcircuits

C-MAC MicroTechnology is a world leader in the supply of microelectronic products and services, with multiple manufacturing facilities throughout North America, Europe and Asia. Extensive design and manufacturing experience in thick film substrates, surface mount hybrids, chip and wire hybrids, multi-chip modules and System-in-a-Package enables them to offer integrated microelectronic solutions to address a range of complex system challenges. Engineering expertise includes extensive circuit design and microelectronic layout skills, supported by state-of-the-art CAD and CAE systems, and offering simulated electronic and thermal performance at the earliest project stages. In addition, considerable experience in silicon design enables C-MAC MicroTechnology to provide cost effective ASIC programmes for a wide variety of electronic packaging products. C-MAC MicroTechnology has established a reputation for providing high reliability microelectronic components for Space integration. The product offering ranges from simple resistor networks to semiconductor die packs, to full custom hybrids, all with Class K screening. An integrated design to engineering package is provided, featuring analog, digital and thermal simulation prior to thick film design realisation. In addition, a separate suite for full custom or FPGA ASIC design requirements is available. We are a major manufacturer of custom hybrids for the military market, with a long history in the design and manufacture of avionics databus products for MIL-STD-1553 and MIL-STD-1760 applications. To achieve the functionality and cost effectiveness required by the military sector, C-MAC MicroTechnology has invested considerably in the development of full and semi-custom ASICs that are at the heart of these products. Many millions of pieces per annum of thick film hybrid circuits are supplied to the Telecommunications market by C-MAC MicroTechnology. Typical applications include central office switching (line cards), enterprise switches and transmission products.

REFERENCES

Albus, J.S.: 2003, 'Engineering of Mind for Enhancing Human Productivity', in: M.C. Roco, W.S. Bainbridge (eds.), *Converging Technologies for Improving Human Performance*, Kluwer, Dordrecht, pp. 281-293.

Altmann, J. & Gubrud, M.A.: 2004, 'Military, Arms Control, and Security Aspects of Nanotechnology', in: D. Baird, A. Nordmann & J. Schummer (eds.), *Discovering the Nanoscale*, Amsterdam: IOS Press, pp. 269-277.

Altmann, J. Military Uses of Microsystem Technologies. Dangers and Preventive Arms Control. Münster: Agenda Verlag, 2001.

Altmann, J.: 2004, 'Military Uses of Nanotechnology: Perspectives and Concerns', *Security Dialogue*, 35 (1): 61-79.

Alves, H., M. Ferreira, U. Koster, and B. Muller. 1996. *Materials Science Forum* 225–227: 769.

Andres, R.P., S. Datta, D.B. Janes, C.P. Kubiak, and R. Reifenberger. 1998. The design, fabrication and electronic properties of self-assembled molecular nanostructures. In *The handbook of nanostructured materials and nanotechnology*, ed. H.S. Nalwa. San Diego: Academic Press.

Andrew Pollack, *Atom by Atom, Scientists Build 'Invisible' Machines of the Future*, The New York Times Science Section, Tuesday, November 26, 1991, page B7.

Anne Eisberg, "Benign Viruses Shine on the Silicon Assembly Line," New York Times, February 12, 2004.

Anonymous: 2002, *Small Wonders, Endless Frontiers: A Review of the National Nanotechnology Initiative*, National Academies Press, Washington, DC.

Antonio Regalado, "Nanotechnology Patents Surge as Companies Vie to Stake Claim," Wall Street Journal, June 18, 2004, p. 1.

Arap W., R. Pasqualini, E. Ruoslahti. 1998. "Cancer Treatment by Targeted Drug Delivery to Tumor Vasculature in a Mouse Model", *Science,* Jan 16; 279 (5349): pp. 377-380.

Arendt, H. The Human Condition, The University of Chicago Press, 1958.

Arias, T.A. and Sohrab Ismail-Beigi. 2000. Novel algebraic formulation of density-functional calculation. *Computer Physics Communications*. In press.

Arriagada, F.J., and K. Osseo-Asave. 1995. *J. Colloid Interface Sci.* 170:8.

Arribart, H.: 2004, 'Les nanomatériaux autres que ceux des technologies de l'information et des communications (TICS)', in: Académie des sciences, Académie des technologies (eds.), *Nanosciences, nanotechnologies*, éditions Tec & Doc, Paris, pp. 361-382.

Arthur C. Clarke, *Profiles of the Future*, Harper & Row, Publishers, New York, 1962.

Asada, M., Y. Mayamoto, and Y. Suematsu. 1986. *IEEE Journ. Quantum Electronics* QE-22(9): 1915-1921.

Atkins, P.: 2003 'The Future of Matter', in: J. Brockman (ed.), *The Next Fifty Years: Science in the First Half of the Twenty-First Century*, Vintage Books, New York, pp.194-205.

Attard, G.S., et al. 1997. *Science* 278:838. Averback, R.S., J. Bernholc, and D.L. Nelson. 1991. *MRS Symposium Proceedings* Vol. 206.

Bai, Q., K.D. Wise, and D.J. Anderson. 2000. A high-yield microassembly structure for three-dimensional microelectrode arrays. *IEEE Trans. Biomed Eng.* 47(3): p. 281-9.

Baird, D.; Nordmann, A. & Schummer, J. (eds.): 2004, *Discovering the Nanoscale*, Amsterdam: IOS Press.

Baker, R.T.K. Synthesis, properties and applications of graphite nanofibers. In *R&D status and trends*, ed. Siegel et al.

Baker, R.T.K. Synthesis, properties and applications of graphite nanofibers. In *R&D status and trends*, ed. Siegel et al.

Ball, P.: 2002, 'Tutorial: Natural Strategies for the Molecular Engineer', *Nanotechnology*, 13, 15-28.

Ball, P.: 2003, 'Nanotechnology in the firing line', *Nanotechweb.org,* 23 December [http://www.nanotechweb.org/articles/society/2/12/1/1].

Barber, Benjamin. 1996. *Jihad vs. McWorld: How Globalism and Tribalism are Reshaping the World* (New York: Ballantine Books).

Baschnagel, J., K. Binder, K. Kremer et al. 2000. Bridging the gap between atomistic and coarse-grained models of polymers: status and perspectives. *Advances in Polymer Science*. Volume 152. Springer- Verlag.

Bass, T.A., "Gene Genie," Wired Magazine, August 1995, p. 114. *Well-informed and well-written survey of the technical developments and people in the new field of DNA computing.*

Bate, R., Frazier, G., Frensley, W., Reed., M., "An Overview of Nanoelectronics," *Texas Instruments Technical Journal*, July-August 1989, pp. 13-20.

Baum, R.; Drexler, K.E.; Smalley, R.: 2003, 'Nanotechnology: Drexler and Smalley Make the Case For and Against 'Molecular Assemblers', *Chemical and Engineering News*, 81, 37-42.

Beck, J.S., J.C. Vartuli, W.J. Roth, M.E. Leonowicz, C.T. Kresge, K.D. Schmitt, C.T.-W. Chu, D.H. Olsen, E.W. Shepard, S.B. McCullen, J.B. Higgins, and J.L. Schlenker. 1992. *J. Am. Chem. Soc.* 114:10834.

Becker, M.F., J.R. Brock, H. Cai, N. Chaudhary, D. Henneke, L. Hilsz, J.W. Keto, J. Lee, W.T. Nichols, and H.D. Glicksman. 1997. Nanoparticles generated by laser ablation. In *Proc. of the Joint NSF-NIST Conf. on Nanoparticles*.

Ben-David, Joseph and Randall Collins, 1966. "Social Factors in the Origins of a New Science: the Case of Psychology." *American Sociological Review.* Vol. 31 pp 451-465.

Bennett, C. H., "Quantum Information and Computation," *Physics Today*, October

Bensaude-Vincent, B.: 2001, 'The Construction of a Discipline: Materials Science in the United States', *Historical Studies in the Physical and Biological Sciences*, 31, 223-248.

Bensaude-Vincent, B.: 2004, 'Two Cultures of Nanotechnology?', *Hyle: International Journal for Philosophy of Chemistry*, 10(2), 65-82.

Bensaude-Vincent, B.; Arribart, B.Y.; Sanchez, C.: 2002, 'Chemists at the School of Nature', *New Journal of Chemistry*, 26, 1-5.

Berger, R., C. Gerber, H.P. Lang, and J.K. Gimzewski. 1996. Micromechanics: a toolbox for femtoscale science: "Towards a laboratory on a tip." *Microelectronic Engineering*, 35:373-9. (International Conference on Micro- and Nanofabrication, Glasgow, U.K., 22- 25 Sept. 1996).

Berkowitz, A.E., J.R. Mitchell, M.J. Carey, A.P. Young, S. Zhang, F.E. Spada, F.T. Parker, A. Hutten, and G. Thomas. 1992. *Phys. Rev. Lett.* 68:3745.

Berndt, C.C., J. Karthikeyan, T. Chraska, and A.H. King. 1997. Plasma spray synthesis of nanozirconia powder. In *Proc. of the Joint NSF-NIST Conf. on Nanoparticles.*

Berube, D.M.: 2004, 'The Rhetoric of Nanotechnology', in: D. Baird, A. Nordmann & J. Schummer (eds.), *Discovering the Nanoscale*, Amsterdam: IOS Press, pp. 173-192.

Bijker, W.E.: 1995a, *Of Bicycles, Bakelite, and Bulbs: Toward a Theory of Sociotechnical Change*, MIT Press, Cambridge, MA.

Bijker, W.E.: 1995b, 'Sociohistorical Technology Studies', in: S. Jasanoff, G.E. Markle, J. Petersen, T. Pinch (eds.), *Handbook of Science and Technology Studies*, Sage, London, pp. 229-256.

Bijker, W.E.; Law, J.: 1992, 'Do Technologies Have Trajectories?', in: W.E. Bijker, J. Law (eds.), *Shaping Technology/Building Society: Studies in Sociotechnical Change*, MIT Press, Cambridge, MA, pp. 1-16.

Bijker, W.E.; Pinch, T.: 1987, 'The Social Construction of Facts and Artifacts: Or How the Sociology of Science and the Sociology of Technology Might Benefit Each Other', in: W.E. Bijker, T.P. Hughes, T. Pinch (eds.), *The Social Construction of Technological Systems: New Directions in the Sociology and History of Technology*, MIT Press, Cambridge, MA, pp. 17-50.

Billings P.R., M.A. Kohn, M. de Cuevas, J. Beckwith, J.S. Alper, M.R. Natowicz. 1992. "Discrimination as a Consequence of Genetic Testing," *American Journal of Human Genetics*, Mar; 50 (3): pp. 476-82. Drews, J. 1996. "Genomic Sciences and the Medicine of Tomorrow," *Nature Biotechnology*, Nov., 14 (11), pp. 1516-1518.

Bimber, B., and Guston, D., eds., *Technology Assessment: The End of OTA*, special issue of *Technology Forecasting and Social Change*, Vol. 54, No. 2 and 3, 1997

Bimber, B.: 1994, 'Three Faces of Technological Determinism', in: M.R. Smith, L. Marx (eds.), *Does Technology Drive History?: The Dilemma of Technological Determinism*, MIT Press, Cambridge, MA, pp. 79-100.

Bimberg, D., N. Kirstaedter, N.N. Ledentsov, Zh.I. Alferov, P.S. Kopev, and V.M. Ustinov. 1998. InGaAs-GaAs quantum-dot lasers. *IEEE J. of Selected Topics in Quant. Electronics* 3:196-205.

Binnig, G., Quate, C.F., and Gerber, Ch. (1986) Atomic force microscope. Phys. Rev. Lett. 56(9), 930-933

Bishop, A.R., and R.G. Nuzzo. 1996. Self-assembled monolayers: Recent developments and applications. *Current Opinion in Colloid & Interface Sci.* 1:127-136.

Bishop, J. et al. 1990. *Surface modified drug nanoparticles.* U.S. Patent application. Docket 61894 Filed 9/17/90.

Black, C.T.; Murray, C.B.; Sandstrom, R.L.; Sun, S.H. 2000 "Spin-dependent tunneling in self-assembled cobalt-nanocrystal superlattices", *Science*, 290, 1131-1134.

Blackman, M. 1994. *An evaluation of the Link Nanotechnology Programme and the National Initiative on Nanotechnology.* Cambridge.

Blumenthal, D. 1992. "Academic-Industry Relationships: Extent, Consequences, and Management," *The Journal of the American Medical Association*, 268:3, December, pp. 3344 -3349.

Blumenthal, D.M., M. Gluck, K.S. Louis, M. Stoto, and D. Wise. 1986. "University-Industry Research Relationships in Biotechnology: Implications for the University," *Science* 232: pp. 1361-1366 (June 13).

BMBF. 1996. "Delphi-Bericht 1995 zur Entwicklung von Wissenschaft und Technik – Mini-Delphi." Bonn, author's translation.

Bockrath, M., D.H. Cobden, P.L. McEuen, N.G. Chopra, A. Zettl, A. Thess, and R.E. Smalley. 1997. "Single-Electron Transport in Ropes of Carbon Nanotubes." *Science*, 275.

Bohn, R., T. Haubold, R Birringer, and H. Gleiter. 1991. *Scripta Metall. Mater.* 25:811.

Bonnemain, B. 1998. "Superparamagnetic agents in magnetic resonance imaging: Physicochemical characteristics and clinical applications - A review", *Journal of Drug Targeting*, 6, 167-174.

Bowes, C.L., A. Malek, and G.A. Ozin. 1996. *Chem. Vap. Deposition* 2: 97.

Brave, Ralph. 2000. "Genome Hell is Revealed in Many Forms." *Baltimore Sun*, July 2, p. 1C.

Brinker, C.J. 1996. *Curr. Opin. Solid State Mater. Sci.* 1:798.

Broderick, D. The Spike, Forge, New York, 2001.

Brooks, L.J.: 2003, *The Institute for the Future: A Site for Mapping, Maintaining, and Transforming Images of the Future*, (Ph.D. dissertation) University of California, San Diego, La Jolla.

Brotzman, R. 1998. Nanoparticle Dispersions. In *R&D status and trends*, ed. Siegel et al. Bu, X., P. Feng, and G.D. Stucky. 1998. Large-cage zeolite structures with multidimensional 12-ring channels. *Science* 278:2080-2085.

Brotzman, R. 1998. Nanoparticle dispersions. In *R&D status and trends*, ed. Siegel et al. Brus, L.E. 1996. Theoretical metastability of semiconductor crystallites in high pressure phases with applications to beta tin structures of silicon. *J. Am. Chem. Soc.* 118:4834-38.

Brown, J. S. and Duguid, P. Don't Count Society Out: A Response to Bill Joy. Paper prepared for the Societal Implications of Nanotechnology Conference, 2000.

Brown, J.S. and P. Daguid. 2001. "Don't Count Society Out: A Response to Bill Joy" in this volume: *Societal Implications of Nanoscience and Nanotechnology.* National Science Foundation, Arlington, VA.

Brown, Lawrence A. 1981. *Innovation Diffusion: A New Perspective*. New York: Methuen.

Brown, N.: 2003, 'Hope Against Hype – Accountability in Biopasts, Presents and Futures', *Science Studies*, 16(2), 3-21.

Brown, N.: 2004, 'Needle on the Real: Technoscience and Poetry at the Limits of Fabrication', in: N.K. Hayles (ed.), *Nanoculture: Implications of the New Technoscience*, Bristol, UK: Intellect Books, pp. 173-190.

Bruchez, M.; Moronne, M.; Gin, P.; Weiss, S.; Alivisatos, A.P. 1998. "Semiconductor nanocrystals as fluorescent biological labels", *Science*, 281, 2013-2016.

Brus, L. 1996. Semiconductor colloids: Individual nanocrystals, opals and porous silicon. *Current Opinion in Colloid & Interface Science* 1:197-201. *Chemical Engineering News*. 1997. Particulate matter health studies to be reanalyzed (August 18):33.

Bueno, O.: 2004, 'The Drexler-Smalley Debated on Nanotechnology: Incommensurability at Work?', *Hyle: International Journal for Philosophy of Chemistry*, 10(2), 83-98.

Bueno, O.: 2004, 'Von Neumann, Self-Reproduction and the Constitution of Nanophenomena', in: D. Baird, A. Nordmann & J. Schummer (eds.), *Discovering the Nanoscale*, Amsterdam: IOS Press, pp. 101-115.

Bush, V.: 1945, *Science, the Endless Frontier*, US Government Printing Office, Washington, DC.

Bush, Vannevar. 1945. *Science, the Endless Frontier*. Washington D.C. U.S. Government Printing Office.

Canguilhem, G.: 1952, *Machine et organisme' in La connaissance de la vie*, Hachette, Paris [quoted from the fourth edition Vrin, Paris, 1971].

Carroll, J.S.: 2001, 'Social Science Research Methods for Assessing Societal Implications of Nanotechnology', in: M.C. Roco and W.S. Bainbridge (eds.), *Societal Implications of Nanoscience and Nanotechnology*, Dordrecht: Kluwer, pp. 188-192.

Carroll, J.S.: 2001, 'Social Science Research Methods for Assessing Societal Implications of Nanotechnology', in: M.C. Roco, W.S. Bainbridge (eds.), *Societal Implications of Nanoscience and Nanotechnology*, National Science Foundation, pp. 188-192.

Carsley, J.E., R. Shaik, W.W. Milligan, and E.C. Aifantis. 1997. In *Chemistry and physics of nanostructures and related non-equilibrium materials*. ed. E. Ma, B. Fultz, R. Shull, J. Morral, and P. Nash. Warrendale, PA: TMS.

Casey, Jr., H.C., and M.B. Panish. 1978. *Heterostructure Lasers, Part A: Fundamental Principles*. (Academic Press. New York). pp. 1-9.

Cetron, M., and Davies, O. 1997. *Probable Tomorrows: How Science and Technology Will Transform Our Lives in the Next Twenty Years* (New York: St. Martin's Press).

Chan, W.C.W. and S. Nie. 1998. Quantum dot bioconjugates for ultrasensitive nonisotopic detection. *Science*. 281, 2016 (25 September).

Chance, B., Mueller, P., DeVault, D. & Powers, L. (1980) *Phys. Today* 33 (10), 32-38.

Chang, K. "Smaller Computer Chips Built Using DNA as Template." New York Times, November 21, 2003.

Chantrell. 1994. *J. Appl. Phys.* 76:6811. Erb, U., G. Palumbo, R. Zugic, and K.T. Aust. 1996. In *Processing and properties of nanocrystalline materials*, ed. C. Suryanarayana, J. Singh, and F.H. Froes. Warrendale, PA: TMS. Gertsman, V.Y., M. Hoffman, H. Gleiter and R. Birringer. 1994. *Acta Metall. Mater*. 42:3539-3544.

Charles B. Musgrave, Jason K. Perry, Ralph C. Merkle, William A. Goddard III, "Theoretical studies of a hydrogen abstraction tool for nanotechnology," *Nanotechnology* 2(1991):187-195; http://www.zyvex.com/nanotech/Habs/Habs.html

Charles B. Musgrave, Jason K. Perry, Ralph C. Merkle, William A. Goddard III, "Theoretical studies of a hydrogen abstraction tool for nanotechnology," *Nanotechnology* 2(1991):187-195; http://www.zyvex.com/nanotech/Habs/Habs.html

Chaudhry, L., *Valley to Bill Joy: 'Zzzzzzz'*, WiredNews, Apr. 5, 2000. Ref: http://piglet.ex.ac.uk/mail/cybersociety.2000/0305.html.

Chen, C.-Y., K. Ishihara, N. Nakabayashi, E. Tamiya, and I. Karube. 1999. Multifunctional biocompatible membrane and its application to fabricate a miniaturized glucose sensor with potential for use *in vivo. Journal of Biomedical Microdevices.* 1, No.2, 155-166.

Chen, J. and N.C. Seeman. 1991. The synthesis from DNA of a molecule with the connectivity of a cube. *Nature* 350:631-633.

Chianelli, R.R. 1998. Synthesis, fundamental properties and applications of nanocrystals, sheets, and fullerenes based on layered transition metal chalcogenides. In *R&D status and trends*, ed. Siegel et al.

Chianelli, R.R., M. Daage, and M.J. Ledoux. 1994. *Advances in Catalysis* 40:177.

Chidsey, E.E.D., and R.W. Murray. 1986. *Science*, 231: 25.

Chondroudis, K. and D.B. Mitzi. 1999. Electroluminescence from an organic-inorganic perovskite incorporating a quaterthiophene dye within lead halide perovskite layers. *Chemistry of Materials.* Vol 11. p. 3028.

Chopra, N.G., R.J. Luyken, K. Cherrey, H. Crespi, M.L. Cohen, S.G. Louie, and A. Zettl. 1995. *Science* 269:966.

Christensen, Clayton M. *The Innovator's Dilemma: When New Technologies Cause Great Firms to Fail*. New York: Harperbusiness, 2000.

Cipra, Barry A. 2000. The best of the 20th century: editors name top 10 algorithms. *SIAM News*. vol. 33(4).

Claeson, T., and Likharev, K.," Single Electronics," *Scientific American,* June 1992, pp. 80-85.

Clasen, R. 1990. *Int. Journal of Glass and Science Technology* 63: 291. Crandall, B.C. and J. Lewis, (eds.). 1992. *Nanotechnology: Research and perspectives*. Cambridge: MIT Press.

Clausen-Schaumann, H.; Seitz, M.; Krautbauer, R.; Gaub, H.E.: 2000, 'Force Spectroscopy with Single Bio-Molecules', *Current Opinion in Chemical Biology*, 4, 524-530.

Clifford Lau of the US Defense Department to Barnaby Feder, "Frontier of Military Technology is the Size of a Molecule," New York Times, April 8, 2003, p. C2.

Coenen, C.: 2004, 'Nanofuturismus: Anmerkungen zu seiner Relevanz, Analyse und Bewertung', *Technikfolgenabschätzung ' Theorie und Praxis*, 13 (2), 78-85.

Colbert, D.T. and R.E. Smalley. 1999. Fullerene nanotubes for molecular electronics. *Trends in Biotechnology*. Vol 17. 46-50.

Collier, C.P., E.W. Wong, M. Belohradsky, F.M. Raymo, J.F. Stoddart, P.J. Kuekes, R.S. Williams, and J.R. Heath. 1999. "Electronically Configurable Molecular-Based Logic Gates." *Science*, 285.

Computing in Science and Engineering, A.I.P and IEEE press, also covered by http://www.siam.org/siamnews/05-00/topten.pdf, Jan/Feb.

Constant, E.W., II: 1980, *The Origins of the Turbojet Revolution*, Johns Hopkins University Press, Baltimore.

Cook, T. K. and Campbell, D. T. *Quasi-Experimentation*. Rand-McNally, 1979 Fischer, C. S. *America Calling: A Social History of the Telephone to 1940*. Berkeley: U. of California, 1992.

Crandall, B.: 1999, 'Molecular Engineering', in: B. Crandall (ed.), *Nanotechnology: Molecular Speculations on Global Abundance*, MIT Press, Cambridge, MA, pp. 1-46.

Crandall, B.C. and Lewis, J., *Nanotechnology: Research and Perspectives*, MIT Press, Cambridge, 1992.

Crommie, M.F.; Lutz, C.P.; Eigler, D.M.; Heller, E.J. 1996. "Quantum interference in 2D atomic-scale structures," *Surface Science*, 362, 864-869.

Crow, M. & Sarewitz, D.: 2001, 'Nanotechnology and Societal Transformation', in: M.C. Roco and W.S. Bainbridge (eds.), *Societal Implications of Nanoscience and Nanotechnology*, Dordrecht: Kluwer, pp. 45-54.

Crow, M. & Sarewitz, D.: 2001, 'Nanotechnology and Societal Transformation', in: M.C. Roco, W.S. Bainbridge (eds.), *Societal Implications of Nanoscience and Nanotechnology*, National Science Foundation, pp. 45-54.

Crow, M.M. 2001, "Harnessing Science to Benefit Society," *The Chronicle of Higher Education, Chronicle.com*. 9 March, 2001: 17 pars. March 8, 2001 (http://chronicle.com/weekly/v471/i26/26b02001.htm).

Crowe, W. J., Statement of Admiral William J. Crowe, Jr., before the House Committee on Government Reform, October 12, 1999. Ref: http://www.house.gov/reform/hearings/healthcare/99.10.12/ crowe.htm.

Csicsery-Ronay, I. Jr.: 1991, 'The SF of Theory: Baudrillard and Haraway', *Science Fiction Studies*, 18, 387-404.

Drell, S. D., A. D. Sofaaer, and G. D. Wilson, Hoover Digest 2000, No. 1, *The Present Threat*.

Drell, S. D., A. D. Sofaaer, and G. D. Wilson, *The New Terror, Facing the Threat of Biological and Chemical Weapons*, Hoover Institution Press, Stanford University (1999).

Dupuy, J.P. "Common Knowledge, Common Sense." Theory and Decision 27 (1989): 37-62.

Dupuy, J.P. "Two Temporalities, Two Rationalities: A New Look at Newcomb's Paradox." in Bourgine, P.&Walliser, B. (eds.), Economics and Cognitive Science. New York: Pergamon. 1992, 191-220.

Dupuy, J.P.: 2000, *The Mechanization of the Mind*, Princeton University Press, Princeton N.J.

Dupuy, J.-P.: 2004, 'Complexity and Uncertainty: A Prudential Approach To Nanotechnology', in: European Commission (Community Health and Consumer Protection): *Nanotechnologies: A Preliminary Risk Analysis on the Basis of a Workshop, Brussels, 1-2 March 2004*, pp. 71-93 (www.europa.eu.int/comm/health/ ph_risk/documents/ ev_20040301_en.pdf).

Einsiedel, E.F.; Goldenberg, L.: 2004, 'Dwarfing the Social? Nanotechnology Lessons from the Biotechnology Front', *Bulletin of Science, Technology & Society*, 24, 28-33.

Elkins, C.: 1979, 'Science Fiction versus Futurology: Dramatic Versus Rational Models', *Science Fiction Studies*, 6, 20-31.

Ellsberg, D. "Risk, Ambiguity and the Savage Axioms." Quarterly Journal of Economics 75 (1961): 643-669.

Ellul, J. 1964. *The Technological Society*. New York: Alfred Knopf. (First published 1954).

Empedocles, S.; Bawendi, M. 1999. "Spectroscopy of single CdSe nanocrystallites," *Accounts of Chemical Research*, 32, 389-396.

Entingh, D., Dunn, A., Glassman, E., Wilson, J. E., Hogan, E. & Damstra, T. (1975) in *Handbook of Psychobiology*, eds. Gazzinga, M. S. & Blakemore, C. (Academic, New York), pp. 201-238.

Esant. 1999. "Economic and Social Aspects of Nanotechnology." First draft report of working group 6 of the Euroconferences on Nanoscience for Nanotechnology. Also at http:// www.nano.org.uk/ ESANT99.htm.

Estermann, M., L.B. McCusker, C. Baerlocher, A. Merrouche, and H. Kessler. 1991. *Nature* 331:698.

Esumi, A., A. Suzuki, N. Aihara, K. Uswi, and K. Torigoe. 1998. *Langmuir*, 14:3157.

ETC: 2003a, *The Big Down: Atom Tech ' Technologies Converging at the Atomic Scale*. Winnipeg, Canada: Action Group on Erosion, Technology and Concentration [www.etcgroup.org/documents/TheBigDown.pdf].

ETC: 2003b, 'No Small Matter II: The Case for a Global Moratorium ' Size Matters!' *Occasional Paper Series* 7(1) [www.etcgroup.org/documents/Occ.Paper_Nanosafety.pdf].

Etzkowitz, H., 2001, 'Nano-Science and Society: Finding a Social Basis for Science Policy', in: M.C. Roco and W.S. Bainbridge (eds.), *Societal Implications of Nanoscience and Nanotechnology*, Dordrecht: Kluwer, pp. 121-128.

Etzkowitz, H., M. Gulbrandsen and J. Levitt. 2000. *Public Venture Capital: Government Funding Sources for Technology Entrepreneurs*. New York: Harcourt.

Etzkowitz, Henry 2001. "The Endless Transition: Finding a Social Basis for Science Policy." *Scipolicy: The Journal of Science and Health Policy*. Vol. 1. No.2, Spring 2001.

Etzkowitz, Henry, Carol Kemelgor and Brian Uzzi. 2000. *Athena Unbound: The Advancement of Women in Science and Technology*. Cambridge: Cambridge University Press.

European Comission: *European Workshop on Social and Economic Research on Nanotechnologies and Nanosciences, Brussels, 14-15 April 2004* [www.stage-research.net/STAGE/PAGES/ Nano.html].

European Commission: 2004, *Converging Technologies: Shaping the Future of European Societies (Report of the High Level Expert Group "Foresighting the New Technology Wave")*, Brussels: European Commission Research [http://europa.eu.int/comm/research/conferences/2004/ntw/index_en.html].

Fedor N. Dzegilenko, Deepak Srivastava, Subhash Saini, "Simulations of carbon nanotube tip assisted mechano-chemical reactions on a diamond surface," *Nanotechnology* 9(December 1998):325-330.

Fedor N. Dzegilenko, Deepak Srivastava, Subhash Saini, "Simulations of carbon nanotube tip assisted mechano-chemical reactions on a diamond surface," *Nanotechnology* 9(December 1998):325-330.

Fennema, O. R. (1973) in *Low-Temperature Preservation of Foods and Living Matter*, eds. Fennema, O. R., Powrie, W. D. & Marth, E. H. (Dekker, New York), pp. 436-475.

Fermi, E., J. Pasta, and S. Ulam. 1974. Studies of nonlinear problems, I. *Los Alamos Report*. LA1940. 1955.

Feyerabend, P.: 1981, *Realism, Rationalism and Scientific Method* (Philosophical Papers, volume 1), Cambridge University Press, Cambridge.

Feynman R. "There's Plenty of Room At the Bottom." Talk given on at the annual meeting of the American Physical Society at the California Institute of Technology, 1959.

Feynman, R. (1961) in *Miniaturization*, ed. Gilbert, H. D. (Reinhold, New York), pp. 282-296.

Feynman, R., "There's Plenty of Room at the Bottom: An invitation to Enter a New Field of Physics," Talk at the Annual Meeting of the American Physical Society, 29 December 1959. Reprinted in Appendix B of Crandall and Lewis.

Feynman, R.: 1960, 'There's Plenty of Room at the Bottom', *Engineering and Science*, 23, 22-36.

Feynman, R.P. 1961. "There is plenty of room at the bottom." In *Miniaturization*. New York: Reinhold.

Fielder, F.A.; Reynolds, G.H.: 1994, 'Legal Problems of Nanotechnology: An Overview', *Southern California Interdisciplinary Law Journal*, 3, 593-629.

Filler, M., C. Mui, G. Wang, C. Musgrave, and S. Bent, "Competition and Selectivity in the Reaction of Nitriles on Ge (100)-2x1," *Journal of the American Chemical Society* 125, 4928-4936 (2003).

Fisher, D. E. and Fisher, M. J. *Tube: The Invention of Television*. Harcourt Brace, 1997.

FitzGerald, Frances. 2000. *Way Out There in the Blue: Reagan, Star Wars, and the End of the Cold War*. New York: Simon & Schuster.

Fleischer, Th.; Decker, M. & Fiedeler, U. (eds.): 2004, *Große Aufmerksamkeit für kleine Welten ' Nanotechnologie und ihre Folgen*, special issue of *Technikfolgenabschätzung ' Theorie und Praxis*, 13 (2), 5-85 [www.itas.fzk.de/tatup/042/inhalt.htm].

Fleming, D. "The Economics of Taking Care: An Evaluation of the Precautionary Principle." in Freestone, D.&Hey, E. (eds.), The Precautionary Principle and International Law. La Haye: Kluwer Law International, 1996.

Foerster, H.v.: 1962, 'Bio-Logic', in: E.E. Bernard & M.R. Kare (eds.), *Biological prototypes and synthetic systems*, Plenum Press, New York, vol. 1., pp. 1-12

Fogelberg, H. & Glimell, H.: 2003, 'Molecular Matters: In Search of the Real Stuff', in: H. Fogelberg & H. Glimell, *Bringing Visibility to the Invisible: Towards A Social Understanding of Nanotechnology*, Göteborg: Göteborg University, pp. 5-32 [www.sts.gu.se/publications/STS_report_6.pdf].

Fogelberg, H. & Glimell, H.: 2003, *Bringing Visibility to the Invisible: Towards A Social Understanding of Nanotechnology*, Göteborg: Göteborg University [www.sts.gu.se/publications/STS_report_6.pdf].

Forest, M.G., Q. Wang, and H. Zhou. 2000a. Exact banded patterns from a Doi-Marrucci-Greco model of nematic liquid crystal polymers. *Physical Review E*. Vol. 61(6). 6665-6672.

Forest, M.G., Q. Wang, and H. Zhou. 2000b. Homogeneous pattern selection and director instabilities of nematic liquid crystal polymers induced by elongational flows. *Physics of Fluids*. Vol. 12(3). 490-498.

Forum Proceedings of Sigma Xi, the Scientific Research Society, *Ethics, Values, and the Promise of Science*, Feb. 25-26, 1993.

Foucault, M. 1977. *Power/Knowledge: Selected Interviews and Other Writings 1972-1977*. New York, Pantheon Books.

Foucault, M.: 1992, *The Archaeology of Knowledge*, Routledge, London.

Fourier, J. 1878. *The Analytical Theory of Heat*. translated by A. Freeman. Cambridge University Press. London and New York.

Francoeur, E.: 1997, 'The Forgotten Tool: The Design and Use of Molecular Models', *Social Studies of Science*, 27, 7-40.

Frankel, M. 1996. "Perception, Reality, and the Political Context of Conflict of Interest in University- Industry Relationships," *Academic Medicine*, 71:12, December, pp. 1297-1304.

Franklin, N.R.; Dai, H.J. 2000. "An enhanced CVD approach to extensive nanotube networks with directionality", *Advanced Materials*, 12, 890-894.

Frazier, G., "An Ideology For Nanoelectronics," in *Concurrent Computations: Algorithms, Architecture, and Technology*, Plenum Press, New York, 1988.

Freitas, L. *Nanomedicine*. Bioscience, 1999.

Frenken, J.W.M. 1998. "Scanning Tunneling Microscopy", section 5.2, A. ten Wolde (ed.), *Nanotechnology: Towards a molecular construction kit*. STT Report #60, The Hague, 289-299.

Friedlander, S.K. 1993. Controlled synthesis of nanosized particles by aerosol processes. *Aerosol Sci. Technol*. 19:527.

Friedlander, S.K. 1998. Synthesis of nanoparticles and their agglomerates: aerosol reactors. In *R&D status and trends*, ed. Siegel et al.

Galison, P.: 1997, *Image and Logic: A Material Culture of Microphysics*, University of Chicago Press, Chicago.

Galison, P.; Stump, D.J. (eds.): 1996, *The Disunity of Science: Boundaries, Contexts, and Power*, Stanford University Press, Stanford.

Gallego-Juárez, J.A. (1989) Piezoelectric ceramics and ultrasonic transducers. J. Phys. E: Sci. Instrum. 22, 804-816

Gates, B.; Park, S.H.; Xia, Y.N. 2000. "Tuning the photonic bandgap properties of crystalline arrays of polystyrene beads by annealing at elevated temperatures", *Advanced Materials*, 12, 653-656.

Geneva Convention, *Protocol for the Prohibition of the Use in War of Asphyxiating, Poisonous or Other Gases, and of Bacteriological Methods of Warfare*, signed at Geneva on June 17,1925.

George D. Skidmore, Eric Parker, Matthew Ellis, Neil Sarkar, Ralph Merkle, "Exponential assembly," paper presented at the 8th Foresight Conference, Washington DC, November 2000; http://www.foresight.org/Conferences/MNT8/Papers/Skidmore/index.html

George M. Whitesides, "The Once and Future Nanomachine," *Scientific American* 285(September 2001):78-83.

Gergen, K. J. Social psychology as history. *Journal of Personality and Social Psychology*, 1973, 26, 309- 20.

Gerlach, N.; Hamilton, S.: 2000, 'Telling the Future, Managing the Present: Business Restructuring Literature as SF', *Science Fiction Studies*, 27, 461-477.

Gieryn, T.: 1999, *Cultural Boundaries of Science*, Univ. of Chicago Press, Chicago.

Gieryn, T.F.: 1999, *Cultural Boundaries of Science: Credibility on the Line*, University of Chicago Press, Chicago.

Gieryn, T.F.; Figert, A.E.: 1986, 'Scientists Protect Their Cognitive Authority: The Status Degradation Ceremony of Sir Cyril Burt', in: G. Böhme, N. Stehr (eds.), *The Knowledge Society*, Reidel, Dordrecht, pp. 67-89.

Gleiter, H. 1989. *Prog. Mater. Sci.* 33:223. Goddard, W.A. 1998. Nanoscale theory and simulation. In *R&D status and trends*, ed. Siegel et al.

Gleiter, H. 1990. *Progress in Materials Science* 33:4. Guinier, A. 1938. *Nature* 142:569; Preston, G.D. 1938. *Nature* 142:570.

Glimell, H.: 2001, 'Challenging Limits ' Excerpts from an Emerging Ethnography of Nano Physicists', in: H. Glimell & O. Johlin (eds.), *The Social Production of Technology: On the Everyday Life with Things*, Göteburg, SE: BAS Publisher, chapter 7, pp. 111-131 (reprinted in: H. Fogelberg & H. Glimell, *Bringing Visibility to the Invisible: Towards A Social Understanding of Nanotechnology*, Göteborg: Göteborg University, pp. 115-139.

Glimell, H.: 2001, 'Dynamics of the Emerging Field of Nanoscience', in: M.C. Roco and W.S. Bainbridge (eds.), *Societal Implications of Nanoscience and Nanotechnology*, Dordrecht: Kluwer, pp. 156-160.

Goddard, W.A. 1998. Nanoscale theory and simulation. In *R&D status and trends*, ed. R. Siegel et al.

Godsell, D.: 2003, *Living Machinery. Bionanotechnology: Lessons from Nature*, Wiley-Liss, New-York.

Goldbeck-Wood, G. and A.H. Windle. 1999. Lattice modeling of nematodynamics. *Rheol. Acta*. Vol. 38. 548-561.

Gorman, M.E.; Groves, J.F. & Shrager, J.: 2004, 'Societal Dimensions of Nanotechnology as a Trading Zone: Results from a Pilot Project', in: D. Baird, A. Nordmann & J. Schummer (eds.), *Discovering the Nanoscale*, Amsterdam: IOS Press, pp. 63-73.

Graham, P.: 2002, 'Hypercapitalism: Language, New Media and Social Perceptions of Value', *Discourse and Society*, 13(2), 227-249.

Gref, Ruxandra, Yoshiharu Minamitake, Maria Teresa Peracchia, Vladimir Trubetskoy, Vladimir Torchilin, Robert Langer. 1994. "Biodegradable Long Circulating Polymeric Nanospheres", *Science*, Vol. 263, 18 Mar, pp 1600-1603.

Griliches, Zvi. 1957. "Hybrid Corn: An Exploration in the Economics of Technological Change," *Econometrica*, 25 (4), pp 501-522.

Griliches, Zvi. 1960. "Hybrid Corn and the Economics of Innovation," *Science*, 132, pp 275-280. Mass, William. 1989. "Mechanical and Organisational Innovation: The Case of the Draper Loom," *Business History Review*, 63 (Winter).

Grinbaum, A. & Dupuy, J.-P.: 2004, 'Living with Uncertainty: Toward the Ongoing Normative Assessment of Nanotechnology', *Techne: Research in Philosophy and Technology*, 8(3) (forthcoming).

Grinbaum, A.: 2004, 'La condition de l'homme moderne et les nanotechnologies', in: G. Nivat (ed.), *Les limites de l'humain. 39èmes Rencontres Internationales de Genève*, L'Age d'Homme, Genève, p. 141.

Gröndahl B., W. Puppe, A. Hoppe, I. Kühne, J.A. Weigl, H.J. Schmitt. 1999. "Rapid Identification of Nine Microorganisms Causing Acute Respiratory Tract Infections by Single-Tube Multiplex Reverse Transcription-PCR: Feasibility Study." *Journal of Clinical Microbiology*, Jan; 37 (1): pp. 1-7.

Grosberg, A.Y. and A.R. Khokhlov. 1994. *Statistical Physics of Macromolecules*, AIP Series in Polymers and Complex Materials, AIP Press, New York.

Gross, M.: 1999, *Travels to the Nanoworld: Miniature Machinery in Nature and Technology*, Plenum, New York.

Grunwald, A.: 2004, 'Ethische Aspekte der Nanotechnologie. Eine Felderkundung', *Technikfolgenabschätzung ' Theorie und Praxis*, 13 (2), 71-78.

Guarnieri, F., M. Fliss, and C. Bancroft. 1996. Making DNA Add. *Science* 273:220-223. Hanes, J., J.L. Cleland, and R. Langer. 1997.

Gubrud, M. A., *Nanotechnology and International Security*, Fifth Foresight Conference on Molecular Nanotechnology, November 5-8, 1997. Ref: http://www.foresight.org/Conferences/MNT05/Papers/Gubrud/.

Gurley, Bill. 2000. Digital music: the real law is Moore's law. *Fortune*. 2 October. page 268. Joy, Bill. 2000. Why the future doesn't need us. *Wired*. April. pp. 238 - 262.

Guthold, M.; Falvo, M.R.; Matthews, W.G.; Paulson, S.; Washburn, S.; Erie, D.A.; Superfine, R.; Brooks, F.P.; Taylor, R.M.: 2000, 'Controlled Manipulation of Molecular Samples with the NanoManipulator', *IEEE-ASME Transactions on Mechatronics*, 5, 189-198.

Hacking, I.: 1983, *Representing and Intervening: Introductory Topics in the Philosophy of Natural Science*, Cambridge University Press, Cambridge, UK.

Hacking, I.: 1992, 'The Self-Vindication of the Laboratory Sciences', in: A. Pickering (ed.), *Science as Practice and Culture*, University of Chicago Press, Chicago, pp. 29-64.

Hadjiconstantinou, N.G. 1999. Combining atomistic and continuum simulations of contact-line motion. *Physical Review E.* Vol. 59(2).2, 2475-2478.

Hadjipanayis, C.G. 1998. Nanostructured magnetic materials. In *R&D Status and Trends*, ed. Siegel et al.

Hamilton, S.: 2003, 'Biotechnology, SF and the Media', *Science Fiction Studies*, 30, 267-282.

Han, W., S. Fam, O. Li, and Y. Hu. 1998. Synthesis of gallium nitride nanorods through a carbon nanotube-confined reaction. *Science* 277:1287-1289.

Hansson, S.O.: 2004, 'Great Uncertainty about Small Things', *Techne: Research in Philosophy and Technology*, 8(3) (forthcoming).

Haraway, D.J.: 1997, *Modest Witness@Second Millenium.FemaleManã Meets OncoMouseä*, Routledge, London and New York.

Harold J. Morowitz, "A Model of Reproduction," *American Scientist* 47(1959):261-263.

Hayes, B., "The Square Root of NOT," *American Scientist*, July-August 1995, pp. 304-308.

Hayles, N.K. (ed.): 2004, *Nanoculture: Implications of the New Technoscience*, Bristol, UK: Intellect Books.

Hayles, N.K.: 2004, 'Connecting the Quantum Dots: Nanotechscience and Culture', in: N.K. Hayles (ed.), *Nanoculture: Implications of the New Technoscience*, Bristol, UK: Intellect Books, pp. 11-26.

Hayward, R.C.; Saville, D. A.; Aksay, I.A. 2000. "Electrophoretic assembly of colloidal crystals with optically tunable micropatterns", *Nature*, 404, 56-59.

Hecht, G., Allen, M.T.: 2001, 'Introduction: Authority, Political Machines, and Technology's History', in: M.T. Allen, G. Hecht (eds.), *Technologies of Power: Essays in Honor of Thomas Parke Hughes and Agatha Chipley Hughes*, MIT Press, Cambridge, MA, pp. 1-23.

Hecht, G.: 1998, *The Radiance of France: Technology, Politics, and National Identity after World War II*, MIT Press, Cambridge, MA.

Heidegger, M.: 1962, *Being and Time*, trans. J. Macquarrie, E. Robinson, Harper & Row, New York.

Heidegger, M.: 1977, *The Question Concerning Technology and Other Essays*, Harper Colophon Books, New York.

Held, R., T. Heinzel, A.P. Studerus, K. Ensslin, and M. Holland. 1997. Semiconductor quantum point contact fabricated by lithography with an atomic force microscope. *Appl. Phys. Lett.* 71:2689-91.

Hellemans, A. 1998. X-rays find new ways to shine. *Science* 277:1214-15.

Helmreich, S.: 1998, *Silicon Second Nature: Culturing Artificial Life in a Digital World*, University of California Press, Berkeley, CA.

Hemerén, G.: 2004, 'Nano Ethics Primer', in: European Commission (Community Health and Consumer Protection): *Nanotechnologies: A Preliminary Risk Analysis on theBasis of a Workshop, Brussels, 1-2 March 2004* (www.europa.eu.int/comm/health/ ph_risk/documents/ ev_20040301_en.pdf), pp. 95-102.

Hendrickson, C.T., A. Horvath, S. Joshi, and L.B. Lave. 1998. Economic input-output models for environmental life-cycle assessment. *Environmental Science & Technology*. p. 32.

Hennig, J.: 2004, 'Changes in the Design of Scanning Tunneling Microscopic Images from 1980 to 1990', *Techne: Research in Philosophy and Technology*, 8(3) (forthcoming).

Hessenbruch, A.: 2004, 'Nanotechnology and the Negotiation of Novelty', in: D. Baird, A. Nordmann & J. Schummer (eds.), *Discovering the Nanoscale*, Amsterdam: IOS Press, pp. 135-144.

Hietpas, P.B., S.D. Gilman, R.A. Lee, M.R. Wood, N. Winograd, and A.G. Ewing. 1996. Development of votammetric methods, capillary electrophoresis and tof sims imaging for constituent analysis of single cells. In *Nanofabrication and biosystems*, ed.

Higgins, R.J. 1997. An economical process for manufacturing of nano-sized powders based on microemulsion-mediated synthesis. In *Proc. of the Joint NSF-NIST Conf. on Nanoparticles*.

Hiruma, K., M. Yazawa, T. Katsoyama, K. Ogawa, K. Haraguchi, M. Koguchi, and H. Kakibayashi. 1995. *J. Appl. Phys*. 77(2):476.

Hoch et al. Ho, S.V., P.W. Sheridan, and E. Krupetsky. 1996. Supported polymeric liquid membranes for removing organics from aqueous solutions.

Hoch, H.C., L.W. Jelinski, and H.G. Craighead, eds. 1996. *Nanofabrication and biosystems*. New York: Cambridge University Press.

Hoh, J.H. and Hansma, P.K. (1992) Atomic force microscopy for high-resolution imaging in cell biology. Trends Cell Bio. 2, 208-213

Homer Jacobson, "On Models of Reproduction," *American Scientist* 46(1958):255-284.

Hopwood, J., and S. Mann. 1997. *Chem. Mater*. 9:1819.

Howland, H.: 1962, Structural, hydraulic and economic aspects of leaf venation and shape ', in: E.E. Bernard & M.R. Kare (eds.), *Biological prototypes and synthetic systems*, Plenum Press, New York, vol. 1, pp. 183-192.

Hoyningen-Huene, P.: 1993, *Reconstructing Scientific Revolutions: Thomas S. Kuhn's Philosophy of Science* (trans. by A. Levin), University of Chicago Press, Chicago.

Huang, J.Y., Y.K. Wu, and H.Q. Ye. 1996. *Acta Mater*. 44:1211. Inoue, A. 1997. Private communication. Inturi, R.B., and Z. Szklavska-Smialowska. 1992. *Corrosion* 48:398. Karch, J., R. Birringer, and H. Gleiter. 1987. *Nature* 330:556.

Hughes, T.P.: 1994, 'Technological Momentum', in: M.R. Smith, L. Marx (eds.), *Does Technology Drive History? The Dilemma of Technological Determinism*, MIT Press, Cambridge, MA, pp. 101-114.

Huijer, M.: 2003, 'Reconsidering Democracy: History of the Human Genome Project', *Science Communication*, 24(4), 479-502.

Hullmann, A. & Meyer, M.: 2003, 'Publications and Patents in Nanotechnology. An overview of previous studies and the state of the art', *Scientometrics*, 58 (3) 507-527.

Ihde, D.: 1991, *Instrumental Realism: The Interface between Philosophy of Science and Philosophy of Technology*, Indiana University Press, Bloomington.

Imae, Y., and T. Atsumi. 1989. T. Na+-driven bacterial flagellar motors: A mini-review. *J. Bioenergetics and Biomembranes* 21:705-716.

J.W. Lyding, K. Hess, G.C. Abeln, D.S. Thompson, J.S. Moore, M.C. Hersam, E.T. Foley, J. Lee, Z. Chen, S.T. Hwang, H. Choi, P.H. Avouris, I.C. Kizilyalli, "UHV-STM nanofabrication and hydrogen/deuterium desorption from silicon surfaces: implications for CMOS technology," *Appl. Surf. Sci.* 130(1998):221-230.

J.W. Lyding, K. Hess, G.C. Abeln, D.S. Thompson, J.S. Moore, M.C. Hersam, E.T. Foley, J. Lee, Z. Chen, S.T. Hwang, H. Choi, P.H. Avouris, I.C. Kizilyalli, "UHV-STM nanofabrication and hydrogen/deuterium desorption from silicon surfaces: implications for CMOS technology," *Appl. Surf. Sci.* 130(1998):221-230.

Jaworek, T., D. Deher, G. Wegner, R.H. Wieringa, and A.J. Schouten. 1998. Electromechanical properties of an ultrathin layer of directionally aligned helical polypeptides. *Science* 279:57-60.

Jena, P., S.N. Khanna, and B.K. Rao. 1996. In *Science and technology of atomically engineered materials*, ed. P. Jena. River Edge, NJ: World Scientific.

Jena, P., S.N. Khanna, and B.K. Rao. 1996. In *Science and technology of atomically engineered materials*, ed. P. Jena. River Edge, NJ: World Scientific.

Johansson, M.: 2003, 'Plenty of room at the bottom: Towards an anthropology of nanoscience', *Anthropology Today*, 19 (No. 6), 3-6.

John Cumings, A. Zettl, "Low-friction nanoscale linear bearing realised from multiwall carbon nanotubes," *Science* 289(28 July 2000):602-604.

John von Neumann, *Theory of Self-Reproducing Automata*, A.W. Burks, ed., University of Illinois Press, Urbana IL, 1966.

Johnson, A.: 2004, 'The End of Pure Science: Science Policy from Bayh-Dole to the NNI', in: D. Baird, A. Nordmann & J. Schummer (eds.), *Discovering the Nanoscale*, Amsterdam: IOS Press, pp. 217-230.

Johnson, D.G. and H. Nissenbaum (eds). 1995. *Computers, Ethics and Social Values*. Prentice Hall:Englewood Cliffs, NJ. pp. 262-393.

Johnson, P.C.: 2003, 'Implications of the Continuum of Bioinformatics', in: M.C. Roco; W.S. Bainbridge (eds.), *Converging Technologies for Improving Human Performance*, Kluwer, Dordrecht, pp. 207-212.

Kalman B., F.D. Lublin. 1999. "The Genetics of Multiple Sclerosis. A Review." *Biomedicine and Pharmacotherapy*, Sep; 53 (8): pp. 358-370.

Kang, J. and C. Musgrave, "A Quantum Chemical Study of the Self-Directed Growth Mechanism of Styrene and Propylene Molecular Nanowires on the Silicon (100) 2x1 Surface," *Journal of Chemical Physics* 116, 9907-9913 (2002).

Kang, J. and C. Musgrave, "The Mechanism of Atomic Layer Deposition of SiO 2 on the Silicon (100)-2x1 Surface Using SiC1 4 and H 2 O as Precursors," *Journal of Applied Physics* 91, 3408-3414 (2002).

Karak, N., and S. Maïti. 1997. Dendritic polymers: A class of novel material. *J. Polym. Mater.*14:105.

Karger, J., and D.M. Ruthven. 1992. *Diffusion in zeolites*. New York: J. Wiley.

Karger, J., and D.M. Ruthven. 1992. *Diffusion in zeolites*. New York: J. Wiley. Krätschmer, W., L.D. Lamb, K. Fostiropoulos, and D.R. Huffman. 1990. *Nature* 347:354.

Kasuga, T., M. Hiramatsu, A. Hoson, T. Sekino, and K. Niihara. 1998. *Langmuir* 14:3160.

Katari, J.E.B., V.L. Colvin, and A.P. Alivisatos. 1994. *J. Phys. Chem.* 98:4109.

Kay, L.: 2000, *Who Wrote the Book of Life? A History of the Genetic Code*, Stanford University Press, Stanford.

Kear, B.H., R.K. Sadangi, and S.C. Liao. 1997. Synthesis of WC/Co/diamond nanocomposites. In *Proc. of the Joint NSF-NIST Conf. on Nanoparticles.*

Keiper, A.: 2003, 'The Nanotechnology Revolution', *The New Atlantis*, 2, 17-34.

Keller, D.J. and Chih-Chung, C. (1992) Imaging steep, high structures by scanning force microscopy with electron beam deposited tips. Surf. Sci. 268, 333-339

Kennedy, P., *The Rise and Fall of the Great Powers, Economic Change and Military Conflict from 1500 to 2000*,Random House (1987).

Kennedy, Paul. *The Rise and Fall of the Great Powers, Economic Change and Military Conflict from 1500 to 2000*. New York: Random House,1987.

Keyssar, Alexander. 1986. *Out of Work: The First Century of Unemployment in Massachusetts* (Cambridge: Cambridge University Press).

Khushf, G.: 2004, 'A Hierarchical Architecture for Nano-scale Science and Technology: Taking Stock of the Claims About Science Made By Advocates of NBIC Convergence', in: D. Baird, A. Nordmann & J. Schummer (eds.), *Discovering the Nanoscale*, Amsterdam: IOS Press, pp. 21-33.

Khushf, G.: 2004, 'Systems Theory and the Ethics of Human Enhancement: A Framework for NBIC Convergence', *Annals of the New York Academy of Sciences*, 1013, 124-149.

Kirschvink, J.L., A. Koyayashi-Kirschvink, and B.J. Woodford. 1992. Magnetite biomineralization in the human brain. *Proc. Nat'l. Acad. Sci.* USA 89:7683-7687.

Kishida, M., T. Fujita, K. Umakoshi, J. Ishiyama, H. Nagata, and K. Wakabayashi. 1995. *Chem. Commun.* 763.

Klein, J.D., et al. 1993. *Chem. Mater.* 5:902. Koch, C.C. 1989. Materials synthesis by mechanical alloying. *Annual Review of Mater. Sci.* 19:121-143.

Kline, R.: 1992, *Steinmetz: Engineer and Socialist*, Johns Hopkins University Press, Baltimore.

Knorr-Cetina, K.: 1992, 'The Couch, the Cathedral, and the Laboratory: On the Relationship between Experiment and Laboratory in Science', in: A. Pickering (ed.), *Science as Practice and Culture*, University of Chicago Press, Chicago, pp. 113-138.

Knorr-Cetina, K.: 1999, *Epistemic Cultures: How the Sciences Make Knowledge*, Harvard University Press, Cambridge, MA.

Koch, C.C. 1998. Bulk behavior. In *R&D status and trends*, ed. Siegel et al. 32 *Evelyn L. Hu and David T. Shaw.*

Kolata, G., "A Vat of DNA May Become a Vast Computer of the Future," N.Y. Times, 11 April 1995, p. C1. *Large, much cited spread at the top of the front page of the Tuesday "Science Times" Section that first brought the Adelman discovery to wide public attention. Beautiful illustrations that explain how it works.*

Kortan, A.R., R. Hull, R.L. Opila, M.G. Bawendi, M.L. Steigerwald, P.J. Carroll, and L.E. Brus. 1990. *J. Am. Chem. Soc.* 112:1327.

Kourilsky, Ph.&Viney, G. Le Principe de précaution. Report to the Prime Minister, Paris, Éditions Odile Jacob, 2000.

Kranakis, E.: 1997, *Constructing a Bridge: An Exploration of Engineering Culture, Design, and Research in Nineteenth-Century France and America*, MIT Press, Cambridge, MA.

Krätschmer, W., L.D. Lamb, K. Fostiropoulos, and D.R. Huffman. 1990. *Nature* 347:354.

Krejchi, M.T., E.D.T. Atkins, A.J. Waddon, M.J. Fournier, T.L. Mason, and D.A. Tirrell. 1994. Chemical sequence control of beta-sheet assembly in macromolecular crystals of periodic proteins. *Science* 265:1427-1432.

Krejchi, M.T., S.J. Cooper, Y. Deguchi, E.D.T. Atkins, M.J. Fournier, T.L. Mason, and D.A. Tirrell. 1997. Crystal structures of chain-folded antiparallel beta-sheet assemblies from sequence-designed periodic polypeptides. *Macromolecules* 30:5012-5024.

Kresge, C.T., M.E. Leonowicz, W.J. Roth, J.C. Vartuli, and J.S. Beck. 1992. *Nature* 359:710.

Kresge, C.T., M.E. Leonowicz, W.J. Roth, J.C. Vartuli, and J.S. Beck. 1992. *Nature* 359:710. Kupperman, A., S. Nadimi, S. Oliver, G. Ozin, J. Garcés, and M. Olken. 1993. *Nature* 365:239.

Kroes, P. & Meijers, A. "The Dual Nature of Technical Artifacts - Presentation of a New Research Programme." Techné 6, 2 (2002): 4 - 8.

Krstic, V., U. Erb, and G. Palumbo. 1993. *Scripta Metall. et Mater*. 29:1501.

Krumhansl, J. A. & Pao, Y. H. (1979) *Phys. Today* 32 (11), 25-32.

Kupperman, A., S. Nadimi, S. Oliver, G. Ozin, J. Garcés, and M. Olken. 1993. *Nature* 365:239.

Kurzweil, R. 1999. *The Age of Spiritual Machines*. Penguin Books.

Kurzweil, R.: 1998, *The Age of Spiritual Machines, How We Will Live, Work and Think in the New Age of Intelligent Machines*, Phoenix, New York.

Kuusi, O. 1999b. *Expertise in the future use of generic technologies*. VATT Research Reports #59, Helsinki.

Kuusi, O. and M. Meyer. 2001. "Technology Generalizations and Leitbilds – The Anticipation of Technological Opportunities." Manuscript submitted to *Technological Forecasting and Social Change*.

Kuusi, O., 1999. "Epistemic value analysis – a tool to evaluate technological options." Personal communication.

Landon, B.: 2004, 'Less is More: Much Less is Much More: The Insistent Allure of Nanotechnology Narratives in Science Fiction', in: K. Hayles (ed.), *NanoCulture*, Intellect Books, Bristol, pp. 59-78.

Laudan, L.: 1984, *Science and Values: The Aims of Science and their Role in Scientific Debate*, University of California Press, Berkeley.

Lave, L.B., C.T. Hendrickson, and F. C. McMichael. 1995b. Environmental implications of electric cars. *Science*. vol. 268: 992-5. May 19.

Lave, L.B., E. Cobas-Flores, C.T. Hendrickson, and F.C. McMichael. 1995a. Using input-output analysis to estimate economy wide discharges. *Environmental Science & Technology*, 29(9).

Lax, Peter D. 1989. The flowering of applied mathematics in America. in *A Century of Mathematics in America-Part II*. edited by Peter Duren et al. American Mathematical Society. 455-466.

Layton, E.T., Jr.: 1971, 'Mirror-Image Twins: The Communities of Science and Technology in 19th-Century America', *Technology and Culture*, 12, 562-580.

Leah Eisenberg, "Medicating Death Row Inmates So They Qualify for Execution," AMA Case in Health Law, Vol. 6, No. 9, September 2004.

Lehn, J.M.: 1995, *Supramolecular Chemistry*, VCH, Weinheim.

Lemann, Nicholas. 1991. *The Promised Land: The Great Black Migration and How it Changed America* (New York: Random House).

Lenhard, J.: 2004, 'Nanoscience and the Janus-Faced Character of Simulations', in: D. Baird, A. Nordmann & J. Schummer (eds.), *Discovering the Nanoscale*, Amsterdam: IOS Press, pp. 93-100.

Lenhard, J.: 2004, 'Nanoscience and the Janus-Faced Character of Simulations', in: D. Baird, A. Nordmann, J. Schummer (eds.), *Discovering the Nanoscale*, IOS Press, Amsterdam, pp. 93-100.

Lent, C.S., Tougaw, P.D., Porod, W., Bernstein, G.H., "Quantum Cellular Automata," *Nanotechnology*, Vol. 4, p. 49, 1993.

Leonard, D., M. Krishnamurthy, C.M. Reaves, S.P. Denbaars, and P.M. Petroff. 1993. Direct formation of quantum-sized dots from uniform coherent islands of InGaAs on GaAs surfaces. *Appl. Phys. Lett.* 63:3203-5.

Leslie-Pelecky, D.L., and R.D. Reike. 1996. *Chem. Mater.* 8:1770.

Lewak, S.E.: 2004, 'What's the Buzz? Tell Me What's A-Happening: Wonder, Nanotechnology, and Alice's Adventures in Wonderland', in: N.K. Hayles (ed.), *Nanoculture: Implications of the New Technoscience*, Bristol, UK: Intellect Books, pp. 201-.

Liang, G., Z. Li, and E. Wang. 1996. *J. Mater. Sci.* Makino, A., A. Inoue, T. Hatanai, and T. Bitoh. 1997. *Materials Science Forum* 235-238: 723.

Lide, D.R., ed. 1993-1994. *CRC Handbook of Chemistry and Physics*, 74th ed.

Lide, D.R., ed. 1993-1994. *CRC Handbook of Chemistry and Physics*, 74th ed. Lin, H.-P., and C.-Y. Mou. 1996. *Science* 273:765.

Lin-Easton, P.C.: 2001, 'It's Time for Environmentalists to Think Small ' Real Small: A Call for the Involvement of Environmental Lawyers in Developing Precautionary Policies Molecular Nanotechnology', *Georgetown International Law Review*, 14, 106-134.

Lipowsky, R. 1991. The conformation of membranes. *Nature*. Vol. 349. 475.

Lipson, H. and J. B. Pollack, Nature, 406, 974-978, (2000). *Automatic design and manufacture of robotic life forms*. See, also, several additional articles on this subject in contemporary issues of Nature.

López, J.: 2003, *Society and Its Metaphors: Language Social Theory and Social Structure*, Continuum, New York and London.

López, J.: 2004, 'Bridging the Gaps: Science Fiction in Nanotechnology', *Hyle: International Journal for Philosophy of Chemistry*, 10(2), 129-152.

López, J.: 2004, 'How Sociology Can Save Bioethics...Maybe', *Sociology of Health and Illness*, 26(7), 1-22.

Lösch, A.: 2004, 'Nanomedicine and Space: Discursive Orders of Mediating Innovations', in: D. Baird, A. Nordmann & J. Schummer (eds.), *Discovering the Nanoscale*, Amsterdam: IOS Press, pp. 193-202.

Lu, H.P.; Xun, L.Y.; Xie, X.S. 1998. "Single-molecule enzymatic dynamics", *Science*, 282, 1877-1882.

Majetich, S.A. and A.C. Canter. 1993. *J. Phys. Chem*. 97:8727.

Malinow, R., Z.F. Mainen, and Y. Hayashi. 2000. LTP mechanisms: from silence to four-lane traffic. *Curr. Opin. Neurobiol*. 10(3): p. 352-7.

Mao, C., W. Sun, and N.C. Seeman. 1997. Construction of Borromean rings from DNA. *Nature* 386:137-138.

Mark A. Reed, James M. Tour, "Computing with molecules," *Scientific American* 282(June 2000):86-93.

Mark, H., Statement of The Honorable Hans Mark, Director, Defense Research and Engineering, before the House Armed Service Committee, October 20, 1999. Ref: http://www.house.gov/hasc/testimony/ 106thcongress/99-10-20mark.htm.

Marshall, K.: 2004, 'Atomizing Risk Technology', in: K. Hayles (ed.), *NanoCulture*, Intellect Books, Bristol, pp. 79-102.

Marshall, K.: 2004, 'Atomizing the Risk Technology', in: N.K. Hayles (ed.), *Nanoculture: Implications of the New Technoscience*, Bristol, UK: Intellect Books, pp. 147-160.

Martin, T.P., N. Malinowski, U. Zimmerman, U. Naher, and H. Schaber. 1993. *J. Chem. Phys.* 99:4210.

Martin, T.P., N. Malinowski, U. Zimmerman, U. Naher, and H. Schaber. 1993. *J. Chem. Phys.* 99:4210.

Martin, T.P., U. Naher, H. Schaber, U. Zimmerman. 1993. *Phys. Rev. Lett*. 70:3079.

Martin, T.P., U. Naher, H. Schaber, U. Zimmerman. 1993. *Phys. Rev. Lett*. 70:3079. Prigogine, I., and S. Rice. 1988. *Advances in chemical physics*, Vol. 70, Parts 1 & 2. New York: J. Wiley.

Marty, A.: 2003, *Testimony before the Committee on Science on HR 766, Nanotechnology Research and Development Act*, US Government Printing Office, Washington, DC.

Matsuki, H., K. Ofuji, N. Chubachi, and S. Nitta. 1996. Signal transmission for implantable medical devices using figure-of-eight coils. *IEEE Trans. on Magnetics.*, vol. 32. No.5, 5121. September.

Matsuki, H., Y. Yamakata, N. Chubachi, S. Nitta, and H. Hashimoto. 1996. Transcutaneous DC-DC converter for totally implantable artificial heart using synchronous rectifier. *IEEE Trans. On Magnetics*. 32, No.5, 5118 (Sept.).

Mayer, S.: 2002, 'From genetic modification to nanotechnology: the dangers of 'sound science", in: T. Gilland (ed.), *Science: Can We Trust the Experts?*, London: Hodder and Stoughton, pp. 1-15.

McCain, L.: 2002, 'Informing Technology Policy Decisions: the US Human Genome Project's Ethical, Legal and Social Implications Programmes as a Critical Case', *Technology and Society*, 24, 111-132.

McCammon, J. A., Gelin, B. R. & Karplus, M. (1977) *Nature (London)* 267, 585-590.

McConnell, H.M. 1996. Light-addressable potentiometric sensor: Applications to drug discovery. In *Nanofabrication and biosystems*, ed. Hoch et al.

McConnell, S.A., ed. 1998. *Pharmaceuticals*. Dun and Bradstreet Industry Reference Handbooks, Gale, Detroit, pp. 30-34.

McCulloch, W.S.: 1962, 'The imitation of one form of life by another – Biomimesis', in: E.E. Bernard & M.R. Kare (eds.), *Biological prototypes and synthetic systems*, Plenum Press, New York, vol. 1., p. 393-97.

McHale, B.: 1987, *Postmodern Fiction*, Methuen, New York & London.

Meadows, D.L., et al. 1972. *The Limits to Growth*, New York: University.

Mehl, R.F., and R.W. Cahn. 1983. Historical development. In *Physical metallurgy*. North Holland. Milligan, W.W., S.A. Hackney, M. Ke, and E.C. Aifantis. 1993. *Nanostructured Materials* 2:267.

Mehta, M.D.: 2002, 'Nanoscience and Nanotechnology: Assessing the Nature of Innovation in These Fields', *Bulletin of Science, Technology & Society*, 22(4), 269-273.

Mehta, M.D.: 2004, 'From Biotechnology to Nanotechnology: What Can We Learn from Earlier Technologies?, *Bulletin of Science, Technology & Society*, 24, 34-39.

Merz, J.L. 1986. "The Optoelectronics Joint Research Laboratory: Light Shed on Cooperative Research in Japan." *Scientific Bulletin* (ONR Far East Office). 11 (4), 1-30 (Oct.-Dec.).

Methods of Molecular Quantum Mechanics, by R. McWeeny, Second Edition, Academic Press 1989.

Meyer, G. and Amer, N.M. (1988) Novel optical approach to atomic force microscopy. Appl. Phys. Lett. 53(12), 1045-1047

Meyer, G. and Amer, N.M. (1990) Simultaneous measurement of lateral and normal forces with an optical-beam-deflection atomic force microscope. Appl. Phys. Lett. 57(20), 2089-2091

Meyer, M. & Kuusi, O.: 2004, 'Nanotechnology: Generalizations in an Interdisciplinary Field of Science and Technology', *Hyle: International Journal for Philosophy of Chemistry*, 10(2), 153-168.

Meyer, M. and O. Persson. 1998. "Nanotechnology - interdisciplinarity, patterns of collaboration and differences in application," *Scientometrics*, XXXXII (2), 195-205.

Meyer, M.: 2000a, 'Does science push technology? Patents citing scientific literature', *Research Policy*, 29, 409-434.

Meyer, M.: 2000b, 'Patent citations in a novel field of technology: What can they tell about interactions of emerging communities of science and technology?', *Scientometrics*, 48, 151-178.

Meyer, M.: 2001a, 'Patent citations in a novel field of technology: An exploration of nano-science and nano-technology', *Scientometrics*, 51, 163-183.

Meyer, M.: 2001b, 'Socio-economic Research on Nanoscale Science and Technology: A European Overview and Illustration', in: M.C. Roco and W.S. Bainbridge (eds.), *Societal Implications of Nanoscience and Nanotechnology*, Dordrecht: Kluwer, pp. 217-241.

Meyer, M.; Persson, O.: 1998, 'Nanotechnology ' interdisciplinarity, patterns of collaboration and differences in application', *Scientometrics*, 47, 195'205.

Michael E.J. Holwill, Peter Satir, "II.4 Generation of Propulsive Forces by Cilia and Flagella," in J. Bereiter-Hahn, O.R. Anderson, W.-E. Reif, eds., *Cytomechanics: The Mechanical Basis of Cell Form and Structure*, Springer-Verlag, Berlin, 1987, pp. 120-130.

Michael Page, Donald W. Brenner, "Hydrogen abstraction from a diamond surface: *Ab initio* quantum chemical study using constrained isobutane as a model," *J. Am. Chem. Soc.* 113(1991):3270-3274.

Michael Page, Donald W. Brenner, "Hydrogen abstraction from a diamond surface: *Ab initio* quantum chemical study using constrained isobutane as a model," *J. Am. Chem. Soc.* 113(1991):3270-3274.

Michalet, X. and D. Bensimon. 1988. Vesicles of toroidal topology: observed morphology and shape transformations. *J. Phys. II*. France. Vol. 5. 598.

Mihail Roco and William Sims Bainbridge, eds., Converging Technologies for Improving Human Performance, NSF/DOC Report, June 2002.

Miksanek, T.: 2001, 'Microscopic Doctors and Molecular Black Bag: Science Fiction's Prescription for Nanotechnology and Medicine', *Literature and Medicine*, 21(1), 55-70.

Milburn, C.: 2002 'Nanotechnology in the Age of Posthuman Engineering: Science Fiction as Science', *Configurations*, 10, 261-295.

Milburn, C.: 2002, 'Nanotechnology in the age of post-human engineering: science fiction as science', *Configurations*, 10, 261-295 [reprinted in: N.K. Hayles (ed.), *Nanoculture: Implications of the New Technoscience*, Bristol, UK: Intellect Books, 2004, pp. 109-130].

Milburn, C.: 2002, 'Nanotechnology in the Age of Posthuman Engineering: Science Fiction as Science', *Configurations*, 10, 261-295.

Milburn, C.: 2004, 'Nano/Splatter: Disintegrating the Postbiological Body', *New Literary History*, 35 (in print).

Minsky, M.: 1995, 'Virtual Molecular Reality', in: Krummenacker, M. & Lewis, J. (eds.), *Prospects in Nanotechology. Proceedings of the 1st general conference on nanotechnology: developments, applications, and opportunities, November 11-14, 1992, Palo-Alto*, John Wiley & Sons, New York, pp. 187-205.

Mirsky, Steve. 2000. "Tantalizing Tubes." *Scientific American*. June.

Misa, T.: 1988, 'How Machines Make History and How Historians (and Others) Help Them to Do So', *Science, Technology, and Human Values*, 13, 308-331.

Mishra, R.S., and A.K. Mukherjee. 1997. Oral presentation at TMS meeting, Indianapolis, Indiana, 16-18 September 1997, to be published in proceedings of Symp. "Mechanical Behavior of Bulk Nano-Materials."

Mishra, R.S., R.Z. Valiev, and A.K. Mukherjee. 1997. *Nanostructured Materials* 9:473. Morris, D.G., and M.A. Morris. 1991. *Acta Metall. Mater.* 39:1763-1779.

Mizrach, S., *Should there be a limit placed on the integration of humans and computers and electronic technology?* Ref: http://www.limmat.ch/koni/texte/cyborg-ethics.html.

Mnyusiwalla, A.; Abdallah, S.D.; Singer, P.A.: 2003, 'Mind the gap: science and ethics in nanotechnology', *Nanotechnology*, 14, R9-R13.

Mody, C.C.M.: 2004, 'Instruments in Training: The Growth of American Probe Microscopy in the 1980s', in: D. Kaiser (ed.), *Pedagogy and the Practice of Science: Producing Physical Scientists, 1800-2000*, Cambridge, MA: MIT Press (forthcoming).

Mody, C.C.M.: 2004, 'Small, but Determined: Technological Determinism in Nanoscience', *Hyle: International Journal for Philosophy of Chemistry*, 10(2), 99-128.

Mody, C.C.M.: 2004, *Crafting the Tools of Knowledge: The Invention, Spread, and Commercialization of Probe Microscopy, 1960-2000*, Ph.D. dissertation, Cornell University.

Molecular engineering: An approach to the development of general capabilities for molecular manipulation, by K. Eric Drexler, Proceedings of the National Academy of Sciences U.S.A. 78(9): 5275-5278, 1981.

Molecular Mechanics, by Ulrich Burkert and Norman L. Allinger, ACS Monograph 177, American Chemical Society, 1982.

Molecular Mechanics: The Art and Science of Parameterization, by J. Phillip Bowen and Norman L. Allinger, page 81-97 of reference.

Montemerlo, M.S., Love, J.C., Opiteck, G.J., Goldhaber, D. J., and Ellenbogen, J.C., "Technologies and Designs for Electronic Nanocomputers," MITRE Technical Report 96W0000044, The MITRE Corporation, McLean, VA, July 1996. For more information, send e-mail to nanotech@mitre.org.

Moor, J.H. & Weckert, J.: 2004, 'Nanoethics: Assessing the Nanoscale From an Ethical Point of View', in: D. Baird, A. Nordmann & J. Schummer (eds.), *Discovering the Nanoscale*, Amsterdam: IOS Press, pp. 301-310.

Moore, J.C., H.M. Jin, O. Kuchner, and F.H. Arnold. 1997. Strategies for the in vitro evolution of protein function: Enzyme evolution by random recombination of improved sequences. *J. Mol. Biol.* 272:336-347.

Moronne, M.M.; Ben Dahan, M.; Bruchez, M.; Hamamoto, D.J.; Weiss, S.; Alivisatos, A.P. 1999. "Nanocrystal phosphors as biological probes for fluorescence microscopy", *Biophysical Journal*, 76, A450-A450.

Morris, D.G., and M.A. Morris. 1997. *Materials Science Forum* 235-238:861. Nagpal, P., and I. Baker. 1990. *Scripta Metall. Mater.* 24:2381.

Moshe Sipper, "Fifty Years of Research on Self-Replication: An Overview," *Artificial Life 4*

(Summer 1998):237-257, in "Special Issue on Self-Replication." See also: Moshe Sipper, "The Artificial Self-Replication Page," http://lslwww.epfl.ch/~moshes/selfrep/

Mui, C. and C. Musgrave, "The Hydroxylation and Oxidation of the Ge(100)-2x1 Surface by Hydrogen Peroxide and Water," *Langmuir,* 20, 7604-7609 (2004).

Mui, C., S. Bent and C. Musgrave, "A Quantum Chemistry Based Statistical Mechanical Model of Hydrogen Deposition from Si(100)-2x1, Ge(100)-2x1, and SiGe Alloy Surfaces," *Journal of Physical Chemistry B,* 108, 18243-18253, (2004).

Mui, C., Y. Widjaja, J. Kang, and C. Musgrave, "Surface Reaction Mechanisms for Atomic Layer Deposition of Silicon Nitride," *Surface Science,* 557, 159-170, (2004).

Mumford, Lewis. 1934. *Technics and Civilization* (New York: Harcourt Brace Jovanovich, Inc.). National Science and Technology Council (NSTC). 1999. *Nanotechnology: Shaping the World Atom by Atom* (Washington, DC: September; available at http://www.nano.gov).

Munn Sanchez, E.: 2004, 'The Expert's Role in Nanoscience and Technology', in: D. Baird, A. Nordmann & J. Schummer (eds.), *Discovering the Nanoscale,* Amsterdam: IOS Press, pp. 257-266.

Murray, C.B.; Kagan, C.R.; Bawendi, M.G. 2000. "Synthesis and characterization of monodisperse nanocrystals and close-packed nanocrystal assemblies", *Annual Review of Materials Science,* 30, 545- 610.

Naughton, G. 1999. The Advanced Tissue Sciences story. *Sci. Am.* 280(4): p. 84-5.

Nauta, K. and R.E. Miller. 1999. Non-equilibrium self-assembly of long chains of polar molecules in superfluid helium. *Science.* Vol. 283. 1895-1897.

New Approaches to Empirical Force Fields, by Uri Dinur and Arnold T. Hagler, pages 99-164, reference.

Nicholas Wade, "A DNA Success Raises Bioterror Concern," New York Times, January 12, 2005.

Nieman, G.W., J.R. Weertman, and R.W. Siegel. 1991a. *Mater. Res. Soc. Symp. Proc.* 206:581-586.

Nomura, M. & Held, W. (1974) in *Ribosomes,* eds. Nomura, M., Tissiers, A. & Lengyel, P. (Cold Spring Harbor Laboratory, Cold Spring Harbor, NY), pp. 193-203.

Nordmann, A. (rapp.) Converging Technologies—Shaping the Future of European Societies, European Commission report, 2004.

Nordmann, A.: 2003, 'Shaping the World Atom by Atom: Eine nanowissenschaftliche WeltBildanalyse', in: A. Grunwald (ed.), *Technikgestaltung zwischen Wunsch und Wirklichkeit,* Berlin: Springer, pp. 191-199.

Nordmann, A.: 2004, 'Molecular Disjunctions: Staking Claims at the Nanoscale', in: D. Baird, A. Nordmann & J. Schummer (eds.), *Discovering the Nanoscale,* Amsterdam: IOS Press, pp. 51-62.

Nordmann, A.: 2004, 'Molecular Disjunctions', in: D. Baird, A. Nordmann, J. Schummer (eds.), *Discovering the Nanoscale,* IOS Press, Amsterdam, pp. 51-62.

Nordmann, A.: 2004, 'Nanotechnology: Convergence and Integration', Presentation at *EuroNanoForum, Trieste, December 10, 2003* (Proceedings in preparation).

Nordmann, A.: 2004, 'Nanotechnology's WorldView: New Space for Old Cosmologies', *IEEE Technology and Society Magazine*, 23 (forthcoming).

Nordmann, A.: 2004, 'Social Imagination for Nanotechnology', in: European Commission (Community Health and Consumer Protection): *Nanotechnologies: A Preliminary Risk Analysis on the Basis of a Workshop, Brussels, 1-2 March 2004*, pp. 111-113 [www.europa.eu.int/comm/health/ph_risk/documents/ev_20040301_en.pdf].

Nordmann, A.: 2004, 'Was ist TechnoWissenschaft? ' Zum Wandel der Wissenschaftskultur am Beispiel von Nanoforschung und Bionik', in: T. Rossmann & C. Tropea (eds.), *Bionik ' Neue Forschungsergebnisse aus Natur-, Ingenieur- und Geisteswissenschaften*, Berlin: Springer, 2004 (forthcoming)

Nowotny, H.; Scott, P.; Gibbons, M.: 2001, *Re-thinking Science*, Polity, Cambridge.

Peng, X.G.; Manna, L.; Yang, W.D.; Wickham, J.; Scher, E.; Kadavanich, A.; Alivisatos, A.P. 2000. "Shape control of CdSe nanocrystals", *Nature*, 404, 59-61.

Perkal M., C. Marks, M.I. Lorber, W.H. Marks. 1992. "A three-year experience with serum anodal trypsinogen as a biochemical marker for rejection in pancreatic allografts. False positives, tissue biopsy, comparison with other markers, and diagnostic strategies." *Transplantation*, Feb; 53 (2): pp. 415-419.

Perspectives on the Computer Revolution, Zenon W. Pylyshyn, Prentice Hall 1970.

Philip Ball, "Synthetic Biology: Starting from Scratch," Nature, 431, pp. 624-626, 7 October 2004.

Philip G. Collins, Michael S. Arnold, Phaedon Avouris, "Engineering carbon nanotubes and nanotube circuits using electrical breakdown," *Science* 292(27 April 2001):706-709.

Philip G. Collins, Phaedon Avouris, "Nanotubes for electronics," *Scientific American* 283(December 2000):62-69.

Philipse, A.P. 1998. "Colloidal Dispersions," section 3.4, A. ten Wolde (ed.), *Nanotechnology: Towards a molecular construction kit*. STT Report #60, The Hague, 171-178.

Phoenix, C.; Drexler, E.: 2004, 'Safe Exponential Manufacturing', *Nanotechnology*, 15, 869-72.

Phoenix, C.; Drexler, E.: 2004, 'Safe Exponential Manufacturing', *Nanotechnology*, 15, 869-872.

Pierce, N.A. and M.B. Giles. 2000. Adjoint reconstruction of superconvergent functionals from PDE approximations. *SIAM Rev.* Vol. 42(2). 247-264.

Pinch, T.: 1986, *Confronting Nature*, Reidel, Dordrecht.

Pinch, T.: 1993, "Testing – One, Two, Three... Testing!': Toward a Sociology of Testing', *Science, Technology, and Human Values*, 18, 25-41.

Pitt, J.C.: 2004, 'The Epistemology of the Very Small', in: D. Baird, A. Nordmann & J. Schummer (eds.), *Discovering the Nanoscale*, Amsterdam: IOS Press, pp. 157-163.

Piveteau, Laurent-Dominique, Anthe S. Zandvliet and Robert Langer. 2001. "Nitroxide and Galactose Grafted Dendrimer: A Specifically Targeted Contrast Agent for MRI and EPR Imaging of the Liver," *in preparation.*

Polanyi, Karl. 1944. *The Great Transformation: The Political and Economic Origins of Our Time* (Boston: Beacon Press, 1957; orig. ed. 1944).

Polanyi, M.: 1962, *Personal Knowledge: Towards a Post-Critical Philosophy*, Harper Torchbooks, New York.

Poldrack, R.A. 2000. Imaging brain plasticity: conceptual and methodological issues — a theoretical review. *Neuroimage*. 12(1): p. 1-13.

Pollock, C.R. 1995. *Fundamentals of Optoelectronics*. Irwin, Inc. pp. 3-5.

Porod, W., C.S. Lent, G.H. Bernstein, A.O. Orlov, I. Amlani, G.L. Snider, and J.L. Merz. 1999. Quantumdot cellular automata: computing with coupled quantum dots. *Int. Journal of Electronics* 86, No.5, 549-590.

pp.140-145.

Pressman, J.: 2004, 'Nano Narrative: A Parable from Electronic Literature', in: N.K. Hayles (ed.), *Nanoculture: Implications of the New Technoscience*, Bristol, UK: Intellect Books, pp. 191-200.

Prigogine, I., and S. Rice. 1988. *Advances in chemical physics*, Vol. 70, Parts 1 & 2. New York: J. Wiley.

Pritchard, M.S. 1996. "Conflicts of Interest: Conceptual and Normative Issues," *Academic Medicine*, 71:12, December, pp. 1305-1313.

Putman, C.A.J., De Grooth, B.G., Van Hulst, N.F., and Greve, J. (1992) A detailed analysis of the optical beam deflection technique for use in atomic force microscopy. J. App. Phys. 72(1), 6-12

Putnam, D., Gentry, C.A., Pack, D.W., Langer, R. 2001. "Polymer-based Gene Delivery with Low Cytotoxicity by a Unique Balance of Side-chain Termini." *Proceedings of the National Academy of Sciences*, U.S.A., Volume 98, Number 3, pages 1200-1205.

Ralph C. Merkle, "Design-ahead for nanotechnology," in Markus Krummenacker, James Lewis, eds., *Prospects in Nanotechnology: Toward Molecular Manufacturing*, John Wiley & Sons, New York, 1995, pp. 23-52.

Ralph C. Merkle, "Molecular building blocks and development strategies for molecular nanotechnology," *Nanotechnology* 11(2000):89-99; http://www.zyvex.com/nanotech/mbb/mbb.html

Ralph C. Merkle, "Self replicating systems and molecular manufacturing," *J. Brit. Interplanet. Soc.* 45(1992):407-413; http://www.zyvex.com/nanotech/selfRepJBIS.html

Ralph C. Merkle, "Self-replicating systems and low cost manufacturing," in M.E. Welland, J.K. Gimzewski, eds., The Ultimate Limits of Fabrication and Measurement, Kluwer, Dordrecht, 1994, pp. 25-32; http://www.zyvex.com/nanotech/selfRepNATO.html

Reinhardt, C.: 2004, 'Chemistry in a Physical Mode: Molecular Spectroscopy and the Emergence of NMR', *Annals of Science*, 61, 1-32.

Reviews in Computational Chemistry, Vol 2, Kenny B. Lipkowitz and Donald B. Boyd, VCH Publishers, 1991.

Rheinberger, H.-J.: 1997, *Toward a History of Epistemic Things: Synthesizing Proteins in the Test Tube*, Stanford University Press, Stanford.

Richard A. Laing, "Automation introspection," *J. Comp. System Sci.* 13(1976):172-183.

Richard A. Laing, "Automaton models of reproduction by self-inspection," *J. Theoret. Biol.* 66(1977):437-456.

Richard A. Laing, "Some alternative reproductive strategies in artificial molecular machines," *J. Theoret. Biol.* 54(1975):63-84.

Richard E. Smalley, "Of chemistry, love, and nanobots," *Scientific American* 285(September 2001):76-77.

Richard P. Feynman, "There's Plenty of Room at the Bottom," *Engineering and Science* (California Institute of Technology), February 1960, pp. 22-36. Reprinted in B.C. Crandall, James Lewis, eds., *Nanotechnology: Research and Perspectives*, MIT Press, 1992. pp. 347-363, and in D.H. Gilbert, ed., *Miniaturization*, Reinhold, New York, 1961, pp. 282-296; http://www.zyvex.com/nanotech/feynman.html

Rietman, E.A.: 2001, 'Drexler hypothesis of a universal assembler is supported not by theoretical arguments alone but by existence proof in the form of biological life', in: *Molecular Engineering of Nanosystems*, Springer, New York & Berlin.

Smith, M.R.; Marx, L. (eds.): 1994, *Does Technology Drive History? The Dilemma of Technological Determinism*, MIT Press, Cambridge, MA.

Star, S.L.; Griesemer, J.R.: 1989, 'Institutional Ecology, 'Translations' and Boundary Objects: Amateurs and Professionals in Berkeley's Museum of Vertebrate Zoology, 1907-1939', *Social Studies of Science*, 19, 387-420.

Steelman, John R. 1947. *Science and Public Policy* (Washington, DC: US Government Printing Office). Stephenson, Neal. 1995. *The Diamond Age, or, Young Lady's Illustrated Primer* (New York: Bantam Books).

Tenner, E.: 2001, 'Nanotechnology and Unintended Consequences', in: M.C. Roco and W.S. Bainbridge (eds.), *Societal Implications of Nanoscience and Nanotechnology*, Dordrecht: Kluwer, pp. 241-246.

Theoretical studies of a hydrogen abstraction tool for nanotechnology, by Charles B. Musgrave, Jason K. Perry, Ralph C. Merkle, and William A. Goddard III, *Nanotechnology*.

Thompson, P.B.: 2001, 'Social Acceptance of Nanotechnology', in: M.C. Roco, W.S. Bainbridge (eds.), *Societal Implications of Nanoscience and Nanotechnology*, online: National Science Foundation, pp. 198-202.

Tighe, T.S., J.M. Worlock, and M.L. Roukes. 1997. Direct thermal conductance measurements on suspended monocrystalline nanostructures. *Appl. Phys. Lett.* 70:2687-9.

Tihamer Toth-Fejel, "Modeling kinematic cellular automata: An approach to self-replication," Environmental Research Institute of Michigan, Ann Arbor, MI, 4 October 2000.

9

Research Methodology in Artificial Intelligence Science

INTRODUCTION

Artificial Intelligence (AI) is the part of computer science concerned with designing intelligent computer systems, that is, computer systems that exhibit the characteristics we associate with intelligence in human behaviour — understanding language, learning, reasoning and solving problems. Artificial Intelligence, or AI for short, is a combination of computer science, physiology, and philosophy. AI is a broad topic, consisting of different fields, from machine vision to expert systems. The element that the fields of AI have in common is the creation of machines that can "think". In order to classify machines as "thinking", it is necessary to define intelligence. To w what degree does intelligence consist of, for example, solving complex problems, or making generalizations and relationships? And what about perception and comprehension? Research into the areas of learning, of language, and of sensory perception have aided scientists in building intelligent machines. One of the most challenging approaches facing experts is building systems that mimic the behavior of the human brain, made up of billions of neurons, and arguably the most complex matter in the universe. Perhaps the best way to gauge the intelligence of a machine is British computer scientist Alan Turing's test. He stated that a computer would deserves to be called intelligent if it could deceive a human into believing that it was human. Artificial Intelligence has come a long way from its early roots, driven by dedicated researchers. The beginnings of AI reach back before electronics, to philosophers and mathematicians such as Boole and others theorizing on principles that were used as the foundation of AI Logic.

AI really began to intrigue researchers with the invention of the computer in 1943. The technology was finally available, or so it seemed, to simulate intelligent behavior. Over the next four decades, despite many stumbling blocks, AI has grown from a dozen researchers, to thousands of engineers and specialists; and from programmes capable of playing checkers, to systems designed to diagnose disease. AI has always been on the pioneering end of computer science. Advanced-level computer languages, as well as computer interfaces and word-processors owe their existence to the research into artificial intelligence. The theory and insights brought about by AI research will set the trend in the future of computing. The products available today are only bits and pieces of what are soon to follow, but they are a movement towards the future of artificial intelligence. The advancements in the quest for artificial intelligence have, and will continue to affect our jobs, our education, and our lives. In an attempt to understand the concept 'Artificial Intelligence' (AI, for short), and also AI as a field of serious, scientific study, we shall look at the following definition of the term in some detail: The concept 'Artificial or Machine Intelligence' may be provisionally defined as 'electronically simulated human intelligence'. This may appear to be a fairly straightforward and easily comprehensible definition you have ever come across, but I must warn you that it is somewhat deceptive. For example, what do you understand by 'human intelligence' (HI, for short), to begin with? I am sure that you believe that it is one of those household words like, 'foolishness', 'sensitiveness', and so on. Synonyms such as 'shrewdness', 'cleverness', brightness of mind', etc., may also occur to you.

The problem, however, is that while HI as it is used here is certainly not incompatible with the sense of 'cleverness' or 'brightness of mind', it is technically wrong and misleading to equate the word with these meanings. For example, even a fool is said to be endowed with HI, if he can spontaneously smile a smile of recognition, say, when he meets a friend after a gap of time. That smile is a sure indication of the fact that the person is capable of adjusting his behaviour to what happens around him and this capability is symptomatic of HI. Again, if an idiot can avoid fire when he sees it, with out having to come into bodily contact with it, he is 'intelligent', too. In other words, HI is that faculty of any normal (that is, not mentally retarded) human being, which accounts for all his actions, speech and behaviour – idiotic as well as informed. Now, when a computer is programmed to simulate this faculty of human beings, it is said to be 'intelligent', and this capability of the machine is 'Artificial or Machine Intelligence'. Let us, in the next place, turn to AI as a field of serious, scientific study. Here it is useful to make a distinction between AI theory and practice.

As theory, AI is the engineering counterpart of 'cognitive science'. Cognitive science is a cover term for 'philosophy', 'linguistics', and 'psychology', each of which claims to account for certain complementary aspects of HI. Philosophy asks and answers questions about 'knowledge', and its nature and expression; it seeks to find out whether all knowledge is realised only in terms of language or whether there is a kind of knowledge which is language independent. The nature and expression of 'meaning' have also been the subject matter of almost any school of philosophical enquiry. Now, no serious discussion of 'intelligence', be it natural or artificial, can afford to ignore the philosophical enquiries concerning 'knowledge' and 'meaning'. As for linguistics, it is the scientific study of language, its structure and function. It aims to relate the world of meanings to the world of sounds on the one hand, and the world of graphic symbols on the other. Our definition of HI laid down that all human actions and behaviour, including 'speech' are symptomatic of intelligence, and speech certainly belongs to the domain of linguistics.

Psychology, to be more precise, cognitive psychology is an equally important constituent of cognitive science, and is very closely related to intelligence. Of particular interest to us, here, are those aspects of psychology which attempt to characterize mental models of objects – captured as facts and reasoning of various kinds. Thus, various aspects of HI are fruitfully studied from the view points of philosophy, linguistics and psychology. The viability of any AI theory directly depends on how effectively it can capture the essential features of HI for purposes of AI modeling and hence the suggested influence of cognitive science on AI theory.

Let us return to our definition of AI once again, before taking up AI practice. There are two parts to this definition: (a) human intelligence, and (b) electronic simulation. Our discussion of AI practice, which primarily stresses on the functional aspects of AI, will center on the electronic simulation part of the definition. Thus, functionally, AI is concerned with the design of a computer to simulate human intelligence. We said that all human action, including speech, is symptomatic of intelligence. In order to determine the scope of electronic simulation of HI, however, a greater understanding of HI than a mere analysis of symptoms becomes necessary. A clearer perception of what constitutes HI than what evidence we have of it is highly relevant to the simulation efforts of AI practice. Knowledge' and 'reasoning' are two important constituents of HI. Knowledge comes to us first or second hand. We see things, feel them, and in appropriate cases, smell and taste them and thereby, KNOW about them first hand. We hear about persons, places, and things; we read about them, and in appropriate cases, see 2-D or 3-D images of them, and thereby KNOW about them second hand.

We develop intellectually perceived artifacts, talk about them, and thereby exchange KNOWLEDGE about them. We come to KNOW about many things without being aware of the process; we also acquire KNOWLEDGE through conscious training, and education. A large body of knowledge is merely inferred knowledge; there is no need for us to have personally experienced this body of knowledge. For example, it is intelligent behaviour to avoid coming into violent, head-on, bodily contact with a fast-moving object, like a speeding bus. The human ability to avoid this collision course is not necessarily based on personally experienced knowledge; nor does this ability come necessarily or conscious education or training; it is, in a large number of cases, based on inference or reasoning. Now, having said that all actions of any normal human being are symptomatic of intelligence, how then do we account for variations in intelligent behaviour among people? Good question. Well, these variations are due to personal styles of behaviour, culturally conditioned styles of behaviour, and even misbehaviour, brought about by intellectual intervention. Thus, at the heart of any intelligent behaviour are knowledge and reasoning, which may be assumed to be located in the human brain. The application of this knowledge in a given situation and the activation of the reasoning capability may be substantially moderated, coloured, or influenced by the personal emotions and the impulses of the individual concerned. We know how the parents' judgment of their children's behaviour is often characterized by a degree of tolerance, which they may not be prepared to show when it comes to judging the behaviour of other children. Our intelligent response in a given situation may be further moderated by 'cultural conditioning'. Thus people belonging to different cultures show different intelligent behaviours. For example, what is morally right and constitutes social justice in culture 'x' may cause a revolt in culture 'y'. Against the background of this discussion of what constitutes HI, let us examine the principal concern of AI practice, namely, electronic simulation of HI, or the design of an intelligent software to programme a computer to behave like a human being to the extent possible.

Main Branches of AI

Logical AI

What a program knows about the world in general the facts of the specific situation in which it must act, and its goals are all represented by sentences of some mathematical logical language. The program decides what to do by inferring that certain actions are appropriate for achieving its goals. The first article proposing this was [McC59]. [McC89] is a more recent summary. [McC96b] lists some of the concepts involved in logical aI. [Sha97] is an important text.

Search

AI programs often examine large numbers of possibilities, e.g. moves in a chess game or inferences by a theorem proving program. Discoveries are continually made about how to do this more efficiently in various domains.

Pattern Recognition

When a program makes observations of some kind, it is often programmed to compare what it sees with a pattern. For example, a vision program may try to match a pattern of eyes and a nose in a scene in order to find a face. More complex patterns, e.g. in a natural language text, in a chess position, or in the history of some event are also studied. These more complex patterns require quite different methods than do the simple patterns that have been studied the most.

Representation

Facts about the world have to be represented in some way. Usually languages of mathematical logic are used.

Inference

From some facts, others can be inferred. Mathematical logical deduction is adequate for some purposes, but new methods of *non-monotonic* inference have been added to logic since the 1970s. The simplest kind of non-monotonic reasoning is default reasoning in which a conclusion is to be inferred by default, but the conclusion can be withdrawn if there is evidence to the contrary. For example, when we hear of a bird, we man infer that it can fly, but this conclusion can be reversed when we hear that it is a penguin. It is the possibility that a conclusion may have to be withdrawn that constitutes the non-monotonic character of the reasoning. Ordinary logical reasoning is monotonic in that the set of conclusions that can the drawn from a set of premises is a monotonic increasing function of the premises. Circumscription is another form of non-monotonic reasoning.

Common Sense Knowledge and Reasoning

This is the area in which AI is farthest from human-level, in spite of the fact that it has been an active research area since the 1950s. While there has been considerable progress, e.g. in developing systems of *non-monotonic reasoning* and theories of action, yet more new ideas are needed. The Cyc system contains a large but spotty collection of common sense facts.

Learning from Experience

Programs do that. The approaches to AI based on *connectionism* and *neural nets* specialize in that. There is also learning of laws expressed in logic. [Mit97] is a comprehensive undergraduate text on machine learning. Programs can only learn what facts or behaviors their formalisms can represent, and unfortunately learning systems are almost all based on very limited abilities to represent information.

Planning

Planning programs start with general facts about the world (especially facts about the effects of actions), facts about the particular situation and a statement of a goal. From these, they generate a strategy for achieving the goal. In the most common cases, the strategy is just a sequence of actions.

Epistemology

This is a study of the kinds of knowledge that are required for solving problems in the world.

Ontology

Ontology is the study of the kinds of things that exist. In AI, the programs and sentences deal with various kinds of objects, and we study what these kinds are and what their basic properties are. Emphasis on ontology begins in the 1990s.

Heuristics

A heuristic is a way of trying to discover something or an idea imbedded in a program. The term is used variously in AI. *Heuristic functions* are used in some approaches to search to measure how far a node in a search tree seems to be from a goal. *Heuristic predicates* that compare two nodes in a search tree to see if one is better than the other, i.e. constitutes an advance toward the goal, may be more useful.

Genetic Programming

Genetic programming is a technique for getting programs to solve a task by mating random Lisp programs and selecting fittest in millions of generations. It is being developed by John Koza's group and here's a tutorial.

Developments in AI

Artificial intelligence (AI) has been one of the most controversial domains of inquiry in computer science since it was first proposed in the 1950s. Defined as the part of computer science concerned with designing systems that exhibit the characteristics associated with human intelligence—understanding language, learning, reasoning, solving problems, and so on (Barr and Feigenbaum, 1981)—the field has attracted researchers because of its ambitious goals and enormous underlying intellectual challenges. The field has been controversial because of its social, ethical, and philosophical implications. Such controversy has affected the funding environment for AI and the objectives of many research programmes. AI research is conducted by a range of scientists and technologists with varying perspectives, interests, and motivations.

Scientists tend to be interested in understanding the underlying basis of intelligence and cognition, some with an emphasis on unraveling the mysteries of human thought and others examining intelligence more broadly. Engineering-oriented researchers, by contrast, are interested in building systems that behave intelligently. Some attempt to build systems using techniques analogous to those used by humans, whereas others apply a range of techniques adopted from fields such as information theory, electrical engineering, statistics, and pattern recognition. Those in the latter category often do not necessarily consider themselves AI researchers, but rather fall into a broader category of researchers interested in machine intelligence. The concept of AI originated in the private sector, but the growth of the field, both intellectually and in the size of the research community, has depended largely on public investments. Public monies have been invested in a range of AI programmes, from fundamental, long-term research into cognition to shorter-term efforts to develop operational systems. Most of the federal support has come from the Defense Advanced Research Projects Agency (DARPA, known during certain periods as ARPA) and other units of the Department of Defense (DOD). Other funding agencies have included the National Institutes of Health, National Science Foundation, and National Aeronautics and Space Administration (NASA), which have pursued AI applications of particular relevance to their missions—health care, scientific research, and space exploration.

This chapter highlights key trends in the development of the field of AI and the important role of federal investments. The sections of this chapter, presented in roughly chronological order, cover the launching of the AI field, the government's initial participation, the pivotal role played by DARPA, the success of speech recognition research, the shift from basic to applied research, and AI in the 1990s. The final section summarizes the lessons to be learned from history. This case study is based largely on published accounts, the scientific and technical literature, reports by the major AI research centers, and interviews conducted with several leaders of AI research centers. (Little information was drawn from the records of the participants in the field, funding agencies, editors and publishers, and other primary sources most valued by professional historians.)

The Private Sector Launches the Field

The origins of AI research are intimately linked with two landmark papers on chess playing by machine. They were written in 1950 by Claude E. Shannon, a mathematician at Bell Laboratories who is widely acknowledged as a principal creator of information theory. In the late 1930s, while still a graduate student, he developed a method for symbolic analysis of switching systems and networks (Shannon, 1938), which provided scientists and engineers with much-improved analytical and conceptual tools. After working at Bell Labs for half a decade, Shannon published a paper on information theory (Shannon, 1948). Shortly thereafter, he published two articles outlining the construction or programming of a computer for playing chess (Shannon, 1950a,b).

Shannon's work inspired a young mathematician, John McCarthy, who, while a research instructor in mathematics at Princeton University, joined Shannon in 1952 in organising a conference on automata studies, largely to promote symbolic modeling and work on the theory of machine intelligence. A year later, Shannon arranged for McCarthy and another future pioneer in AI, Marvin Minsky, then a graduate student in mathematics at Princeton and a participant in the 1952 conference, to work with him at Bell Laboratories during 1953.

By 1955, McCarthy believed that the theory of machine intelligence was sufficiently advanced, and that related work involved such a critical mass of researchers, that rapid progress could be promoted by a concentrated summer seminar at Dartmouth University, where he was then an assistant professor of mathematics. He approached the Rockefeller Foundation's Warren Weaver, also a mathematician and a promoter of cutting-edge science, as well as Shannon's collaborator on information theory. Weaver and his colleague Robert S. Morison, director for Biological and Medical Research, were initially skeptical (Weaver, 1955). Morison pushed McCarthy and Shannon to widen the range of participants and made other suggestions. McCarthy and Shannon responded with a widened proposal that heeded much of Morison's advice. They brought in Minsky and a well-known industrial researcher, Nathaniel Rochester of IBM, as co-principal investigators for the proposal, submitted in September 1955. In the proposal, the four researchers declared that the summer study was "to proceed on the basis of the conjecture that every aspect of learning or any other feature of intelligence can in principle be so precisely described that a machine can be made to simulate it." They sought to bring a number of U.S. scholars to Dartmouth to create a research agenda for AI and begin actual work on it. In spite of Morison's skepticism, the Rockefeller Foundation agreed to fund this summer project with a grant of $7,500 (Rhind, 1955), primarily to cover summer salaries and expenses of the academic participants. Researchers from industry would be compensated by their respective firms.

Although most accounts of AI history focus on McCarthy's entrepreneurship, the role of Shannon—an intellectual leader from industry—is also critical. Without his participation, McCarthy would not have commanded the attention he received from the Rockefeller Foundation. Shannon also had considerable influence on Marvin Minsky. The title of Minsky's 1954 doctoral dissertation was "Neural Nets and the Brain Model Problem."

The role of IBM is similarly important. Nathan Rochester was a strong supporter of the AI concept, and he and his IBM colleagues who attended the 1956 Dartmouth workshop contributed to the early research in the field. After the workshop IBM welcomed McCarthy to its research laboratories, in large part because of IBM's previous work in AI and because "IBM looked like a good bet to pursue artificial intelligence research vigorously" in the future. Rochester was a visiting professor at the Massachusetts Institute of Technology (MIT) during 1958-1959, and he unquestionably helped McCarthy with the development of LISP, an important list-processing language. Rochester also apparently lent his support to the creation in 1958 of the MIT Artificial Intelligence Project (Rochester and Gelertner, 1958). Yet, in spite of the early activity of Rochester and other IBM researchers, the corporation's interest in AI cooled. Although work continued on computer-based checkers and chess, an internal report prepared about 1960 took a strong position against broad support for AI. Thus, the activities surrounding the Dartmouth workshop were, at the outset, linked with the cutting-edge research at a leading private research laboratory (AT&T Bell Laboratories) and a rapidly emerging industrial giant (IBM). Researchers at Bell Laboratories and IBM nurtured the earliest work in AI and gave young academic researchers like McCarthy and Minsky credibility that might otherwise have been lacking. Moreover, the Dartmouth summer research project in AI was funded by private philanthropy and by industry, not by government. The same is true for much of the research that led up to the summer project.

The federal government's initial involvement in AI research was manifested in the work of Herbert Simon and Allen Newell, who attended the 1956 Dartmouth workshop to report on "complex information processing." Trained in political science and economics at the University of Chicago, Simon had moved to Carnegie Institute of Technology in 1946 and was instrumental in the founding and early research of the Graduate School of Industrial Administration (GSIA). Funded heavily by the Ford Foundation and the Office of Naval Research (ONR), and the Air Force, GSIA was the pioneer in bringing quantitative behavioral social sciences research (including operations research) into graduate management education. Because of his innovative work in human decision making, Simon became, in March 1951, a consultant to the RAND Corporation, the pioneering think tank established by the Air Force shortly after World War II. At RAND, where he spent several summers carrying out collaborative research, Simon encountered Newell, a mathematician who helped to conceive and develop the Systems Research Laboratory, which was spun out of RAND as the System Development Corporation in 1957. In 1955, Simon and Newell began a long collaboration on the simulation of human thought, which by the summer of 1956 had resulted in their fundamental work (with RAND computer programr J.C. Shaw) on the Logic Theorist, a computer programme capable of proving theorems found in the *Principia* of Bertrand Russell and Alfred North Whitehead (Newell and Simon, 1956).

This programme is regarded by many as the first successful AI programme, and the language it used, IPL2, is recognized as the first significant list-processing language. As programmed by Simon, Newell, and Shaw, a computer simulated human intelligence, solving a problem in logic in much the same way as would a skilled logician. In this sense, the machine demonstrated artificial intelligence. The project was funded almost entirely by the Air Force through Project RAND, and much of the computer programming was done at RAND on an Air Force-funded computer (the Johnniac, named after RAND consultant John von Neumann, the creator of the basic architecture for digital electronic computers).

Newell's collaboration with Simon took him to Carnegie Tech, where, in 1957, he completed the institution's first doctoral dissertation in AI, "Information Processing: A New Technique for the Behavioral Sciences." Its thrust was clearly driven by the agenda laid out by the architects of GSIA. As Newell later stressed, his work with Simon (and that of Simon's several other AI students at GSIA) reflected the larger agenda of GSIA, even though most of this work was funded by the Air Force and ONR until the early 1960s. All of this work concentrated on the formal modeling of decision making and problem solving.

Simon and Newell developed another well-known AI programme as a sequel to Logic Theorist—the General Problem Solver (GPS), first run in 1957 and developed further in subsequent years. Their work on GPS, like that on Logic Theorist, was characterized by its use of heuristics (i.e., efficient but fallible rules of thumb) as the means to simulate human cognitive processes (Newell et al., 1959). The GPS was capable of solving an array of problems that challenge human intelligence (an important accomplishment in and of itself), but, most significantly, it solved these problems by simulating the way a human being would solve them. These overall research efforts at GSIA, including the doctoral research of Simon's students—all funded principally by Air Force and ONR money—remained modest in scale compared to those at Carnegie Tech after 1962.

Also modest were the efforts at MIT, where McCarthy and Minsky established the Artificial Intelligence Project in September 1957. This effort was funded principally through a word-of-mouth agreement with Jerome Wiesner, then director of MIT's military-funded Research Laboratory in Electronics (RLE). In exchange for "a room, two programrs, a secretary and a keypunch [machine]," the two assistant professors of mathematics agreed, according to McCarthy, to "undertake the supervision of some of the six mathematics graduate students that RLE had undertaken to support."

The research efforts at Carnegie Tech (which became Carnegie Mellon University [CMU] in 1967), RAND, and MIT, although limited, yielded outstanding results in a short time. Simon and Newell showed that computers could demonstrate human-like behavior in certain well-defined tasks. Substantial progress was also made by McCarthy, with his pioneering development of LISP, and Minsky, who formalized heuristic processes and other means of reasoning, including pattern recognition. Previously, computers had been used principally to crunch numbers, and the tools for such tasks were primitive. The AI researchers found ways to represent logical formulas, carry out proofs, conduct plans, and manipulate such objects. Buoyed by their successes, researchers at both institutions projected bold visions—which, as the research was communicated to the public, became magnified into excessive claims—about the future of the new field of AI and what computers might ultimately achieve.

DARPA'S Pivotal Role

The establishment in 1962 of ARPA's Information Processing Techniques Office (IPTO) radically changed the scale of research in AI, propelling it from a collection of small projects into a large-scale, high-profile domain. From the 1960s through the 1990s, DARPA provided the bulk of the nation's support for AI research and thus helped to legitimize AI as an important field of inquiry and influence the scope of related research. Over time, the nature of DARPA's support changed radically—from an emphasis on fundamental research at a limited number of centers of excellence to more broad-based support for applied research tied to military applications—both reflecting and motivating changes in the field of AI itself.

The early academic centers were MIT and Carnegie Tech. Following John McCarthy's move to Stanford in 1963 to create the Stanford Artificial Intelligence Laboratory (SAIL), IPTO worked a similar transformation of AI research at Stanford by making it the third center of excellence in AI. Indeed, the IPTO increased Stanford's allocation in 1965, allowing it to upgrade its computing capabilities and to launch five major team projects in AI research. Commenting in 1984 about how AI-related research at Carnegie Tech migrated out of GSIA into what became an autonomous department (and later a college) of CMU, Newell (1984) captured the transformation wrought by IPTO:

> ... the DARPA support of AI and computer science is a remarkable story of the nurturing of a new scientific field. Not only with MIT, Stanford and CMU, which are now seen as the main DARPA-supported university computer-science research environments, but with other universities as well... DARPA began to build excellence in information processing in whatever fashion we thought best.... The DARPA effort, or anything similar, had not been in our wildest imaginings....

Another center of excellence—the Stanford Research Institute's (SRI's) Artificial Intelligence Center—emerged a bit later (in 1966), with Charles Rosen at the command. It focused on developing "automatons capable of gathering, processing, and transmitting information in a hostile environment" (Nilsson, 1984). Soon, SRI committed itself to the development of an AI-driven robot, Shakey, as a means to achieve its objective. Shakey's development necessitated extensive basic research in several domains, including planning, natural-language processing, and machine vision. SRI's achievements in these areas (e.g., the STRIPS planning system and work in machine vision) have endured, but changes in the funder's expectations for this research exposed SRI's AI programme to substantial criticism in spite of these real achievements.

Under J.C.R. Licklider, Ivan Sutherland, and Robert Taylor, DARPA continued to invest in AI research at CMU, MIT, Stanford, and SRI and, to a lesser extent, other institutions. Licklider (1964) asserted that AI was central to DARPA's mission because it was a key to the development of advanced command-and-control systems. Artificial intelligence was a broad category for Licklider (and his immediate successors), who "supported work in problem solving, natural language processing, pattern recognition, heuristic programming, automatic theorem proving, graphics, and intelligent automata. Various problems relating to human-machine communication—tablets, graphic systems, hand-eye coordination—were all pursued with IPTO support" (Norberg and O'Neill, 1996). These categories were sufficiently broad that researchers like McCarthy, Minsky, and Newell could view their institutions' research, during the first 10 to 15 years of DARPA's AI funding, as essentially unfettered by immediate applications. Moreover, as work in one problem domain spilled over into others easily and naturally, researchers could attack problems from multiple perspectives. Thus, AI was ideally suited to graduate education, and enrollments at each of the AI centers grew rapidly during the first decade of DARPA funding. DARPA's early support launched a golden age of AI research and rapidly advanced the emergence of a formal discipline. Much of DARPA's funding for AI was contained in larger programme initiatives. Licklider considered AI a part of his general charter of Computers, Command, and Control. Project MAC, a project on time-shared computing at MIT, allocated roughly one-third of its $2.3 million annual budget to AI research, with few specific objectives.

Success in Speech Recognition

The history of speech recognition systems illustrates several themes common to AI research more generally: the long time periods between the initial research and development of successful products, and the interactions between AI researchers and the broader community of researchers in machine intelligence. Many capabilities of today's speech-recognition systems derive from the early work of statisticians, electrical engineers, information theorists, and pattern-recognition researchers. Another key theme is the complementary nature of government and industry funding. Industry supported work in speech recognition at least as far back as the 1950s, when researchers at Bell Laboratories worked on systems for recognizing individual spoken digits "zero" through "nine." Research in the area was boosted tremendously by DARPA in the 1970s. DARPA established the Speech Understanding Research (SUR) programme to develop a computer system that could understand continuous speech. Lawrence Roberts initiated this project in 1971 while he was director of IPTO, against the advice of a National Academy of Sciences committee. Roberts wanted a system that could handle a vocabulary of 10,000 English words spoken by anyone. His advisory board, which included Allen Newell and J.C.R. Licklider, issued a report calling for an objective of 1,000 words spoken in a quiet room by a limited number of people, using a restricted subject vocabulary (Newell et al., 1971).

Roberts committed $3 million per year for 5 years, with the intention of pursuing a 5-year follow-on project. Major SUR project groups were established at CMU, SRI, MIT's Lincoln Laboratory, Systems Development Corporation (SDC), and Bolt, Beranek, and Newman (BBN). Smaller contracts were awarded to a few other institutions. Five years later, SUR products were demonstrated. CMU researchers demonstrated two systems, HARPY and HEARSAY-I, and BBN developed Hear What I Mean (HWIM). The system developed cooperatively by SRI and SDC was never tested (Green, 1988). The system that came the closest to satisfying the original project goals—and may have exceeded the benchmarks—was HARPY, but controversy arose within DARPA and the AI community about the way the tests were handled. Full details regarding the testing of system performance had not been worked out at the outset of the SUR programme. As a result, some researchers—including DARPA research managers—believed that the SUR programme had failed to meet its objectives. DARPA terminated the programme without funding the follow-on. Nevertheless, industry groups, including those at IBM, continued to invest in this research area and made important contributions to the development of continuous speech recognition methods.

DARPA began funding speech recognition research on a large scale again in 1984 as part of the Strategic Computing Programme (discussed later in this chapter) and continued funding research in this area well into the late 1990s. Many of the same institutions that had been part of the SUR programme, including CMU, BBN, SRI, and MIT, participated in the new initiatives. Firms such as IBM and Dragon Systems also participated. As a result of the controversy over SUR testing, evaluation methods and criteria for these programmes were carefully prescribed though mutual agreements between DARPA managers and the funded researchers. Some researchers have hailed this development and praised DARPA's role in benchmarking speech-recognition technology, not only for research purposes but also for the commercial market. By holding annual system evaluations on carefully designed tasks and test materials, DARPA and the National Bureau of Standards (later the National Institute of Standards and Technology) led the standards-definition process, drawing the participation of not only government contractors but also industry and university groups from around the world, such as AT&T, Cambridge University (of the United Kingdom), and LIMSI (of France). The overall effect was the rapid adoption of the most successful techniques by every participant and quick migration of those techniques into products and services. Although it resulted in quick diffusion of successful techniques, this approach may also have narrowed the scope of approaches taken. Critics have seen this as symptomatic of a profound change in DARPA's philosophy that has reduced the emphasis on basic research. DARPA's funding of research on understanding speech has been extremely important. First, it pushed the research frontiers of speech recognition and AI more generally. HEARSAY-II is particularly notable for the way it parsed information into independent knowledge sources, which in turn interacted with each other through a common database that CMU researchers labeled a "blackboard" (Englemore et al., 1988). This blackboard method of information processing proved to be a significant advance in AI. Moreover, although early speech-recognition researchers appeared overly ambitious in incorporating syntax and semantics into their systems, others have recently begun to adopt this approach to improve statistically based speech-recognition technology. Perhaps more important, the results of this research have been incorporated into the products of established companies, such as IBM and BBN, as well as start-ups such as Nuance Communications (an SRI spinoff) and Dragon Systems. Microsoft Corporation, too, is incorporating speech recognition technology into its operating system (DARPA, 1997; McClain, 1998). The leading commercial speech-recognition programme on the market today, the Dragon Systems software, traces its roots directly back to the work done at CMU between 1971 and 1975 as part of SUR. The DRAGON programme developed in CMU's SUR project (the predecessor of the HARPY program) pioneered the use of techniques borrowed from mathematics and statistics (hidden Markov models) to recognize continuous speech (Baker, 1975). According to some scholars, the adoption of hidden Markov models by CMU's research team owes much to activities outside the AI field, such as research by engineers and statisticians with an interest in machine intelligence. Other examples of commercial success abound. Charles Schwab and Company adopted DARPA technology to develop its VoiceBroker system, which provides stock quotes over the telephone. The system can recognize the names of 13,000 different securities as well as major regional U.S. accents. On the military side, DARPA provided translingual communication devices for use in Bosnia. These devices translated spoken English phrases into corresponding Serbo-Croatian or Russian phrases. The total market for these new personal-use voice recognition technologies is expected to reach about $4 billion in 2001 (Garfinkel, 1998).

Shift to Applied Research Increases Investment

Although most founders of the AI field continued to pursue basic questions of human and machine intelligence, some of their students and other second-generation researchers began to seek ways to use AI methods and approaches to tackle real-world problems. Their initiatives were important, not only in their own right, but also because they were indicative of a gradual but significant change in the funding environment toward more applied realms of research. The development of expert systems, such as DENDRAL at SAIL, provides but one example of this trend. The promotion in 1969 of Lawrence Roberts to director of IPTO also contributed to a perceived tightening of AI research. Under Roberts, IPTO developed a formal AI programme, which in turn was divided into formal subprogrammes (Norberg and O'Neill, 1996). The line-item budget of AI research inevitably led to greater scrutiny owing to reporting mechanisms and the need to justify programmes to the DOD, the Administration, and the U.S. Congress. Consequently, researchers began to believe that they were being boxed in by IPTO and DARPA, and to a certain extent they were. The flow of DARPA's AI research money to CMU, MIT, and Stanford University did not cease or even diminish much, but the demand grew for interim reports and more tangible results. External developments reinforced this shift. The most important was the passage of the Mansfield Amendment in 1969. Passed during the Vietnam War amid growing public concern about the "military-industrial complex" and the domination of U.S. academic science by the military, the Mansfield Amendment restricted the DOD to supporting basic research that was of "direct and apparent" utility to specific military functions and operations. It brought about a swift decline in some of the military's support for basic research, often driving it toward the applied realm. Roberts and his successors now had to justify AI research programmes on the basis of immediate utility to the military mission. The move toward relevance spawned dissatisfaction among both the established pioneers of the AI field and its outside skeptics. Another external development provided further impetus for change. In 1973, at the request of the British Scientific Research Council, Sir James Lighthill, the Lucasian Professor of Applied Mathematics at Cambridge University and a Fellow of the Royal Society of London, produced a survey that expressed considerable skepticism about AI in general and research domains in particular. Despite having no expertise in AI himself, Lighthill suggested that any particular successes in AI had stemmed from modeling efforts in more traditional disciplines, not from AI per se. He singled out robotics research for especially sharp criticism. The Lighthill report raised questions about AI research funding in the United States and led DOD to establish a panel to assess DARPA's AI programme. Known as the American Study Group, the panel (which included some of AI's major research figures) raised some of the same questions as did Lighthill's report and served to inform George Heilmeier, a former research manager from RCA Corporation who was then assistant director of Defense R&D and later became director of DARPA. The Lighthill report and its U.S. equivalent led to a shifting of DARPA funds out of robotics research (hurting institutions such as SRI that had committed heavily to the area) and toward "mission-oriented direct research, rather than basic undirected research" (Fleck, 1982). As a result of these forces, DARPA's emphasis on relevance in AI research grew during the late 1970s and 1980s. Despite the disgruntlement among some scientists, the changes led to increased funding—although not directly to widespread commercial success—for AI research. A magnet for these monies was the Strategic Computing Programme (SCP), announced in 1983 (DARPA, 1983). DARPA committed $1 billion over the planned 10-year course of the programme. The four main goals of the SCP were as follows:

- Advance machine intelligence technology and high-performance computing, including speech recognition and understanding, natural-language computer interfaces, vision comprehension systems, and advanced expert systems development, and to do so by providing significant increases in computer performance, through parallel-computer architectures, software, and supporting microelectronics;
- Transfer technology from DARPA-sponsored university research efforts to the defense industry through competitive research contracts, with industry and universities jointly participating;
- Develop more new scientists in AI and high-performance computing through increased funding of graduate student research in these areas; and
- Provide the supporting research infrastructure for AI research through advanced networking, new microcircuit fabrication facilities, advanced emulation facilities, and advanced symbolic processors (Kahn, 1988).

To achieve these goals, DARPA established three specific applications as R&D objectives: a pilot's associate for the Air Force, an autonomous land vehicle for the Army, and an aircraft battle management system for the Navy. The applications were intended to spark the military services' interest in developing AI technology based on fundamental research. The SCP differed from some other large-scale national efforts in that its goals were extremely ambitious, requiring fundamental advances in the underlying technology. (By contrast, efforts such as the Apollo space programme were principally engineering projects drawing from an established scientific base [Office of Technology Assessment, 1985]). The SCP also differed from earlier large AI programmes in that some 60 percent of its funds were committed to industry. However, of the 30 prime contractors for the SCP involved in software or AI research, more than 20 were established defense contractors (Goldstein, 1992). The SCP significantly boosted overall federal funding for AI research but also altered its character. Between 1984 and 1988, total federal funding for AI research, excluding the SCP, tripled from $57 million to $159 million. With support for the SCP included, federal funding increased from $106 million to $274 million. Because the SCP was budgeted as an applied programme, it tipped the balance of federal funding toward applied research. Although DARPA's funding for basic AI research doubled from roughly $20 million to $40 million during this same period, the DOD's overall role in basic AI research declined. Meanwhile, it continued to play the dominant role in supporting applied research in AI. Although budget categorizations for programmes such as the SCP are somewhat arbitrary and subject to political influence, researchers noted a change in DARPA's funding style. The SCP also attracted a tremendous amount of industry investment and venture capital to AI research and development. Firms developing and selling expert systems entered the market, often basing their systems on the LISP machines developed by the AI community. Several new firms entered the market to design, make, and sell the very expensive LISP machines.

Yet the rapid development of engineering workstations, especially those of Sun Microsystems, Inc., soon undermined the LISP machine industry. This segment of the market, which was clearly tied to the SCP, collapsed. Even with the development of expert-system shells to run on less-costly machines, doubts began to arise about the capabilities and flexibility of expert systems; this doubt hampered the commercialization of AI. In addition, commercial contractors had difficulty meeting the high-profile milestones of the major SCP projects because of difficulties with either the AI technologies themselves or their incorporation into larger systems. Such problems undermined the emergence of a clearly identifiable AI industry and contributed to a shift in emphasis in high-performance computing, away from AI and toward other grand challenges, such as weather modeling and prediction and scientific visualization.

Lessons from History

As this case study demonstrates, federal funding is critical in establishing new disciplines because it can sustain long-term, high-risk research areas and nurture a critical mass of technical and human resources. DARPA helped legitimize the AI field and served as the major source of research funds beginning in the 1960s. It created centers of excellence that evolved into today's major computer science research centers. This support was particularly critical given that some objectives took much longer to realise than was originally anticipated. A diversity of approaches to research problems can be critical to the development of practical tools. A prime example is the field of speech recognition, in which the most effective products to date have used techniques borrowed from the mathematics and statistics communities rather than more traditional AI techniques. This outcome could not have been predicted and demonstrates the importance of supporting competing approaches, even those outside the mainstream. Federal funding has promoted innovation in commercial products such as expert systems, the establishment of new companies, the growth of billion-dollar markets for technologies such as speech recognition, and the development of valuable military applications. AI technologies often enhance the performance of the larger systems into which they are increasingly incorporated. There is a creative tension between fundamental research and attempts to create functional devices. Original attempts to design intelligent, thinking machines motivated fundamental work that created a base of knowledge. Initial advances achieved through research were not sufficient to produce, by themselves, commercial products, but they could be integrated with other components and exploited in different applications. Efforts to apply AI technology often failed initally because they uncovered technical problems that had not yet been adequately addressed. Applications were fed back into the research process, thus motivating inquiries into new areas.

Artificial Intelligence in the 1990s

Despite the commercial difficulties associated with the Strategic Computing Programme, the AI-driven advances in rule-based reasoning systems (i.e., expert systems) and their successors—many of which were initiated with DARPA funding in the 1960s and 1970s—proved to be extremely valuable for the emerging national information infrastructure and electronic commerce. These advances, including probabilistic reasoning systems and Bayesian networks, natural language processing, and knowledge representation, brought AI out of the laboratory and into the marketplace. Paradoxically, the major commercial successes of AI research applications are mostly hidden from view today because they are embedded in larger software systems. None of these systems has demonstrated general human intelligence, but many have contributed to commercial and military objectives.

An example is the Lumiere project initiated at Microsoft Research in 1993. Lumiere monitors a computer user's actions to determine when assistance may be needed. It continuously follows the user's goals and tasks as software programmes run, using Bayesian networks to generate a probability distribution over topic areas that might pose difficulties and calculating the probability that the user will not mind being bothered with assistance. Lumiere forms the basis of the "office assistant" that monitors the behavior of users of Microsoft's Office 97 and assists them with applications. Lumiere is based on earlier work on probabilistic models of user goals to support the display of customized information to pilots of commercial aircraft, as well as user modeling for display control for flight engineers at NASA's Mission Control Center. These earlier projects, sponsored by the NASA-Ames Research Center and NASA's Johnson Space Center, were undertaken while some of the Lumiere researchers were students at Stanford University.

Patent trends suggest that AI technology is being incorporated into growing numbers of commercial products. The number of patents in AI, expert systems, and neural networks jumped from fewer than 20 in 1988 to more than 120 in 1996, and the number of patents citing patents in these areas grew from about 140 to almost 800.[29] The number of AI-related patents (including patents in AI, expert systems, neural networks, intelligent systems, adaptive agents, and adaptive systems) issued annually in the United States increased exponentially from approximately 100 in 1985 to more than 900 in 1996. Changes in the U.S. Patent and Trademark Office's rules on the patentability of algorithms have no doubt contributed to this growth, as has the increased commercial value of AI technology. The vast majority of these patents are held by private firms, including large manufacturers of electronics and computers, as well as major users of information technology. These data indicate that AI technology is likely to be embedded in larger systems, from computers to cars to manufacturing lines, rather than used as stand-alone products. A central problem confronting the wider commercialization of AI today revolves around integration. Both the software and the hardware developed by the AI research community were so advanced that their integration into older, more conservative computer and organisational systems proved to be an enormous challenge.

As one observer has noted, "Because AI was a leading-edge technology, it arrived in this world too early. As a consequence, the AI application community had to ride many waves of technological quick fixes and fads.... Many of these integration problems are now being addressed head on by a broad community of information technologists using Internet-based frameworks such as CORBA [common object request broker architecture] and the World Wide Web" (Shrobe, 1996). The rapid development of computer hardware and software, the networking of information systems, and the need to make these systems function smoothly and intelligently are leading to wide diffusion of AI knowledge and technology across the infrastructure of the information age. Federal funding reflects these changes. Meanwhile, much of the knowledge acquired through AI research over the years is now being brought to bear on real-world problems and applications while also being deepened and broadened. The economic and social benefits are enormous. Technologies such as expert systems, natural-language processing, and computer vision are now used in a range of applications, such as decision aids, planning tools, speech-recognition systems, pattern recognition, knowledge representation, and computer-controlled robots. AI technologies help industry diagnose machine failures, design new products, and plan, simulate, and schedule production. They help scientists search large databases and decode DNA sequences, and they help doctors make more-informed decisions about diagnosis and treatment of particular ailments. AI technologies also make the larger systems into which they are incorporated easier to use and more productive. These benefits are relatively easy to identify, but measuring them is difficult.

Federal investments in AI have produced a number of notable results, some envisioned by the founders of the field and others probably not even imagined. Without question, DARPA's generous, enduring funding of various aspects of AI research created a scientific research discipline that meets the standard criteria of discipline formation laid out by sociologists of science. At least three major academic centers of excellence and several other significant centers were established, and they produced a large number of graduates with Ph.D.s who diffused AI research to other research universities, cross-pollinated the major research centers, and moved AI methods into commercial markets.

Application of AI

In the quest to create intelligent machines, the field of Artificial Intelligence has split into several different approaches based on the opinions about the most promising methods and theories. These rivaling theories have lead researchers in one of two basic approaches; bottom-up and top-down. Bottom-up theorists believe the best way to achieve artificial intelligence is to build electronic replicas of the human brain's complex network of neurons, while the top-down approach attempts to mimic the brain's behavior with computer programmes.

Neural Networks and Parallel Computation

The human brain is made up of a web of billions of cells called neurons, and understanding its complexities is seen as one of the last frontiers in scientific research. It is the aim of AI researchers who prefer this bottom-up approach to construct electronic circuits that act as neurons do in the human brain. Although much of the working of the brain remains unknown, the complex network of neurons is what gives humans intelligent characteristics. By itself, a neuron is not intelligent, but when grouped together, neurons are able to pass electrical signals through networks.

Research has shown that a signal received by a neuron travels through the dendrite region, and down the axon. Separating nerve cells is a gap called the synapse. In order for the signal to be transferred to the next neuron, the signal must be converted from electrical to chemical energy. The signal can then be received by the next neuron and processed. Warren McCulloch after completing medical school at Yale, along with Walter Pitts a mathematician proposed a hypothesis to explain the fundamentals of how neural networks made the brain work. Based on experiments with neurons, McCulloch and Pitts showed that neurons might be considered devices for processing binary numbers. An important back of mathematic logic, binary numbers (represented as 1's and 0's or true and false) were also the basis of the electronic computer. This link is the basis of computer-simulated neural networks, also know as Parallel computing.

A century earlier the true / false nature of binary numbers was theorized in 1854 by George Boole in his postulates concerning the Laws of Thought. Boole's principles make up what is known as Boolean algebra, the collection of logic concerning AND, OR, NOT operands. Boole also assumed that the human mind works according to these laws, it performs logical operations that could be reasoned. Ninety years later, Claude Shannon applied Boole's principles in circuits, the blueprint for electronic computers. Boole's contribution to the future of computing and Artificial Intelligence was immeasurable, and his logic is the basis of neural networks. McCulloch and Pitts, using Boole's principles, wrote a paper on neural network theory. The thesis dealt with how the networks of connected neurons could perform logical operations. It also stated that, one the level of a single neuron, the release or failure to release an impulse was the basis by which the brain makes true / false decisions. Using the idea of feedback theory, they described the loop which existed between the senses — > brain —> muscles, and likewise concluded that Memory could be defined as the signals in a closed loop of neurons. Although we now know that logic in the brain occurs at a level higher then McCulloch and Pitts theorized, their contributions were important to AI because they showed how the firing of signals between connected neurons could cause the brains to make decisions. McCulloch and Pitt's theory is the basis of the artificial neural network theory. Using this theory, McCulloch and Pitts then designed electronic replicas of neural networks, to show how electronic networks could generate logical processes. They also stated that neural networks may, in the future, be able to learn, and recognize patterns. The results of their research and two of Weiner's books served to increase enthusiasm, and laboratories of computer simulated neurons were set up across the country. Two major factors have inhibited the development of full scale neural networks. Because of the expense of constructing a machine to simulate neurons, it was expensive even to construct neural networks with the number of neurons in an ant.

Although the cost of components have decreased, the computer would have to grow thousands of times larger to be on the scale of the human brain. The second factor is current computer architecture. The standard Von Neuman computer, the architecture of nearly all computers, lacks an adequate number of pathways between components. Researchers are now developing alternate architectures for use with neural networks. Even with these inhibiting factors, artificial neural networks have presented some impressive results. Frank Rosenblatt, experimenting with computer simulated networks, was able to create a machine that could mimic the human thinking process, and recognize letters. But, with new top-down methods becoming popular, parallel computing was put on hold. Now neural networks are making a return, and some researchers believe that with new computer architectures, parallel computing and the bottom-up theory will be a driving factor in creating artificial intelligence.

Chess

AI-based game playing programmes combine intelligence with entertainment. On game with strong AI ties is chess. World-champion chess playing programmes can see ahead twenty plus moves in advance for each move they make. In addition, the programmes have an ability to get progressably better over time because of the ability to learn. Chess programmes do not play chess as humans do. In three minutes, Deep Thought (a master program) considers 126 million moves, while human chessmaster on average considers less than 2 moves. Herbert Simon suggested that human chess masters are familiar with favorable board positions, and the relationship with thousands of pieces in small areas. Computers on the other hand, do not take hunches into account. The next move comes from exhaustive searches into all moves, and the consequences of the moves based on prior learning. Chess programmes, running on Cray super computers have attained a rating of 2600 (senior master), in the range of Gary Kasparov, the Russian world champion.

Frames

On method that many programmes use to represent knowledge are frames. Pioneered by Marvin Minsky, frame theory revolves around packets of information. For example, say the situation was a birthday party. A computer could call on its birthday frame, and use the information contained in the frame, to apply to the situation. The computer knows that there is usually cake and presents because of the information contained in the knowledge frame. Frames can also overlap, or contain sub-frames. The use of frames also allows the computer to add knowledge. Although not embraced by all AI developers, frames have been used in comprehension programmes such as Sam. This page touched on some of the main methods used to create intelligence. These approaches have been applied to a variety of programmes. As we progress in the development of Artificial Intelligence, other theories will be available, in addition to building on today's methods.

Top Down Approaches; Expert Systems

Because of the large storage capacity of computers, expert systems had the potential to interpret statistics, in order to formulate rules. An expert system works much like a detective solves a mystery. Using the information, and logic or rules, an expert system can solve the problem. For example it the expert system was designed to distinguish birds it may have the following: Charts like these represent the logic of expert systems. Using a similar set of rules, experts can have a variety of applications. With improved interfacing, computers may begin to find a larger place in society. We have been studying this issue of AI application for quite some time now and know all the terms and facts. But what we all really need to know is what can we do to get our hands on some AI today. How can we as individuals use our own technology? We hope to discuss this in depth (but as briefly as possible) so that you the consumer can use AI as it is intended.

First, we should be prepared for a change. Our conservative ways stand in the way of progress. AI is a new step that is very helpful to the society. Machines can do jobs that require detailed instructions followed and mental alertness. AI with its learning capabilities can accomplish those tasks but only if the worlds conservatives are ready to change and allow this to be a possibility. It makes us think about how early man finally accepted the wheel as a good invention, not something taking away from its heritage or tradition. Secondly, we must be prepared to learn about the capabilities of AI. The more use we get out of the machines the less work is required by us. In turn less injuries and stress to human beings. Human beings are a species that learn by trying, and we must be prepared to give AI a chance seeing AI as a blessing, not an inhibition. Finally, we need to be prepared for the worst of AI. Something as revolutionary as AI is sure to have many kinks to work out. There is always that fear that if AI is learning based, will machines learn that being rich and successful is a good thing, then wage war against economic powers and famous people? There are so many things that can go wrong with a new system so we must be as prepared as we can be for this new technology. However, even though the fear of the machines are there, their capabilities are infinite Whatever we teach AI, they will suggest in the future if a positive outcome arrives from it. AI are like children that need to be taught to be kind, well mannered, and intelligent. If they are to make important decisions, they should be wise. We as citizens need to make sure AI programrs are keeping things on the level. We should be sure they are doing the job correctly, so that no future accidents occur.

AIAI Teaching Computers Computers

Does this sound a little Redundant? Or maybe a little redundant? Well just sit back and let me explain. The Artificial Intelligence Applications Institute has many project that they are working on to make their computers learn how to operate themselves with less human input. To have more functionality with less input is an operation for AI technology. I will discuss just two of these projects: AUSDA and EGRESS. AUSDA is a programme which will exam software to see if it is capable of handling the tasks you need performed. If it isn't able or isn't reliable AUSDA will instruct you on finding alternative software which would better suit your needs. According to AIAI, the software will try to provide solutions to problems like "identifying the root causes of incidents in which the use of computer software is involved, studying different software development approaches, and identifying aspects of these which are relevant to those root causes producing guidelines for using and improving the development approaches studied, and providing support in the integration of these approaches, so that they can be better used for the development and maintenance of safety critical software."

Sure, for the computer buffs this programme is a definitely good news. But what about the average person who think the mouse is just the computers foot pedal?

Where do they fit into computer technology. Well don't worry guys, because us nerds are looking out for you too! Just ask AIAI what they have for you and it turns up the EGRESS is right down your alley. This is a programme which is studying human reactions to accidents. It is trying to make a model of how peoples reactions in panic moments save lives. Although it seems like in tough situations humans would fall apart and have no idea what to do, it is in fact the opposite. Quick Decisions are usually made and are effective but not flawless. These computer models will help rescuers make smart decisions in time of need. AI can't be positive all the time but can suggest actions which we can act out and therefor lead to safe rescues.

So AIAI is teaching computers to be better computers and better people. AI technology will never replace man but can be an extension of our body which allows us to make more rational decisions faster. And with Institutes like AIAI- we continue each stay to step forward into progress.

Apple Computers may not have ever been considered as the state of art in Artificial Intelligence, but a second look should be given. Not only are today's PC's becoming more powerful but AI influence is showing up in them. From Macros to Voice Recognition technology, PC's are becoming our talking buddies. Who else would go surfing with you on short notice- even if it is the net. Who else would care to tell you that you have a business appointment scheduled at 8:35 and 28 seconds and would notify you about it every minute till you told it to shut up. Even with all the abuse we give today's PC's they still plug away to make us happy. We use PC's more not because they do more or are faster but because they are getting so much easier to use. And their ease of use comes from their use of AI.

All Power Macintoshes come with Speech Recognition. That's right- you tell the computer to do what you want without it having to learn your voice. This implication of AI in Personal computers is still very crude but it does work given the correct conditions to work in and a clear voice. Not to mention the requirement of at least 16Mgs of RAM for quick use. Also Apple's Newton and other hand held note pads have Script recognition. Cursive or Print can be recognized by these notepad sized devices. With the pen that accompanies your silicon note pad you can write a little note to yourself which magically changes into computer text if desired. No more complaining about sloppy written reports if your computer can read your handwriting. If it can't read it though- perhaps in the future, you can correct it by dictating your letters instead. Macros provide a huge stress relief as your computer does faster what you could do more tediously. Macros are old but they are to an extent, Intelligent. You have taught the computer to do something only by doing it once. In businesses, many times applications are upgraded. But the files must be converted. All of the businesses records but be changed into the new software's type. Macros save the work of conversion of hundred of files by a human by teaching the computer to mimic the actions of the programr. Thus teaching the computer a task that it can repeat whenever ordered to do so. AI is all around us all but get ready for a change. But don't think the change will be harder on us because AI has been developed to make our lives easier.

The Scope of Expert Systems

As stated in the 'approaches' section, an expert system is able to do the work of a professional. Moreover, a computer system can be trained quickly, has virtually no operating cost, never forgets what it learns, never calls in sick, retires, or goes on vacation. Beyond those, intelligent computers can consider a large amount of information that may not be considered by humans. But to what extent should these systems replace human experts? Or, should they at all? For example, some people once considered an intelligent computer as a possible substitute for human control over nuclear weapons, citing that a computer could respond more quickly to a threat. And many AI developers were afraid of the possibility of programmes like Eliza, the psychiatrist and the bond that humans were making with the computer. We cannot, however, over look the benefits of having a computer expert. Forecasting the weather, for example, relies on many variables, and a computer expert can more accurately pool all of its knowledge. Still a computer cannot rely on the hunches of a human expert, which are sometimes necessary in predicting an outcome. In conclusion, in some fields such as forecasting weather or finding bugs in computer software, expert systems are sometimes more accurate than humans. But for other fields, such as medicine, computers aiding doctors will be beneficial, but the human doctor should not be replaced. Expert systems have the power and range to aid to benefit, and in some cases replace humans, and computer experts, if used with discretion, will benefit human kind.

Main Types

Game Playing

You can buy machines that can play master level chess for a few hundred dollars. There is some AI in them, but they play well against people mainly through brute force computation—looking at hundreds of thousands of positions. To beat a world champion by brute force and known reliable heuristics requires being able to look at 200 million positions per second.

Speech Recognition

In the 1990s, computer speech recognition reached a practical level for limited purposes. Thus United Airlines has replaced its keyboard tree for flight information by a system using speech recognition of flight numbers and city names. It is quite convenient. On the the other hand, while it is possible to instruct some computers using speech, most users have gone back to the keyboard and the mouse as still more convenient.

Understanding Natural Language

Just getting a sequence of words into a computer is not enough. Parsing sentences is not enough either. The computer has to be provided with an understanding of the domain the text is about, and this is presently possible only for very limited domains.

Computer Vision

The world is composed of three-dimensional objects, but the inputs to the human eye and computers' TV cameras are two dimensional. Some useful programs can work solely in two dimensions, but full computer vision requires partial three-dimensional information that is not just a set of two-dimensional views. At present there are only limited ways of representing three-dimensional information directly, and they are not as good as what humans evidently use.

Expert Systems

A "knowledge engineer" interviews experts in a certain domain and tries to embody their knowledge in a computer program for carrying out some task. How well this works depends on whether the intellectual mechanisms required for the task are within the present state of AI. When this turned out not to be so, there were many disappointing results. One of the first expert systems was MYCIN in 1974, which diagnosed bacterial infections of the blood and suggested treatments. It did better than medical students or practicing doctors, provided its limitations were observed. Namely, its ontology included bacteria, symptoms, and treatments and did not include patients, doctors, hospitals, death, recovery, and events occurring in time. Its interactions depended on a single patient being considered. Since the experts consulted by the knowledge engineers knew about patients, doctors, death, recovery, etc., it is clear that the knowledge engineers forced what the experts told them into a predetermined framework. In the present state of AI, this has to be true. The usefulness of current expert systems depends on their users having common sense.

Heuristic Classification

One of the most feasible kinds of expert system given the present knowledge of AI is to put some information in one of a fixed set of categories using several sources of information. An example is advising whether to accept a proposed credit card purchase. Information is available about the owner of the credit card, his record of payment and also about the item he is buying and about the establishment from which he is buying it (e.g., about whether there have been previous credit card frauds at this establishment).

Histroy

Evidence of Artificial Intelligence folklore can be traced back to ancient Egypt, but with the development of the electronic computer in 1941, the technology finally became available to create machine intelligence. The term artificial intelligence was first coined in 1956, at the Dartmouth conference, and since then Artificial Intelligence has expanded because of the theories and principles developed by its dedicated researchers. Through its short modern history, advancement in the fields of AI have been slower than first estimated, progress continues to be made. From its birth 4 decades ago, there have been a variety of AI programmes, and they have impacted other technological advancements.

The Beginnings of AI

Although the computer provided the technology necessary for AI, it was not until the early 1950's that the link between human intelligence and machines was really observed. Norbert Wiener was one of the first Americans to make observations on the principle of feedback theory feedback theory. The most familiar example of feedback theory is the thermostat: It controls the temperature of an environment by gathering the actual temperature of the house, comparing it to the desired temperature, and responding by turning the heat up or down. What was so important about his research into feedback loops was that Wiener theorized that all intelligent behavior was the result of feedback mechanisms. Mechanisms that could possibly be simulated by machines. This discovery influenced much of early development of AI. In late 1955, Newell and Simon developed *The Logic Theorist*, considered by many to be the first AI programme. The programme, representing each problem as a tree model, would attempt to solve it by selecting the branch that would most likely result in the correct conclusion. The impact that the logic theorist made on both the public and the field of AI has made it a crucial stepping stone in developing the AI field. In 1956 John McCarthy regarded as the father of AI, organised a conference to draw the talent and expertise of others interested in machine intelligence for a month of brainstorming. He invited them to Vermont for "The Dartmouth summer research project on artificial intelligence." From that point on, because of McCarthy, the field would be known as Artificial intelligence. Although not a huge success, (explain) the Dartmouth conference did bring together the founders in AI, and served to lay the groundwork for the future of AI research.

The Era of the Computer

In 1941 an invention revolutionized every aspect of the storage and processing of information. That invention, developed in both the US and Germany was the electronic computer. The first computers required large, separate air-conditioned rooms, and were a programrs nightmare, involving the separate configuration of thousands of wires to even get a programme running. The 1949 innovation, the stored programme computer, made the job of entering a programme easier, and advancements in computer theory lead to computer science, and eventually Artificial intelligence. With the invention of an electronic means of processing data, came a medium that made AI possible.

Knowledge Expansion

In the seven years after the conference, AI began to pick up momentum. Although the field was still undefined, ideas formed at the conference were re-examined, and built upon. Centers for AI research began forming at Carnegie Mellon and MIT, and a new challenges were faced: further research was placed upon creating systems that could efficiently solve problems, by limiting the search, such as the Logic Theorist. And second, making systems that could learn by themselves. In 1957, the first version of a new programme The *General Problem Solver*(GPS) was tested. The programme developed by the same pair which developed the Logic Theorist. The GPS was an extension of Wiener's feedback principle, and was capable of solving a greater extent of common sense problems. A couple of years after the GPS, IBM contracted a team to research artificial intelligence. Herbert Gelerneter spent 3 years working on a programme for solving geometry theorems. While more programmes were being produced, McCarthy was busy developing a major breakthrough in AI history. In 1958 McCarthy announced his new development; the LISP language, which is still used today. LISP stands for LISt Processing, and was soon adopted as the language of choice among most AI developers. In 1963 MIT received a 2.2 million dollar grant from the United States government to be used in researching Machine-Aided Cognition (artificial intelligence). The grant by the Department of Defense's Advanced research projects Agency (ARPA), to ensure that the US would stay ahead of the Soviet Union in technological advancements. The project served to increase the pace of development in AI research, by drawing computer scientists from around the world, and continues funding.

The Multitude of Programmes

The next few years showed a multitude of programmes, one notably was SHRDLU. SHRDLU was part of the microworlds project, which consisted of research and programming in small worlds (such as with a limited number of geometric shapes). The MIT researchers headed by Marvin Minsky, demonstrated that when confined to a small subject matter, computer programmes could solve spatial problems and logic problems. Other programmes which appeared during the late 1960's were STUDENT, which could solve algebra story problems, and SIR which could understand simple English sentences. The result of these programmes was a refinement in language comprehension and logic. Another advancement in the 1970's was the advent of the expert system. Expert systems predict the probability of a solution under set conditions. For example: Because of the large storage capacity of computers at the time, expert systems had the potential to interpret statistics, to formulate rules. And the applications in the market place were extensive, and over the course of ten years, expert systems had been introduced to forecast the stock market, aiding doctors with the ability to diagnose disease, and instruct miners to promising mineral locations. This was made possible because of the systems ability to store conditional rules, and a storage of information.

During the 1970's Many new methods in the development of AI were tested, notably Minsky's frames theory. Also David Marr proposed new theories about machine vision, for example, how it would be possible to distinguish an image based on the shading of an image, basic information on shapes, color, edges, and texture. With analysis of this information, frames of what an image might be could then be referenced. another development during this time was the PROLOGUE language. The language was proposed for In 1972, During the 1980's AI was moving at a faster pace, and further into the corporate sector. In 1986, US sales of AI-related hardware and software surged to $425 million. Expert systems in particular demand because of their efficiency. Companies such as Digital Electronics were using XCON, an expert system designed to programme the large VAX computers. DuPont, General Motors, and Boeing relied heavily on expert systems Indeed to keep up with the demand for the computer experts, companies such as Teknowledge and Intellicorp specializing in creating software to aid in producing expert systems formed. Other expert systems were designed to find and correct flaws in existing expert systems.

The Transition from Lab to Life

The impact of the computer technology, AI included was felt. No longer was the computer technology just part of a select few researchers in laboratories. The personal computer made its debut along with many technological magazines. Such foundations as the American Association for Artificial Intelligence also started. There was also, with the demand for AI development, a push for researchers to join private companies. 150 companies such as DEC which employed its AI research group of 700 personnel, spend $1 billion on internal AI groups.

Other fields of AI also made there way into the marketplace during the 1980's. One in particular was the machine vision field. The work by Minsky and Marr were now the foundation for the cameras and computers on assembly lines, performing quality control. Although crude, these systems could distinguish differences shapes in objects using black and white differences. By 1985 over a hundred companies offered machine vision systems in the US, and sales totaled $80 million.

The 1980's were not totally good for the AI industry. In 1986-87 the demand in AI systems decreased, and the industry lost almost a half of a billion dollars. Companies such as Teknowledge and Intellicorp together lost more than $6 million, about a third of there total earnings. The large losses convinced many research leaders to cut back funding. Another disappointment was the so called "smart truck" financed by the Defense Advanced Research Projects Agency. The projects goal was to develop a robot that could perform many battlefield tasks. In 1989, due to project setbacks and unlikely success, the Pentagon cut funding for the project. Despite these discouraging events, AI slowly recovered. New technology in Japan was being developed. Fuzzy logic, first pioneered in the US has the unique ability to make decisions under uncertain conditions. Also neural networks were being reconsidered as possible ways of achieving Artificial Intelligence. The 1980's introduced to its place in the corporate marketplace, and showed the technology had real life uses, ensuring it would be a key in the 21st century.

AI Put to the Test

The military put AI based hardware to the test of war during Desert Storm. AI-based technologies were used in missile systems, heads-up-displays, and other advancements. AI has also made the transition to the home. With the popularity of the AI computer growing, the interest of the public has also grown. Applications for the Apple Macintosh and IBM compatible computer, such as voice and character recognition have become available. Also AI technology has made steadying camcorders simple using fuzzy logic. With a greater demand for AI-related technology, new advancements are becoming available. Inevitably Artificial Intelligence has, and will continue to affecting our lives. To what extent can we *ever* expect to be able to create a machine which is 'truly' intelligent? In trying to answer this question we have taken quite a long digression; first to describe what is known about the way the brain is constructed and operates, second to explore one of the most promising new approaches to machine intelligence — computers modeled after the brain or neural networks. We have debated both the successes and deficiencies of these network models and in the light of what we have discovered we must now return and try to answer the question. We first discuss how we might recognize intelligence in a machine and to what extent neural networks and other approaches to AI can be said to have met this criteria. Finally, we conclude by describing current research directions.

Consciousness, Nets and the Future

What do we mean when we say someone is intelligent? Is it that they are, for example, very good at mathematics or translating foreign languages? These people are certainly good at understanding and manipulating abstract concepts. But what about poets, novelists and musicians? They are clearly intelligent because they are creative. Indeed, intelligence is visible in almost every form of human activity — ability to adapt, learn new skills, form complex relationships and societies. Much of this appears to be unique to humans (at least on Earth) and differentiates us from all other species. We might say that all of these aspects of our lives and behavior can be attributed to the fact that we are *conscious*.

Unfortunately, there is no precise, widely agreed upon definition of the word *consciousness*. However, most of us have an intuitive sense of what is meant by the term. Consciousness, or cognition, is a sort of awareness — of self, of interaction with the world, of thought processes taking place, and of our ability to at least partially control these processes. We also associate consciousness with an inner voice that expresses our high level, deliberate, thoughts, as well as intentionality and emotion. It seems doubtful whether true intelligence can ever arise in the absence of consciousness. Perhaps, one might take the view that intelligent behavior is the outward sign of a conscious being. If so any machine which could display human-like intelligence qualities could be said to be conscious.

This point of view was taken by Alan Turing, who in 1950 invented a test whose result could be used to determine whether, in any practical sense, a machine could be said to be conscious. The test is quite simple. You enter a room and encounter two terminals: one terminal connects with a computer, and the other interfaces with a person who types responses. The goal of the test is for you to determine which terminal is connected with the computer. You are allowed to ask questions, make assertions, question feelings and motivations for as long as you wish. If you fail to determine which terminal is communicating with the computer or guess that the computer is the human, the computer has passed the test and can be said to be 'conscious'. Turing invented his test at a time when it was thought that mind-like computers might be only fifty years away. A whole new science was born with the aim of producing such intelligent machines — the subject of artificial intelligence or AI. In fact that has not happened — initial efforts to create computers with mind-like reasoning have failed miserably. Many researchers now believe that part of the reason for this failure was that traditional computers function in a way very different from the brain and that the key to true intelligent machines lies in understanding in detail the functioning of the brain and emulating this with artificial neural networks.

Needless to say this view is not held by all — some philosophers maintain that the phenomenon of consciousness cannot be ascribed to purely physical processes (the cooperative firing of networks of neurons) and is *in principle* inaccessible to arbitrarily advanced scientific assault. This is the traditional mind/matter split advocated by the seventeenth century philosophers. There is a famous argument due to John Searle in 1980 which attempts to rebut the Turing test as a way of assessing consciousness. In his argument one imagines a non-Chinese speaking person sitting in a room with a long list of rules for translating strings of Chinese characters into new strings of Chinese characters. When a string of characters is slipped under the door, the person consults the rules and slips back an appropriate response under the door. If the incoming strings actually represented questions (like a Turing test), then a particularly clever and exhaustive set of rules could conceivably allow the person in the room to produce outgoing strings that furnished answers to the questions. From the point of view of a person outside, the room would seem to contain an intelligent person who is responding to the questions. But yet the person in the room has no understanding of the content of these questions — he or she is merely acting out a set of rules, translating one set of random symbols into another. In other words while we could possibly programme a machine to mock up the effects of intelligence it would never be truly conscious. While this criticism may be applied to the old style of AI (rule based AI systems rather similar to Searle's Chinese room exist and have met with some success — they are called expert systems), it not clear that it truly applies to neural network based AI, since there is no real concept of a set of rules determining a response. Consciousness is not envisaged as arising out of a machine obeying a set of rules but as some as yet ill-defined property of the natural functioning of billions of neurons.

Building a Human Computer

Of course, it is entirely possible that the processes involved in brain function are so complex as to make an effective understanding of them impossible in any practical sense. Then AI is rendered a practical impossibility. At the other end of the spectrum a few scientists hold that there is nothing all that special about consciousness and that any machine packed with enough intelligence will automatically acquire consciousness along the way.

Our investigations seem to indicate that existing neural networks exhibit many promising features. They do not require sophisticated rule sets to be programmed in to them but function by a dynamic pattern matching mechanism much closer at least in spirit to the firing activities found in the brain. But at present the models are somewhat stupid — they can only learn by a slow procedure which requires constant supervision from an external 'teacher'. Their decision making ability is strictly limited. While this has not stopped them being becoming a very useful tool in many areas, it effectively prevents them from tackling more complicated problems and ultimately from acquiring any degree of true intelligence. What is really needed is a way of allowing the internode connections to change with time not according to some scheme determined by the external teacher but as a response to the node firings activated by input patterns. After all this is essentially what happens in the brain — the neurons and their connections self-organise into a structure which, considered as a whole, is capable of very sophisticated functions.

The network must select its own output patterns and connection strengths dynamically. Furthermore, the association between input and output must be useful — the network must be able to make decisions as a consequence of its firings. If it is to store memories, it must first be able to see whether a new input is close to an old memory, or really new. If the latter it must be able to store the new pattern without destroying the old. It must be able to focus only on certain types of pattern and screen out the rest in order to perform useful tasks. Ultimately, it must be able to make complex decisions by a succession of hierarchical pattern association steps. Rather surprisingly, there are new neural network models (for example the Kohonen network and Steven Grossberg's 1987 ART network) which attempt with some success to satisfy some of these criteria. These networks learn by a 'competitive' process in which nodes on the hidden layers compete to represent the input image in such a way that the final representation of the input pattern is localized on a single winning unit. The way this happens is that when an image is presented to the network, some node on the hidden layer will respond most strongly to the image.

The connections to the this node are then progressively strengthened in such a way as to increase the node's response to this input pattern whilst the connections of all the other nodes are adjusted to minimize their response. Thus a given type of image can be made to excite only one hidden layer neuron. A new image can then be made to activate another node and so on. In this way different generic features of an input pattern can be handled by different hidden layer nodes. Furthermore, the hidden layer nodes can also have connections between each other which can be arranged in such a way that nodes that are strongly connected within the layer respond to similar images. For example, after such a competitive learning process one hidden layer node might respond to ellipses of a certain size, whilst one of its 'neighbors' (those to which it has a strong intra-layer connection) might respond to ellipses of the same size but rotated through some angle. This method of learning requires no 'teacher' and is typically much faster than the supervised methods we discussed below. It also bears some resemblance to the learning mechanisms exhibited by certain types of neuron. It can also be more powerful in its classification capabilities — to use our old example, it may capable of spotting a triangle whatever its size, orientation and position in the input pattern plane. Such models are still in their infancy, however the fields of science on which they are based — neuroscience and complex systems are advancing very rapidly and it is almost certain that great strides will be made in the near future in our understanding of artificial neural networks. Whether that increased understanding will be sufficient to build an intelligent machine — a computer with a mind — is an impossible question to answer at this time.

REFERENCES

Adelson-Velsky, G. M., Arlazarov, V. L., and Donskoy, M. V. (1975). Some methods of controlling the tree search in chess programmes. *Artificial Intelligence*, 6(4):361-371.

Agmon, S. (1954). The relaxation method for linear inequalities. *Canadian Journal of Mathematics*, 6(3):382-392.

Agre, P. E. and Chapman, D. (1987). Pengi: an implementation of a theory of activity. In *Proceedings of the Tenth International Joint Conference on Artificial Intelligence (IJCAI-87)*, pages 268-272, Milan, Italy. Morgan Kaufmann.

Allen, J. F., Hendler, J., and Tate, A., editors (1990a). *Readings in Planning*. Morgan Kaufmann, San Mateo, California.

Allen, J. F., Kautz, H. A., Pelavin, R., and Tenenberg, J., editors (1990b). *Reasoning about Plans*. Morgan Kaufmann, San Mateo, California. .

Almuallim, H. and Dietterich, T. G. (1991). Learning with many irrelevant features. In *Proceedings of the Ninth National Conference on Artificial Intelligence (AAAI-91)*, volume 2, pages 547-552, Anaheim, California. AAAI Press.

Armstrong, D. M. (1968). *A Materialist Theory of the Mind*. Routledge and Kegan Paul, London.

Baum, L. E. and Petrie, T. (1966). Statistical inference for probabilistic functions of finite state Markov chains. *Annals of Mathematical Statistics*, 41.

Berliner, H. J. and Ebeling, C. (1989). Pattern knowledge and search: The SUPREM architecture. *Artificial Intelligence*, 38(2):161-198.

Berliner, H. J. and Goetsch, G. (1984). A quantitative study of search methods and the effect of constraint satisfaction. Technical Report CMU-CS-84-187, Computer Science Department, Carnegie-Mellon University, Pittsburgh, Pennsylvania.

Brachman, R. J., Fikes, R. E., and Levesque, H. J. (1983). Krypton: A functional approach to knowledge representation. *Computer*, 16(10):67-73.

Brachman, R. J. and Levesque, H. J., editors (1985). *Readings in Knowledge Representation*. Morgan Kaufmann, San Mateo, California.

Bransford, J. and Johnson, M. K. (1973). Consideration of some problems in comprehension. In Chase, W. G., editor, *Visual Information Processing*. Academic Press.

Bratko, I. (1986). *Prolog Programming for Artificial Intelligence*. Addison-Wesley, Reading, Massachusetts, first edition.

Bratko, I. (1990). *Prolog Programming for Artificial Intelligence*. Addison-Wesley, Reading, Massachusetts, second edition.

Bratman, M. E. (1987). *Intention, Plans, and Practical Reason*. Harvard University Press, Cambridge, Massachusetts.

Bratman, M. E. (1992). Planning and the stability of intention. *Minds and Machines*, 2(1):1-16.

Brelaz, D. (1979). New methods to color the vertices of a graph. *Communications of the Association for Computing Machinery*, 22(4):251-256.

Bresnan, J. (1982). *The Mental Representation of Grammatical Relations*. MIT Press, Cambridge, Massachusetts.

Chung, K. L. (1979). *Elementary Probability Theory with Stochastic Processes*. Springer-Verlag, Berlin, third edition.

Church, A. (1936). A note on the Entscheidungsproblem. *Journal of Symbolic Logic*, 1:40-41 and 101-102.

Church, A. (1941). *The Calculi of Lambda-Conversion*. Princeton University Press, Princeton, New Jersey.

Church, K. (1988). A stochastic parts programme and noun phrase parser for unrestricted texts. In *Proceedings of the Second Conference on Applied Natural Language Processing*, Austin, Texas.

Church, K. and Patil, R. (1982). Coping with syntactic ambiguity or how to put the block in the box on the table. *American Journal of Computational Linguistics*, 8(3-4):139-149.

Churchland, P. M. (1979). *Scientific Realism and the Plasticity of Mind*. Cambridge University Press, Cambridge.

Churchland, P. M. and Churchland, P. S. (1982). Functionalism, qualia, and intentionality. In Biro, J. I. and Shahan, R. W., editors, *Mind, Brain and Function: Essays in the Philosophy of Mind*, pages 121-145. University of Oklahoma Press, Norman, Oklahoma.

Churchland, P. S. (1986). *Neurophilosophy: Toward a Unified Science of the Mind- Brain*. MIT Press, Cambridge, Massachusetts.

Clark, K. L. (1978). Negation as failure. In Gallaire, H. and Minker, J., editors, *Logic and Data Bases*, pages 293-322. Plenum, New York.

Clark, K. L. and Gregory, S. (1986). PARLOG: parallel programming in logic. *ACM Transactions on Programming Languages*, 8:1-49.

Clark, R. (1992). The selection of syntactic knowledge. *Language Acquisition*, 2(2):83-149.

Clarke, M. R. B., editor (1977). *Advances in Computer Chess 1*. Edinburgh University Press, Edinburgh, Scotland.

Clocksin, W. F. and Mellish, C. S. (1987). *Programming in Prolog*. Springer-Verlag, Berlin, third revised and extended edition.

de Dombal, F. T., Staniland, J. R., and Clamp, S. E. (1981). Geographical variation in disease presentation. *Medical Decision Making*, 1:59-69.

Dempster, A., Laird, N., and Rubin, D. (1977). Maximum likelihood from incomplete data via the EM algorithm. *Journal of the Royal Statistical Society*, 39 (Series B):1-38.

Dempster, A. P. (1968). A generalization of Bayesian inference. *Journal of the Royal Statistical Society*, 30 (Series B):205-247.

Doyle, J. (1983). What is rational psychology? Toward a modern mental philosophy. *AI Magazine*, 4(3):50-53.

Doyle, J. and Patil, R. S. (1991). Two theses of knowledge representation: language restrictions, taxonomic classification, and the utility of representation services. *Artificial Intelligence*, 48(3):261-297.

Drabble, B. (1990). Mission scheduling for spacecraft: Diaries of T- SCHED. In *Expert Planning Systems*, pages 76-81. Institute of Electrical Engineers.

Draper, D., Hanks, S., and Weld, D. (1994). Probabilistic planning with information gathering and contingent execution. In *Proceedings 2nd AIPS*.

Dreyfus, H. L. (1972). *What Computers Can't Do: A Critique of Artificial Reason*. Harper and Row, New York, first edition.

Dreyfus, H. L. (1979). *What Computers Can't Do: The Limits of Artificial Intelligence*. Harper and Row, New York, revised edition.

Dreyfus, H. L. (1992). *What Computers Still Can't Do: A Critique of Artificial Reason*. MIT Press, Cambridge, Massachusetts.

Dreyfus, H. L. and Dreyfus, S. E. (1986). *Mind over Machine: The Power of Human Intuition and Expertise in the Era of the Computer*. Blackwell, Oxford. With Tom Athanasiou.

Dreyfus, S. E. (1969). An appraisal of some shortest-paths algorithms. *Operations Research*, 17:395-412.

Dubois, D. and Prade, H. (1994). A survey of belief revision and updating rules in various uncertainty models. *International Journal of Intelligent Systems*, 9(1):61-100.

Fikes, R. E. and Nilsson, N. J. (1993). STRIPS, a retrospective. *Artificial Intelligence*, 59(1-2):227-232.

Fu, K.-S. and Booth, T. L. (1986b). Grammatical inference: Introduction and survey-part ii. *IEEE Transactions on Pattern Analysis and Machine Intelligence*, PAMI-8(3):360-375. Reprinted from IEEE Trans. on Systems, Man, and Cybernetics, Vol. SMC-5, No. 4, January 1975.

Gelernter, H., Hansen, J. R., and Gerberich, C. L. (1960). A FORTRAN-compiled list processing language. *Journal of the Association for Computing Machinery*, 7(2):87-101.

Heckerman, D. (1991). *Probabilistic Similarity Networks*. MIT Press, Cambridge, Massachusetts.

Hsu, F.-H., Anantharaman, T. S., Campbell, M. S., and Nowatzyk, A. (1990). A grandmaster chess machine. *Scientific American*, 263(4):44-50.

Hsu, K., Brady, D., and Psaltis, D. (1988). Experimental demonstration of optical neural computers. In Anderson, D. Z., editor, *Neural Information Processing Systems, Denver 1987*, pages 377-386, Denver, Colorado. American Institute of Physics.

Huang, T., Koller, D., Malik, J., Ogasawara, G., Rao, B., Russell, S., and Weber, J. (1994). Automatic symbolic traffic scene analysis using belief networks. In *Proceedings of the Twelfth National Conference on Artificial Intelligence (AAAI-94)*, pages 966-972, Seattle, Washington. AAAI Press.

Joshi, A. (1985). Tree-adjoining grammars: How much context sensitivity is required to provide reasonable structural descriptions. In Dowty, D., Karttunen, L., and Zwicky, A., editors, *Natural Language Parsing*. Cambridge University Press.

Joshi, A., Webber, B., and Sag, I. (1981). *Elements of Discourse Understanding*. Cambridge University Press.

Kolmogorov, A. N. (1941). Interpolation und extrapolation von stationaren zufalligen folgen. *Bulletin of the Academy of Sciences of the USSR*, Ser. Math. 5:3-14.

Lowe, D. G. (1987). Three-dimensional object recognition from single two-dimensional images. *Artificial Intelligence*, 31:355-395.

Moore, J. and Newell, A. (1973). How can Merlin understand? In Gregg, L., editor, *Knowledge and Cognition*. Lawrence Erlbaum Associates, Potomac, Maryland.

Moore, R. C. (1980). Reasoning about knowledge and action. Artificial Intelligence Center Technical Note 191, SRI International, Menlo Park, California.

Nicholson, A. E. and Brady, J. M. (1992). The data association problem when monitoring robot vehicles using dynamic belief networks. In *ECAI 92: 10th European Conference on Artificial Intelligence Proceedings*, pages 689-693, Vienna, Austria. Wiley.

Olesen, K. G. (1993). Causal probabilistic networks with both discrete and continuous variables. *IEEE Transactions on Pattern Analysis and Machine Intelligence (PAMI)*, 15(3):275-279.

Pereira, F. (1983). Logic for natural language analysis. Technical Note 275, SRI International.

Pereira, F. C. N. and Shieber, S. M. (1987). *Prolog and Natural-Language Analysis*. Center for the Study of Language and Information (CSLI), Stanford, California.

Pohl, I. (1977). Practical and theoretical considerations in heuristic search algorithms. In Elcock, E. W. and Michie, D., editors, *Machine Intelligence 8*, pages 55-72. Ellis Horwood, Chichester, England.

Putnam, H. (1963). 'Degree of confirmation' and inductive logic. In Schilpp, P. A., editor, *The Philosophy of Rudolf Carnap*. Open Court, La Salle, Illinois.

Quinlan, J. R. and Cameron-Jones, R. M. (1993). FOIL: a midterm report. In Brazdil, P. B., editor, *European Conference on Machine Learning Proceedings (ECML-93)*, pages 3-20, Vienna. Springer-Verlag.

Reiter, R. (1991). The frame problem in the situation calculus: a simple solution (sometimes) and a completeness result for goal regression. In Lifschitz, V., editor, *Artificial Intelligence and Mathematical Theory of Computation: Papers in Honor of John McCarthy*, pages 359-380. Academic Press, New York.

Rissanen, J. (1984). Universal coding, information, prediction, and estimation. *IEEE Transactions on Information Theory*, IT-30(4):629-636.

Ritchie, G. D. and Hanna, F. K. (1984). AM: a case study in AI methodology. *Artificial Intelligence*, 23(3):249-268.

Rivest, R. L. (1987). Learning decision lists. *Machine Learning*, 2(3):229-246.

Rorty, R. (1965). Mind-body identity, privacy, and categories. *Review of Metaphysics*, 19(1):24-54.

Rosenblatt, F. (1957). The perceptron: A perceiving and recognizing automaton. Report 85-460-1, Project PARA, Cornell Aeronautical Laboratory.

Stockman, G. (1979). A minimax algorithm better than alpha-beta? *Artificial Intelligence*, 12(2):179-196.

Stolcke, A. (1993). An efficient probabilistic context-free parsing algorithm that computes prefix probabilities. Report TR-93-065, ICSI, Berkeley.

10

Research Methodology in Nuclear Science

INTRODUCTION

Nuclear technology is technology that involves the reactions of atomic nuclei. It has found applications from smoke detectors to nuclear reactors, and from gun sights to nuclear weapons. There is a great deal of public concern about its possible implications, and every application of nuclear technology is reviewed with care. In 1896, Henri Becquerel was investigating phosphorescence in uranium salts when he discovered a new phenomenon which came to be called radioactivity. He, Pierre Curie and Maria Sklodowska-Curie began investigating the phenomenon. In the process they isolated the element radium, which is highly radioactive. They discovered that radioactive materials produce intense, penetrating rays of several distinct sorts, which they called alpha rays, beta rays and gamma rays. Some of these kinds of radiation could pass through ordinary matter, and all of them could cause damage in large amounts — all the early researchers received various radiation burns, much like sunburn, and thought little of it. The new phenomenon of radioactivity was seized upon by the manufacturers of quack medicine (as had the discoveries of electricity and magnetism, earlier), and any number of patent medicines and treatments involving radioactivity were put forward.

Gradually it came to be realised that the radiation produced by radioactive decay was ionizing radiation, and that quantities too small to burn presented a severe long-term hazard. Many of the scientists working on radioactivity died of cancer as a result of their exposure. Radioactive patent medicines mostly disappeared, but other applications of radioactive materials persisted, such as the use of radium salts to produce glowing dials on meters. As the atom came to be better understood, the nature of radioactivity became clearer: some atomic nuclei are unstable, and they can decay, releasing energy (in the form of gamma rays, high-energy photons) and nuclear fragments (alpha particles, a pair of protons and a pair of neutrons, and beta particles, high-energy electrons). During World War II, nuclear reactions were sufficiently well understood that all the factions began to see the possibility of constructing a nuclear weapon. Nuclear reactions release far more energy per reaction than chemical reactions, so if large numbers of reactions could be induced to occur at once, tremendous amounts of energy could be released. The British and the Americans set up the Manhattan Project under the direction of Robert Oppenheimer to build such a device. Radioactivity is generally a slow and difficult process to control, and is unsuited to building a weapon. However, other nuclear reactions are possible. In particular, a sufficiently unstable nucleus can undergo nuclear fission, breaking into two smaller nuclei and releasing energy and some fast neutrons. This neutron could, if captured by another nucleus, cause that nucleus to undergo fission as well. The process could then continue in a nuclear chain reaction. Such a chain reaction could release a vast amount of energy in a short amount of time. The design of a nuclear weapon is more complicated than it might seem — it is quite difficult to ensure that such a chain reaction consumes a significant fraction of the fuel before the device flies apart. The construction of a nuclear weapon is also more difficult than it might seem, as no naturally-occurring substance is sufficiently unstable for this process to occur. One isotope of uranium, namely uranium-235, is naturally-occurring and sufficiently unstable, but it is always found mixed with the stable isotope uranium-238.

Thus a complicated and difficult process of isotope separation must be performed to obtain uranium-235. Alternatively, the element plutonium possesses an isotope that is sufficiently unstable for this process to be usable. Plutonium does not occur naturally, so it must be manufactured in a nuclear reactor. Ultimately, the Manhattan Project manufactured nuclear weapons based on each of these. The first atomic bomb was detonated in a test code-named "Trinity", at Alamogordo on July 16, 1945. After much debate on the morality of using such a horrifying weapon, two bombs were dropped on the Japanese cities Hiroshima and Nagasaki, and the Japanese surrender followed shortly. The nations that could afford to began nuclear weapons programmes, developing ever more destructive bombs in an arms race to obtain what they called a nuclear deterrent. Throughout the Cold War, the opposing powers had huge nuclear arsenals, sufficient to kill hundreds of millions of people.

Generations of people grew up under the shadow of nuclear devastation. However, the tremendous energy release in the detonation of a nuclear weapon also suggested the possibility of a new energy source. Nuclear power plants were built to generate household electric power. Nuclear submarines were built, able to travel at speed while submerged for months at a time. Nuclear ships were built, primarily aircraft carriers, although a few icebreakers were built. Research projects were started into the possibility of nuclear-powered aircraft and nuclear thermal rockets. The first generations of nuclear reactors were built to produce power; nuclear safety was a secondary consideration. However, as more nuclear reactors were built, it became clear that they were complex devices in which failures were extremely dangerous. Early safety features were primarily concerned with the exposure of operators to intense radiation. However, it was gradually realised that the release of radioactive material into the environment, called radioactive contamination, was also potentially serious. Radioactive isotopes of common elements are chemically very similar to non-radioactive isotopes, so the human body may take up the radioactive materials and deposit them in the bones, thyroid, lungs, or elsewhere. The radioactive materials then decay in place, often leading to cancer. The Three Mile Island incident, coupled with the release of the disaster film *The China Syndrome* greatly impacted the public's perception of nuclear power. Many human factors engineering improvements were made to American power plants in the wake of Three Mile Island's partial meltdown. The Chernobyl accident in 1986 further sracered the public of nuclear power. While design differences between the RBMK reactor used at Chernobyl and most western reactors virtually eliminate the possibility of such an accident occurring outside of the former Soviet Union, it is only recently that the general public in the United States has started to embrace nuclear energy.

The vast majority of everyday phenomena do not involve nuclear reactions. Most everyday phenomena only involve gravity and electromagnetism. Of the fundamental forces of nature, these are the weakest, but the strong nuclear force and the weak nuclear force are essentially short-range forces so they do not play a role outside the atomic nucleus. Atomic nuclei are generally kept apart because they contain positive electrical charges and therefore repel each other, so in ordinary circumstances they cannot meet. Most natural nuclear reactions fall under the heading of radioactive decay, where a nucleus is unstable and decays after a random interval. The most common processes by which this can occur are alpha decay, beta decay, and gamma decay. Under suitable circumstances, a large unstable nucleus can break into two smaller nuclei, undergoing nuclear fission. Nuclear fission normally releases fast neutrons. If these neutrons are captured by a suitable nucleus, they can trigger fission as well, leading to a chain reaction. A mass of radioactive material large enough (and in a suitable configuration) is called a critical mass. When a neutron is captured by a suitable nucleus, fission may occur immediately, or the nucleus may persist in an unstable state for a short time. If there are enough immediate decays to carry on the chain reaction, the mass is said to be prompt critical, and the energy release will grow rapidly and uncontrollably, usually leading to an explosion.

However, if the mass is critical only when the delayed neutrons are included, the reaction can be controlled, for example by the introduction or removal of neutron absorbers. This is what allows nuclear reactors to be built. Fast neutrons are not easily captured by nuclei; they must be slowed (slow neutrons), generally by collision with the nuclei of a neutron moderator, before they can be easily captured. If nuclei are forced to collide, they can undergo nuclear fusion. This process may release or absorb energy. When the resulting nucleus is lighter than that of iron, energy is normally released; when the nucleus is heavier than that of iron, energy is generally absorbed.

This process of fusion occurs in stars, and is the way all elements heavier than helium were produced. Because of the very strong repulsion of nuclei, fusion is difficult to achieve in a controlled fashion. Fusion bombs obtain their enormous destructive power from fusion, but obtaining controlled fusion power has so far proved elusive. Controlled fusion can be achieved in particle accelerators; this is how many synthetic elements were produced. The Farnsworth-Hirsch Fusor is a device which can produce controlled fusion (and which can be built as a high-school science project), albeit at a net energy loss. It is sold commercially as a neutron source.

What is Atomic Nucleus?

The nucleus of an atom is the very small dense region, of postive charge, in its centre consisting of nucleons (protons and neutrons). The size (diameter) of the nucleus is in the range of 1.6 fm (for a proton in light hydrogen) to about 15 fm (10^{-15} m) (for the heaviest atoms, such as uranium). These dimensions are much smaller than the size of the atom itself by a factor of about 23,000 (uranium) to about 145,000 (hydrogen). Almost all of the mass in an atom is made up from the protons and neutrons with almost no contribution from the electrons. The etymology of the term nucleus is from 1704 meaning "kernel of a nut". In 1844, Michael Faraday used the term to refer to the "central point of an atom". The modern atomic meaning was proposed by Ernest Rutherford in 1912. The adoption of the term "nucleus" to atomic theory, however, was not immediate. In 1916, for example, Gilbert Lewis stated, in his famous article *The Atom and the Molecule*, that "the atom is composed of the *kernel* and an outer atom or *shell*".

Overview

Nuclear Makeup

The nucleus of an atom consists of protons and neutrons (two types of baryons) bound by the strong force. These baryons are further composed of sub-atomic fundamental particles known as quarks.

Isotopes

The isotope of an atom is determined by the number of neutrons in the nucleus. Different isotopes of the same element have very similar chemical properties because chemical reactions depend almost entirely on the number of electrons that an atom has. Different isotopes in a sample of a particular chemical can be separated by using a centrifuge or by using a mass spectrometer. The first method is used in producing enriched uranium from a sample of regular uranium, and the second is used in carbon dating. The number of protons and neutrons together determine the nuclide (type of nucleus). Protons and neutrons have nearly equal masses, and their combined number, the mass number, is approximately equal to the atomic mass of an atom. The combined mass of the electrons is very small in comparison to the mass of the nucleus, since protons and neutrons weigh roughly 2000 times more than electrons.

Nuclear Decay

If a nucleus has too few or too many neutrons it may be unstable, and will decay after some period of time. For example, nitrogen-16 atoms (7 protons, 9 neutrons) beta decay to oxygen-16 atoms (8 protons, 8 neutrons) within a few seconds of being created. In this decay a neutron in the nitrogen nucleus is turned into a proton and an electron by the weak nuclear force. The element of the atom changes because while it previously had seven protons (which makes it nitrogen) it now has eight (which makes it oxygen). Many elements have multiple isotopes which are stable for weeks, years, or even billions of years.

Nucleus Size

The radius of a nucleon (neutron or proton) is of the order of 1 fm = (10^{-15} m). The nuclear radius is less than 0.005 % (1/20,000th) of the radius of the atom. Thus the density of the nucleus is more than 10 trillion times that of the atom as a whole. One solid cubic millimeter of nuclear material, if compressed together, would have a mass of around 200,000 tonnes. Neutron stars are composed of such material.

History

The discovery of the electron was the first indication that the atom had internal structure. At the turn of the 20th century the accepted model of the atom was J. J. Thomson's "plum pudding" model in which the atom was a large positively charged ball with small negatively charged electrons embedded inside of it. By the turn of the century physicists had also discovered three types of radiation coming from atoms, which they named alpha, beta, and gamma radiation. Experiments in 1911 by Lise Meitner and Otto Hahn, and by James Chadwick in 1914 discovered that the beta decay spectrum was continuous rather than discrete. That is, electrons were ejected from the atom with a range of energies, rather than the discrete amounts of energies that were observed in gamma and alpha decays. This was a problem for nuclear physics at the time, because it indicated that energy was not conserved in these decays. The problem would later lead to the discovery of the neutrino.

Around the same time that this was happening (1909) Ernest Rutherford performed a remarkable experiment in which Hans Geiger and Ernest Marsden under his supervision fired alpha particles (helium nuclei) at a thin film of gold foil. The plum pudding model predicted that the alpha particles should come out of the foil with their trajectories being at most slightly bent. He was shocked to discover that a few particles were scattered through large angles, even completely backwards in some cases. The discovery, beginning with Rutherford's analysis of the data in 1911, eventually led to the Rutherford model of the atom, in which the atom has a very small, very dense nucleus consisting of heavy positively charged particles with embedded electrons in order to balance out the charge. As an example, in this model nitrogen-14 consisted of a nucleus with 14 protons and 7 electrons, and the nucleus was surrounded by 7 more orbiting electrons.

The Rutherford model worked quite well until studies of nuclear spin were carried out by Franco Rasetti at the California Institute of Technology in 1929. By 1925 it was known that protons and electrons had a spin of 1/2, and in the Rutherford model of nitrogen-14 the 14 protons and six of the electrons should have paired up to cancel each others spin, and the final electron should have left the nucleus with a spin of 1/2. Rasetti discovered, however, that nitrogen-14 has a spin of one.

In 1930 Wolfgang Pauli was unable to attend a meeting in Tübingen, and instead sent a famous letter with the classic introduction "Dear Radioactive Ladies and Gentlemen". In his letter Pauli suggested that perhaps there was a third particle in the nucleus which he named the "neutron". He suggested that it was very light (lighter than an electron), had no charge, and that it did not readily interact with matter (which is why it hadn't yet been detected). This desperate way out solved both the problem of energy conservation and the spin of nitrogen-14, the first because Pauli's "neutron" was carrying away the extra energy and the second because an extra "neutron" paired off with the electron in the nitrogen-14 nucleus giving it spin one. Pauli's "neutron" was renamed the neutrino (Italian for little neutral one) by Enrico Fermi in 1931, and after about thirty years it was finally demonstrated that a neutrino really is emitted during beta decay.

In 1932 Chadwick realised that radiation that had been observed by Walther Bothe, Herbert Becker, Irène and Frédéric Joliot-Curie was actually due to a massive particle that he called the neutron. In the same year Dmitrij Iwanenko suggested that neutrons were in fact spin 1/2 particles and that the nucleus contained neutrons and that there were no electrons in it, and Francis Perrin suggested that neutrinos were not nuclear particles but were created during beta decay. To cap the year off, Fermi submitted a theory of the neutrino to *Nature* (which the editors rejected for being "too remote from reality"). Fermi continued working on his theory and published a paper in 1934 which placed the neutrino on solid theoretical footing. In the same year Hideki Yukawa proposed the first significant theory of the strong force to explain how the nucleus holds together.

With Fermi and Yukawa's papers the modern model of the atom was complete. The center of the atom contains a tight ball of neutrons and protons, which is held together by the strong nuclear force. Unstable nuclei may undergo alpha decay, in which they emit an energetic helium nucleus, or beta decay, in which they eject an electron (or positron). After one of these decays the resultant nucleus may be left in an excited state, and in this case it decays to its ground state by emitting high energy photons (gamma decay).

The study of the strong and weak nuclear forces led physicists to collide nuclei and electrons at ever higher energies. This research became the science of particle physics, the crown jewel of which is the standard model of particle physics which unifies the strong, weak, and electromagnetic forces.

Nuclear Fusion

When two light nuclei come into very close contact with each other it is possible for the strong force to fuse the two together. It takes a great deal of energy to push the nuclei close enough together for the strong force to have an effect, so the process of nuclear fusion can only take place at very high temperatures or high densities. Once the nuclei are close enough together the strong force overcomes their electromagnetic repulsion and squishes them into a new nucleus. A very large amount of energy is released when light nuclei fuse together because the binding energy per nucleon increases with mass number up until nickel-62. Stars like our sun are powered by the fusion of four protons into a helium nucleus, two positrons, and two neutrinos. The *uncontrolled* fusion of hydrogen into helium is known as a thermonuclear weapon. Research to find an economically viable method of using energy from a *controlled* fusion reaction is currently being undertaken by various research establishments.

Nuclear Fission

For nuclei heavier than nickel-62 the binding energy per nucleon decreases with the mass number. It is therefore possible for energy to be released if a heavy nucleus breaks apart into two lighter ones. This splitting of atoms is known as nuclear fission. The process of alpha decay may be thought of as a special type of spontaneous nuclear fission. This process produces a highly asymmetrical fission because the four particles which make up the alpha particle are especially tightly bound to each other, making production of this nucleus in fission particularly likely. For certain of the heaviest nuclei which produce neutrons on fission, and which also easily absorb neutrons to initiate fission, a self-igniting type of neutron-initiated fission can be obtained, in a so-called chain reaction. [Chain reactions were known in chemistry before physics, and in fact many familiar processes like fires and chemical explosions are chemical chain reactions].

The fission or "nuclear" chain-reaction, using fission-produced neutrons, is the source of energy for nuclear power plants and fission type nuclear bombs such as the two that the United States used against Hiroshima and Nagasaki at the end of World War II. Heavy nuclei such as uranium and thorium may undergo spontaneous nuclear fission, but they are much more likely to undergo decay by alpha decay. For a neutron-initiated chain-reaction to occur, there must be a critical mass of the element present in a certain space under certain conditions (these conditions slow and conserve neutrons for the reactions). There is one known example of a natural nuclear fission reactor, which was active in two regions of Oklo, Gabon, Africa, over 1.5 billion years ago. Measurements of natural neutrino emission have demonstrated at around half of the heat emanating from the earth's core, results from radioactive decay. However, it is not known if any of this results from fission chain-reactions.

Nuclear Physics

A heavy nucleus can contain hundreds of nucleons (neutrons and protons), which means that to some approximation it can be treated as a classical system, rather than a quantum-mechanical one. In the resulting liquid-drop model, the nucleus has an energy which arises partly from surface tension and partly from electrical repulsion of the protons. The liquid-drop model is able to reproduce many features of nuclei, including the general trend of binding energy with respect to mass number, as well as the phenomenon of nuclear fission. Superimposed on this classical picture, however, are quantum-mechanical effects, which can be described using the nuclear shell model, developed in large part by Maria Goeppert-Mayer. Nuclei with certain numbers of neutrons and protons (the magic numbers 2, 8, 20, 50, 82, 126,...) are particularly stable, because their shells are filled.

Much of current research in nuclear physics relates to the study of nuclei under extreme conditions such as high spin and excitation energy. Nuclei may also have extreme shapes (similar to that of American footballs) or extreme neutron-to-proton ratios. Experimenters can create such nuclei using artificially induced fusion or nucleon transfer reactions, employing ion beams from an accelerator. Beams with even higher energies can be used to create nuclei at very high temperatures, and there are signs that these experiments have produced a phase transition from normal nuclear matter to a new state, the quark-gluon plasma, in which the quarks mingle with one another, rather than being segregated in triplets as they are in neutrons and protons.

Production of Heavy Elements

As the Universe cooled after the big bang [never happened] it eventually became possible for particles as we know them to exist. The most common particles created in the big bang which are still easily observable to us today were protons (hydrogen) and electrons (in equal numbers). Some heavier elements were created as the protons collided with each other, but most of the heavy elements we see today were created inside of stars during a series of fusion stages, such as the proton-proton chain, the CNO cycle and the triple-alpha process. Progressively heavier elements are created during the evolution of a star. Since the binding energy per nucleon peaks around iron, energy is only released in fusion processes occurring below this point. Since the creation of heavier nuclei by fusion costs energy, nature resorts to the process of neutron capture. Neutrons (due to their lack of charge) are readily absorbed by a nucleus. The heavy elements are created by either a slow neutron capture process (the so-called *s* process) or by the rapid, or *r* process. The *s* process occurs in thermally pulsing stars (called AGB, or asymptotic giant branch stars) and takes hundreds to thousands of years to reach the heaviest elements of lead and bismuth. The *r* process is thought to occur in supernova explosions due to the fact that the conditions of high temperature, high neutron flux and ejected matter are present. These stellar conditions make the successive neutron captures very fast, involving very neutron-rich species which then beta-decay to heavier elements, especially at the so-called waiting points that correspond to more stable nuclides with closed neutron shells (magic numbers). The *r* process duration is typically in the range of a few seconds.

Addressing the Nuclear Reactor

A nuclear reactor is a device in which nuclear chain reactions are initiated, controlled, and sustained at a steady rate (as opposed to a nuclear bomb, in which the chain reaction occurs in a fraction of a second and is completely uncontrolled).

Nuclear reactors are used for many purposes. The most significant current use is for the generation of electrical power. Research reactors are used for radioisotope production and for beamline experiments with free neutrons. Historically, the first use of nuclear reactors was the production of weapons grade plutonium for nuclear weapons. Another military use is submarine / ship propulsion (Though this involves a much smaller nuclear reactor than the one used in a nuclear power plant).

Currently all commercial nuclear reactors are based on nuclear fission, and are considered by some to be a safe and pollution-free method of generating electricity. Conversely, some consider nuclear reactors problematic for their potential safety and health risks. Fusion power is an experimental technology based on nuclear fusion instead of fission. There are other devices in which nuclear reactions occur in a controlled fashion, including radioisotope thermoelectric generators and atomic batteries, which generate heat and power by exploiting passive radioactive decay, as well as Farnsworth-Hirsch fusors, in which controlled nuclear fusion is used to produce neutron radiation.

Applications

- Nuclear power:
  - Heat for electricity generation
  - Heat for domestic and industrial heating
  - Hydrogen production
  - Desalination
- Nuclear propulsion:
  - Nuclear marine propulsion
  - Proposed nuclear thermal rockets
  - Proposed nuclear pulse propulsion rockets
- Transmutation of elements:
  - Production of plutonium, often for use in nuclear weapons
  - Creating various radioactive isotopes, such as americium for use in smoke detectors, and cobalt-60, molybdenum-99 and others, used for imaging and medical treatment
- Research applications including:
  - Providing a source of neutron and positron radiation (e.g. Neutron Activation Analysis and Potassium-argon dating)
  - Development of nuclear technology

History

Although mankind has only tamed nuclear power recently, the first nuclear reactors were naturally occurring. Fifteen natural fission reactors have so far been found in three separate ore deposits at the Oklo mine in Gabon, West Africa. First discovered in 1972 by French physicist Francis Perrin, they are collectively known as the Oklo Fossil Reactors. These reactors ran for approximately 150 million years, averaging 100 kW of power output during that time. Also, stars rely on nuclear fusion for their output of heat, light and other radiations. The concept of a natural nuclear reactor was theorized as early as 1956 by Paul Kuroda at the University of Arkansas

Enrico Fermi and Leó Szilárd, while both were at the University of Chicago, were the first to build a nuclear pile and demonstrate a controlled chain reaction on December 2, 1942. In 1955 they shared U.S. Patent 2,708,656 for the nuclear reactor.

The first nuclear reactors were used to generate plutonium for nuclear weapons. Additional reactors were used in the navy to propel submarines and aircraft carriers. In the mid-1950s, both the Soviet Union and western countries were expanding their nuclear research to include non-military uses of the atom. However, as with the military programme, much of the non-military work was done in secret.

On December 20, 1951, electric power from a nuclear powered generator was produced for the first time at Experimental Breeder Reactor-I (EBR-1) located near Arco, Idaho. On June 26, 1954, at 5:30 pm, the world's first nuclear power plant to generate electricity began operations at Obninsk, Kaluga Oblast, USSR. It produced 5 megawatts, enough to power 2,000 homes.

The world's first commercial scale nuclear power station, Calder Hall, in England, began generation on October 17, 1956. Another early power reactor was the Shippingport Reactor in Pennsylvania (1957).

Even before the 1979 Three Mile Island accident, new orders for nuclear plants in the U.S. had ceased for economic reasons primarily related to greatly extended construction times. As of 2004, no new nuclear plants have been ordered in the USA since 1978, although that may change by 2010.

Unlike the Three Mile Island accident, the 1986 Chernobyl accident did not increase regulations affecting Western reactors. This was because the Chernobyl reactors were known to be an unsafe design, using the RBMK, without containment buildings and operated unsafely, and the West had little to learn from them. There was however political fallout: Italy held a referendum the next year in 1987, the results of which led to a shutdown of the country's four nuclear power plants.

The Chernobyl accident raised awareness about the possible geographical range of nuclear events, with contamination from Ukraine easily crossing national borders and spreading over significant parts of Europe. As a consequence an international organisation to promote safety awareness and professional development on operators in nuclear facilities was created: WANO; World Association of Nuclear Operators.

In 1992 the Turkey Point Nuclear Generating Station in Florida, USA, was hit directly by Hurricane Andrew. Over $90 million of damage was done, largely to a water tank and to a smokestack of one of the fossil-fueled units on-site, but the containment buildings were undamaged.

The first organisation to develop utilitarian nuclear power, the U.S. Navy, is the only organisation worldwide with a totally clean record. This is perhaps because of the stringent demands of Admiral Hyman G. Rickover, who was the driving force behind nuclear marine propulsion. The U.S. Navy has operated more nuclear reactors than any other entity, other than the Soviet Navy, with no publicly known major incidents. Two U.S. nuclear submarines, USS *Scorpion* and *Thresher*, have been lost at sea, though for reasons not related to their reactors, and their wrecks are situated such that the risk of nuclear pollution is considered low.

Nuclear Power in Electricity Production

Nuclear power from a reactor is typically utilized to produce electricity. The production of electricity is usually accomplished by somewhat standard methods that involve using heat from the nuclear reaction to power steam turbines. Nuclear power is attractive in that relatively small amounts of fuel are used to produce vast amounts of energy with no or much smaller production of free pollutants, such as greenhouse gas.

Nuclear power is controversial since it produces radioactive waste and runs the risk of nuclear meltdown. Such events, though unlikely with proper precautions, are typically viewed as catastrophic and can produce far reaching detrimental effects, such as widespread radiation contamination. Modern reactor designs and the relatively low enrichment of nuclear reactor fuel make it essentially impossible for a nuclear explosion to occur (the Chernobyl accident was neither a modern reactor design nor was it a nuclear explosion).

The Future of the Industry

As of 2006, Watts Bar 1, which came on-line in 1997, was the last U.S. commercial nuclear reactor to go on-line. This is often quoted as evidence of a successful worldwide campaign for nuclear power phase-out. However, political resistance to nuclear power has only ever been successful in parts of Europe, in New Zealand, in the Philippines, and in the United States. Even in the US and throughout Europe, investment in research and in the nuclear fuel cycle has continued, and some experts predict that electricity shortages, fossil fuel price increases and concern over greenhouse gas emissions will renew the demand for nuclear power plants.

Many countries remain active in developing nuclear power, including Japan, China and India, all actively developing both fast and thermal technology, South Korea and the United States, developing thermal technology only, and South Africa and China, developing versions of the Pebble Bed Modular Reactor (PBMR). Finland and France actively pursue nuclear programmes; Finland has a new European Pressurized Reactor under construction by Areva. Japan has an active nuclear construction programme with new units brought on-line in 2005. In the U.S., three consortia responded in 2004 to the U.S. Department of Energy's solicitation under the Nuclear Power 2010 Programme and were awarded matching funds — the Energy Policy Act of 2005 authorized subsidies for up to six new reactors, and authorized the Department of Energy to build a reactor based on the Generation IV Very-High-Temperature Reactor concept to produce both electricity and hydrogen. As of the early 21st century, nuclear power is of particular interest to both China and India to serve their rapidly growing economies — both are developing fast breeder reactors. See also future energy development. In the energy policy of the United Kingdom it is recognized that there is a likely future energy supply shortfall, which may have to be filled by either new nuclear plant construction or maintaining existing plants beyond their programmed lifetime.

On September 22, 2005 it was announced that two sites in the U.S. had been selected to receive new power reactors (exclusive of the new power reactor scheduled for INL) — see Nuclear Power 2010 Programme.

It is possible that the first new nuclear power plant to be built in the United States since the 1970s may be installed in the remote town of Galena, Alaska. The town's City Council approved the idea, and Toshiba proposed to install its model 4S "nuclear battery" in Galena free of charge as a test.

Types of reactors

A number of reactor technologies have been developed. Fission reactors can be divided roughly into two classes, depending on the energy of the neutrons that are used to sustain the fission chain reaction.

- *Thermal (slow) reactors* use slow or thermal neutrons. These are characterized by having moderating materials which are intended to slow the neutrons until they approach the average kinetic energy of the surrounding particles, that is, until they are *thermalized*. Thermal neutrons have a far higher probability of fissioning U-235, and a lower probability of capture by U-238 than the faster neutrons that result from fission do. As well as the moderator, thermal reactors have fuel (fissionable material), containments, pressure vessels, shielding, and instrumentation to monitor and control the reactor's systems. Most power reactors are of this type, and the first plutonium production reactors were thermal reactors using graphite as the moderator. Some thermal power reactors are more thermalised than others; Graphite (ex. Russian RBMK reactors) and heavy water moderated plants (e.g. Canadian CANDU reactors) tend to be more thoroughly thermalised than PWRs and BWRs, which use light water (normal water) as the moderator (due to the extra thermalization, these types can use natural uranium/unenriched fuel).

- *Fast reactors* use fast neutrons to sustain the fission chain reaction, and are characterized by the lack of moderating material. They require highly enriched fuel (sometimes weapons-grade), or plutonium in order to reduce the amount of U-238 that would otherwise capture fast neutrons. Some are capable of producing more fuel than they consume, usually by converting U-238 to Pu-239. Some early power stations were fast reactors, as are some Russian naval propulsion units, and construction of prototypes is continuing, but overall the class has not achieved the success of thermal reactors in any application. An example of this type of reactor is the Fast Breeder Reactor (FBR).

Thermal power reactors can again be divided into three types, depending on whether they use pressurised fuel channels, a large pressure vessel, or gas cooling.

- Pressure vessels holding steam heated by the reactor are used by most commercial and naval reactors. The pressure vessel serves to balance out pressure transients in the primary loops which occur with reactor power changes. The pressure vessel also serves a small role as a primary coolant make-up source. The pressure vessel is almost always lined up to the reactor and is only isolated from the reactor for special maintenance or testing.
- Pressurised channels are used by the RBMK and CANDU reactors. Channel-type reactors can be refuelled under load, which has advantages discussed under CANDU reactor.
- Gas-cooled reactors are cooled by a circulating inert gas, usually helium, but nitrogen and carbon dioxide have also been used. Utilisation of the heat varies, depending on the reactor. Some reactors run hot enough that the gas can directly power a gas turbine. Older designs usually run the gas through a heat exchanger to make steam for a steam turbine. The pebble bed reactor uses a gas-cooled design.

Since water serves as a moderator, it cannot be used as a coolant in a fast reactor. Most designs for fast power reactors have been cooled by liquid metal, usually molten sodium. They have also been of two types, called pool and loop reactors. Further detail on the classification of Nuclear reactors can be found at Classification of Nuclear Reactors.

Current Families of Reactors

- Pool-type reactor
- Pressurized water reactor (PWR)
- Boiling water reactor (BWR)
- Fast breeder reactor (FBR)
- Pressurized Heavy Water Reactor (PHWR) or CANDU
- United States Naval reactor

Obsolete Types Still in Service

- Magnox reactor
- Advanced gas-cooled Reactor (AGR)
- Light water cooled graphite moderated reactor (RBMK)

Other Types of Reactors

- Aqueous Homogeneous Reactor
- Liquid Fluoride Reactor

Advanced Reactors

More than a dozen advanced reactor designs are in various stages of development. Some are evolutionary from the PWR, BWR and PHWR designs above, some are more radical departures. The former include the Advanced Boiling Water Reactor (ABWR), two of which are now operating with others are under construction, and the planned passively safe ESBWR and AP1000 units. The best-known radical new design is the Pebble Bed Modular Reactor (PBMR), a High Temperature Gas Cooled Reactor (HTGCR). The Clean And Environmentally Safe Advanced Reactor (CAESAR) is a nuclear reactor concept that uses steam as a moderator — this design is still in development. Possible designs of subcritical reactors exist on the drawing board, notably the energy amplifier, awaiting political support and funding. Some, such as the Integral Fast Reactor (IFR), have been cancelled due to a political climate unfavorable to nuclear power.

Generation IV Reactors

Even more-advanced reactors are also on the drawing boards. These are the Generation IV reactors, which are divided into seven overall design classes.

- Gas cooled fast reactor
- Lead cooled fast reactor

- Molten salt reactor
- Sodium-cooled fast reactor
- Supercritical water reactor
- Very high temperature reactor
- Fission fragment reactor

Nuclear Fuel Cycle

Thermal reactors generally depend on refined and enriched uranium. Some nuclear reactors can operate with a mixture of plutonium and uranium. The process by which uranium ore is mined, processed, enriched, used, possibly reprocessed and disposed of is known as the nuclear fuel cycle.

Uranium is sampled and mined as other metals are, via open-pit mining or leach mining. Raw uranium ore found in the United States ranges from 0.05% to 0.3% uranium oxide. Uranium ore is not rare; the largest probable resources, extractable at a cost of US$80 per kilogram or cheaper, are located in Australia, Kazakhstan, Canada, South Africa, Brazil, Namibia, Russia, and the United States.

The raw ore is then milled, where it is ground and chemically leached. The resulting powder of natural uranium oxide is called "yellowcake". The yellowcake powder is then converted to uranium hexafluoride to prepare for enrichment.

Under 1% of the uranium found in nature is the easily fissionable U-235 isotope and as a result most reactor designs require enriched fuel. Enrichment involves increasing the percentage of U-235 and is usually done by means of gaseous diffusion or gas centrifuge. The enriched result is then converted into uranium dioxide powder, which is pressed and fired onto pellet form. These pellets are stacked into tubes which are then sealed and called fuel rods. Many of these fuel rods are used in each nuclear reactor.

Most BWR and PWR commercial reactors use uranium enriched to about 4% U-235, many research reactors use highly enriched, or weapons grade uranium, while some commercial reactors with a high neutron economy do not require the fuel to be enriched at all.

Fueling of Nuclear Reactors

The amount of energy in the reservoir of nuclear fuel is frequently expressed in terms of "full-power days," which is the number of 24-hour periods (days) a reactor is scheduled for operation at full power output for the generation of heat energy. The number of full-power days in a reactor's operating cycle (between refueling outage times) is related to the amount of fissile uranium-235 (U-235) contained in the fuel assemblies at the beginning of the cycle. A higher percentage of U-235 in the core at the beginning of a cycle will permit the reactor to be run for a greater number of full-power days.

At the end of the operating cycle, the fuel in some of the assemblies is "spent," and is discharged and replaced with new (fresh) fuel assemblies. Although in practice, it is the buildup of reaction poisons in nuclear fuel that determines the lifetime of nuclear fuel in a reactor; long before all possible fissions have taken place, the buildup of long-lived neutron absorbing fission products damps out the chain reaction. The fraction of the reactor's fuel core replaced during refueling is typically one-fourth for a boiling-water reactor and one-third for a pressurized-water reactor.

Not all reactors need to be shut down for refueling; for example, pebble bed reactors, RBMK reactors, molten salt reactors, Magnox and CANDU reactors allow fuel to be shifted through the reactor while it is running. In a CANDU reactor, this also allows individual fuel elements to be moved about within the reactor core to places that are best suited to the amount of U-235 in the fuel element.

The amount of energy extracted from nuclear fuel is called its "burn up," which is expressed in terms of the heat energy produced per initial unit of fuel weight. Burn up is commonly expressed as megawatt days thermal per metric ton of initial heavy metal.

Waste Management

The final stage of the nuclear fuel cycle is the management of the still highly radioactive, "spent" fuel, which constitutes the most problematic component of the nuclear waste stream. After fifty years of nuclear power the question of how to deal with this material remains fraught with safety concerns and technical problems, and one of the most important lines of criticism of the industry is based on the long-term risks and costs associated with dealing with the waste.

Management of the spent fuel can include various combinations of storage, reprocessing, and disposal. In practice storage has been the primary modality so far. Typically the spent fuel rods are stored in a pool of water which is usually located on-site. The water provides both cooling for the still-decaying uranium, and shielding from the continuing radioactivity. After a few decades some on-site storage involves moving the now cooler, less radioactive fuel to a dry-storage facility, where the fuel is stored in steel and concrete containers which are monitored carefully.

Another, more permanent method of disposal of high-level nuclear waste calls for the material to be buried deep underground in certain geological formations. The Canadian government, for example, is seriously considering this method of disposal, known as the *Deep Geological Disposal* concept. Under the current plan, a vault is to be dug 500 to 1000 meters below ground, under the Canadian Shield, one of the most stable landforms on the planet. The vaults are to be dug inside geological formations known as *batholiths*, formed about a billion years ago. The used fuel bundles will be encased in a corrosion-resistant container, and further surrounded by a layer of *buffer material*, possibly of a special kind of clay (bentonite clay). The case itself is designed to last for thousands of years, while the clay would further slow the corrosion rates of the container. The batholiths themselves are chosen for their low ground-water movement rates, geological stability, and low economic value.

The Finnish government has already started building a vault to store nuclear waste 500 to 1000 meters below ground, not far from the nuclear plant at Olkiluoto. Storing high level nuclear waste above ground for a century or so is considered appropriate by many scientists. This allows for the material to be more easily observed and any problems detected and managed, while the decay over this time period significantly reduces the level of radioactivity and the associated harmful effects to the container material. It is also considered likely that over the next century newer materials will be developed which will not break down as quickly when exposed to a high neutron flux thus increasing the longevity of the container once it is permanently buried.

Reprocessing is attractive in principle because (1) it can recycle nuclear fuel and (2) it can prepare the waste material for disposal. Considerable experience with reprocessing in France however, has indicated that a one way fuel cycle based on extracting and processing fresh supplies of uranium and storing the spent fuel is more economical than reprocessing, not the least because in the process of plutonium extraction, the volume of high-level liquid radioactive waste increases about 17-fold.

Natural Nuclear Reactors

A natural nuclear fission reactor can occur under certain circumstances that mimic the conditions in a constructed reactor. The only known natural nuclear reactor formed 2 billion years ago in Oklo, Gabon, Africa. Such reactors can no longer form on Earth: radioactive decay over this immense time span has reduced the proportion of U-235 in naturally occurring uranium to below the amount required to sustain a chain reaction.

The natural nuclear reactors formed when a uranium-rich mineral deposit became inundated with groundwater that acted as a neutron moderator, and a strong chain reaction took place. The water moderator would boil away as the reaction increased, slowing it back down again and preventing a meltdown. The fission reaction was sustained for hundreds of thousands of years.

These natural reactors are extensively studied by scientists interested in geologic radioactive waste disposal. They offer a case study of how radioactive isotopes migrate through the earth's crust. This is a significant area of controversy as opponents of geologic waste disposal fear that isotopes from stored waste could end up in water supplies or be carried into the environment.

AECL's Leading Edge Nuclear Technology: A Case Study

AECL's Nuclear Laboratories Business Unit has lead responsibility for the company's Technology programme. This programme supports all aspects of the CANDU reactor – from reactor design and engineering, to project management and construction of CANDU nuclear power plants, to the provision of a wide range of nuclear-related services, including the management of nuclear wastes. Our research and development (R&D) activities are centred on eight key technologies: safety; software performance; physics and fuel; fuel channels; components and systems; hydrogen and heavy water; environmental emissions and health physics; and control and information.

- AECL R&D and its Programmes
- Chalk River Laboratories
- Whiteshell Laboratories
- Research Reactors

Products and Services

CANDU Services offers high-quality products and services to help CANDU utilities maximize their station performance. Our CANDU Services team provides customers with the benefits of our unparalleled knowledge of CANDU technology, our global experience and our extensive project management expertise. Our large suite of advanced engineering tools is key to every project we undertake. AECL's Dedicated Isotope Facilities (DIF), is a one-of-a-kind project designed solely to provide a long-term, secure supply of medical isotopes world-wide. Global Nuclear Products and Services (GNP) provides a select group of quality products and services for Pressurized Water Reactors (PWRs), Boiling Water Reactors (BWRs) and other energy sectors that require our expertise.

Nuclear Technology Education Consortium: An Example

A new concept in postgraduate-level training for the nuclear sector has been developed by a strong consortium of UK universities and HE institutions. The breadth and format of the training is designed to meet the UK's projected nuclear skills requirements in decommissioning and clean-up, reactor technology, fusion and nuclear medicine. The consortium is the Nuclear Technology Education Consortium

(NTEC) and comprises the Universities of Birmingham, Lancaster, Leeds, Liverpool, Manchester and Sheffield, City University, London, HMS Sultan, Imperial College London, UHI Millennium Institute & Westlakes Research Institute. Together these institutions represent more than 90% of the nuclear postgraduate teaching expertise residing in the UK's universities and research institutes. NTEC thus provides a one-stop shop for a range of postgraduate training in Nuclear Science & Technology which is unparalleled in the UK. The structure and content of the programme, which leads to qualifications up to Master's level in Nuclear Science & Technology, was established following extensive consultations with the UK nuclear sector, including industry, regulators, MoD, NDA, Government Departments and the Cogent Sector Skills Council. Initially the modules will be delivered by direct teaching but will subsequently also be available in a distance learning format to provide greater choice for students. This development is scheduled to commence in 2007, when the syllabus will also be broadened to provide pathways in nuclear fusion and nuclear medical technologies.. Modules will generally be delivered on the campus of the providing institution. Students seeking a postgraduate qualification will register with the university of their choice and visit other members of the consortium to attend their selected modules. The programme is coordinated by the Dalton Nuclear Institute at The University of Manchester. This programme has been approved by the Institution of Mechanical Engineers.

Energy from Nuclear Fusion

Two nuclei combine into one nucleus with the emission of a nucleon is called nuclear fusion. The most familiar and promising nuclear fusion is the conversion of a deuterium and a tritium into helium with the release of a neutron and a lot of energy. There are many more possibilities and examples of fusion described in this Chapter.

Nuclear fusion reactions provide energy in the sun, which provides us with almost all the energy. Energy from the sun drives the weather, causing rain, snow, wind, and heat. Solar energy makes plants grow, and energy stored in plants sustains animal lives. The sun is 150 million kilometers far, even light leaving the sun surface takes more than 4 minutes to reach the Earth, yet it plays a dominant role in our lives. Indirectly, nuclear fusion energy is part of our daily live. So, understanding what is going on in the sun is important.

On the other hand, raising living standard also raises energy consumption. Increase in energy demand causes all kinds of problems, some of which are social and political ones. Nuclear fission energy was once thought to be a big help, and solve all the problems, but fission technologies are now not acceptable to the public because of the big risks associated with them. The big hope is now fusion technologies.

Therefore, it is important for us to understand the basic science of nuclear fusion.

Nuclear Fusion

The sun is one of the billions of stars out there in the universe. Only a small number of stars are visible to the unaided eyes, and the Sun is one of them. The next nearest star, the Alpha or Proxima Centauri is 4.3 light years from the Sun, whereas the Sirius and the Procyon are 8.6 and 11.4 light years from the Sun respectively. Stars generally occur in pairs and multiple systems (clusters), but the Sun and a few others exist singly. The mean distance between the Sun and the Earth is usually called an Astronomical Unit (AU), which is 149,597,870.7 km (one hundred fifty million kilometers), 4.3 light minutes. The mass of the Sun is 333,000 times that of the Earth, whereas its radius is 109 times that of the Earth. The sun is a big nuclear fusion reactor, which lasts a long time, unlike a fusion bomb.

From Stars to Hydrogen Bombs

A bright sunny day makes everybody happy. The Sun dominates our lives in more than one way, but most people do not know the basic facts about the sun or about stars in general.

- How do stars produce energy?

  What is the power output of the Sun?

  What is the solar power received on Earth?

  Is it possible to produce a small Sun on Earth?

The Sun emits 3.86´10^{26} watts of power, almost 8 kwatt/cm^2 at the surface. Just outside the atmosphere, the planet Earth receives 0.14 watt/cm^2, and this value is known as the solar constant, a value important for the design of solar energy supply for spaceships. Solar power reaching the sea level varies with weather conditions. Thermonuclear reactions (another term for nuclear fusion) provide energy in the Sun, proposed Atkinson and Houtermans in 1929. George Gamow, a young physicist, learned the proposal. At a meeting in Leningrad, he reported that if hydrogen a gas were heated to very high temperatures, nuclear fusion would take place, releasing energy like a sun. He was immediately offered the privilege of research for making a sun for the Communist party, but he did not accept. He is very well known for having written extensively on many aspects of science. He also worked in the Manhattan project. The theory on the energy production of the Sun continued to develop, but difficulties in heating and holding the hot hydrogen gas are the barriers for creating another sun on Earth. Fortunately, particle accelerators have been available for the investigation of fusion reactions. Using accelerated protons and deuterons, exothermic nuclear fusion has been confirmed. The fusion of ^1H, deuterium, tritium, helium, carbon *etc.* releases a large amount of energy. In 1939, Hans A. Bethe (1906-) proposed that stars derive their energy from fusion reactions. During World War II, the idea emerged that a fission explosion could heat up and compress light nuclides to initiate a chain of nuclear fusion (Teller, 1981). Soon after the war Teller, Bethe and others directed the Manhattan project, and they utilized fission explosions to heat and to compress mixtures of deuterium and tritium. They made hydrogen bombs (or thermonuclear bombs) and the first one was tested on November 1, 1952 (code name "Mike") on a Pacific coral atoll. Thereafter, USSR, Britain, China, and France have tested similar bombs. A single bomb can completely destroy a large city. Nearly all scientists who have worked on these weapons campaign for the banning of further nuclear tests.

Nuclear Fusion

Nuclear fusion is a special nuclear reaction, opposite to fission. Like all nuclear reactions, the reactants are converted to different elements.

- What is nuclear fusion?

  How can fusion reactions be studied?

  What are the best fusion materials?

  What fusion reaction is the most promising for a fusion reactor to supply energy for the future?

In nuclear fusion, two light nuclei combine into one, but some nucleons may also be released. For fusion to take place, two bare nuclei must approach each other within a short distance of 10^{-15} m so that the strong force becomes sufficiently effective to hold the nucleons together. Both nuclei are positively charged, and the nuclei must move at high speed to overcome the Coulomb barrier before fusion takes place. In other words, fusion reactions requires very high temperatures.

Particle accelerators invented for the study of nuclear reactions are also wonderful machines for the study of fusion. Accelerated H, D, T, ^{3}He, ^{4}He particles are used to bombard targets of these nuclides. These experiments enable us to learn the nature of fusion. Probabilities of fusion reactions are quantitatively defined as the cross sections. Cross sections for various fusion reactions at various temperatures are different and their variations with temperatures are also different.

Well known plots of these cross sections have been given by Post (1970), and an approximate sketch is illustrated here. The higher the effective fusion cross section, the higher the probability of fusion. In general, the probability increases with kinetic energy of the nuclei. Kinetic energies of nuclei are proportional to their temperatures in K. Note that the vertical axis (effective fusion cross section) is a log scale, and the divisions increase by a factor of 10, whereas the horizontal (kinetic energy or temperature) scale is a linear scale. All things considered, the most favorable reaction is

$$^{2}D + ^{3}T \rightarrow ^{4}He + n.$$

The D + T fusion has the highest cross section at any temperature, and the cross section is high even at moderate temperatures. Thus, thermal nuclear fusion of deuterium and tritium is the reaction of choice for fusion research and thermonuclear bombs. Neutrons are also products of the nuclear fusion reactions.

At the temperature corresponding to a kinetic energy of 40 keV, the cross sections for D-^{3}He, D-D, and D-T fusions are 0.3, 3, and 700 mb respectively. Thus, the probability of D-T fusion is 230 times that of the D-D fusion reaction, and 2300 times that of the D-^{3}He fusion reaction.

Fusion Energy

In the earlier discussion of binding energy, it has been pointed out that fusion of light nuclei into nuclides with a combined mass will release energy.

- How can fusion energy be estimated?

  How much energy is released in a fusion reaction?

  Which reaction releases the most energy?

  All things considered, which is the most promising reaction for application?

The energy released from nuclear fusion reactions is at the expense of the mass. The mass differences before and after the fusion reactions are converted to energy. Thus, the energy of fusion Q can be treated as part of the reaction equation. For example, the equation of D-T fusion is,

$$D + T \rightarrow ^{4}He + n + Q$$

and if the mass excess (in MeV) is available, the equation based on mass and energy conservation is

$$13.136 + 14.950 = 2.425 + 8.071 + Q$$

Therefore, Q = 17.6 MeV

The actual masses rather than mass excesses can also be used, and the calculation is just as easy. The four commonly studied fusion reactions and their Q values are:

$$D + T \rightarrow ^{4}He + n + 17.6 \text{ MeV}$$

$$D + {}^{3}He \rightarrow {}^{4}He + p + 18.4\ MeV$$
$$D + D \rightarrow {}^{3}He + n + 3.3\ MeV$$
$$D + D \rightarrow {}^{3}T + p + 4.0\ MeV$$

Reactions D + T and D + ^{3}He releases similar amounts of energy, but the D + T reaction takes place at much lower temperatures and is particularly interesting from an engineering point of view.

The release of (*Q*) 17.6 MeV per ^{4}He nuclide formed is impressive when this quantity is translated to 1.7×10^{12} J per mole (or 1.7×10^{9} kJ or 4.1×10^{8} kcal for about 4 g) of He formed. This is a very large amount of energy! However, due to the fact that both the reactants and products are at very high temperature, a considerable amount of fusion energy is released as short-wave X- or γ-ray radiation.

It is interesting to note that products (^{3}He and ^{3}T) of the D-D fusion reactions are fuels in the D + T and D + ^{3}He reactions. Thus, ^{3}He and ^{3}T are intermediates of ^{2}D fusion, and protons are also products of these reactions. Taking all the reactions into consideration, no radioactive ash or waste is produced. For this reason, fusion is considered to be clean and environmental friendly for energy generation. However, the product neutrons will cause the structural material to be radioactive due to neutron capture reactions.

The fusion of hydrogen only takes place at very high temperatures, and the energy released is 1.44 MeV, not counting the decay energy of the neutron,

$$H + H \rightarrow D + \beta^{+} + n + 1.44\ MeV$$

Another hypothetical reaction has not been observed, but it releases 23.85 MeV per ^{4}He formed.

$$D + D \rightarrow {}^{4}He + 23.85\ MeV.$$

Can this hypothetical reaction take place under peculiar conditions? Although speculative in nature, the fusion of deuterium to form a ^{4}He nucleus would be an attractive reaction to have.

Although deuterium contains much more energy than helium, the fusion rate is so slow that natural nuclear fusion has been observed at ordinary temperatures and pressures on the planet Earth. Furthermore, electrostatic repulsion of atomic electrons keeps the nuclei more than 10^{-10} m apart. Recall that nuclei have to approach each other at 10^{-15} m in order to from a compound nucleus before a fusion is possible. Even when electrons are stripped, the bare nuclei have to overcome the Coulomb repulsive force before a compound nucleus can be formed.

Plasma and Nuclear Fusion

Energy sources from fossil fuel are limited. Fission nuclear reactors have been supplying energy for decades but the problems of radioactive waste disposal have become apparent in the 1970s. In the 1980s, public faith on fission reactors has been further eroded by reactor accidents. These problems remain unsolved, and the future of fission reactors seems uncertain. Thus, the relative "clean" energy source from fusion is now very attractive. For fusion reactions to take place, D and T mixtures have to be heated to 10 million degrees. At these temperatures, the mixture is a plasma, no longer a gas.

Plasma

At some high temperatures, molecules dissociate to atoms. Further increases in temperature cause atoms to lose electrons. The soup or mixture of positive and negative charged particles is called plasma. A plasma is a macroscopically neutral collection of charged particles. Vigorous thermal motion at high temperature causes the ions (bare nuclei) to collide, approaching each other within short distances (10^{-15} m), at which compound nuclei formation is possible. Thus, studies of plasmas are parts of fusion research.

- How are plasmas formed?

  What is plasma?

  What are in the plasmas?

  What are the properties of plasma?

  What is the distribution of particle energy in plasma?

  How do charged particles move in a magnetic field (the dynamo effect)?

  What is a magnetic bottle?

  How a torus confines plasma?

Plato considered earth, water, air and fire *primal substances*. They correspond to the four states of matter: solid, liquid, gas and plasma. Under ordinary terrestrial conditions, the plasma state is rare, but in the universe, plasma is the dominate (99%) state of all matter. Matters on the Sun, the stars, solar wind, cosmic dust, and white dwarfs are all in the plasma state. Closer to home, fire, the electric spark, the ionosphere, and the northern lights (aurora) are phenomena related to plasma (Frank-Kamenetskii, 1972).

Plasmas are formed by heating, electric discharge, injection of hot ions, and ionization by high pressure. Alkali metals such as sodium, potassium, and cesium have low ionization potentials, and they form plasmas at about 3,000 K. The characteristic yellow light of sodium does not give drivers a glare making them suitable for street illumination. Other substances, which do not ionize easily, require a higher temperature. Some material become plasma at 10,000 K.

No container can hold plasma and still withstand an external application of heat. Plasmas at temperature in the order of 100,000 K correspond to energies of a few eV, and these are called cold plasmas. No nuclear fusion takes place in cold plasmas. Plasmas with energy of greater than 100 eV are called hot plasmas. At these temperatures, nuclear fusion takes place.

Plasmas are heated by alternating electric field much like the action of a transformer, but at a much higher frequency. The electric fields are applied by induction coils, and these methods have raised temperatures of plasma to 10 million K or higher, hotter than the gas at the surface of the Sun. Ion densities greater than 10^{19} particles per cubic meter have also been achieved. One of the major objectives of plasma research is to achieve high temperatures while maintaining a high particle density for an extended period of time. Properties of plasma are studied to achieve these goals.

In terms of properties of plasmas, the motion of individual particles and the collective motion of plasma as a whole are particularly interesting. For individual particles, speeds of ions are very similar to those of a gas. These speeds follow the Maxwell-Boltzmann distribution, as opposed to the normal or bell shape distribution. The mean kinetic energy of all nuclei in the plasma is $(^3/_2)\ k\ T$ ($k = 1.38062 \times 10^{-23}$ J/deg is the Boltzmann constant). At certain temperatures, a fraction of the particles have high energy and they collide with one another like molecules in a gas. Only high-energy nuclei will overcome the Coulomb barrier, and approach each other to a short distance (10^{-15} m) for the formation of a compound nucleus. At low particle densities, particle collisions are relatively infrequent. The motion of individual particles follows the electro-magnetic rules. Electric and magnetic fields are used to control the movements of charged particles in the plasma.

Motions of charged particles in a magnetic field have been well studied. In general, moving charges will generate magnetic fields. Therefore their motions are affected by a magnetic field. In a magnetic field, charged particles move in a spiral fashion. Positive and negative particles move in opposite directions. Their motions are depicted here. Plasmas consist of electrons and nuclei. They spiral in different directions. Knowing the details of their motion enable us to design magnetic fields to direct their motion. However, particles in plasma move collectively as a neutral fluid.

Plasma is a neutral fluid with charged particles. As a whole, it is strongly diamagnetic. That is the strength of a magnetic field over the region occupied by plasma is greatly reduced, if not entirely canceled. The magnetic field outside the region occupied by plasma is stronger, and the difference in magnetic field develops a magnetic pressure that keeps the plasma together. Thus, plasma can be confined by magnetic walls, pushed by magnetic pistons (moving magnetic fields), and confined by magnetic traps. An arrangement of magnetic mirrors for the confinement of plasma is shown here. A magnetic field is generated by the currents in the coils. The plasma can not flow out of the strong magnetic field, and it is confined in a volume by magnetic field. Such an arrangement is often called a magnetic bottle.

During the early stage of investigation, plasma confined by magnetic walls was found to be unstable. Compressed plasma showed signs of fusion. Due to possible weapon development in fusion, both the U.S. and the former U.S.S.R. conducted their fusion research in secrecy. However, both sides experienced the same plasma instability when confined by magnetic fields. The instability is caused by the counter movements of the positive particles and the negatively charge electrons. Scientists in the U.S. were able to persuade the American authorities to declassify the whole magnetic bottle approach, and world-wide free information exchange took place at the second Atoms for Peace Conference held in 1958, (Teller, 1981).

After having learned that plasmas confined in magnetic bottles were unstable, researchers turned to magnetic enclosure with a configuration of a toroid or torus. Such a device is a closed magnetic trap. The plasma is confined in a ring shape in the center of the toroid. Current flows along the windings to create a magnetic field in the ring.

Traditional electric power generation has been to convert heat sources to steam, which drives turbines of generators. Thus far, fossilized fuel and fission nuclear reactors are used to generate heat for electric power generation. Nuclear fusion of deuterium plasma would release a large amount of energy in the form kinetic energy of the products, neutrons and helium atoms (Rose and Clark, 1961). Great amounts of heat are produced. Converting heat into electric power is inefficient. It has been suggested to use the dynamo effect for power generation involving plasma. When plasma moves perpendicular to a magnetic field, an electromotive force, according to Faraday's law, is generated in a direction perpendicular to both the direction of flow of the plasma and the magnetic field. This dynamo effect can be used to drive a current in an external circuit connected to electrodes in the plasma, and thus electrical power may be produced directly from fusion energy without having to convert heat to mechanical and then to electric energy. This method is called magnetohydro-dynamic (MHD) power generation.

The motor effect or the inverse dynamo effect may be used to accelerate plasma.

Thermonuclear Bombs

Thermonuclear or hydrogen bombs explode with enormous power using uncontrolled self-sustaining chain fusion reactions. Deuterium and tritium, under extremely high temperatures, form helium providing the energy.

A typical thermonuclear bomb liberates more than 10^{17} J per explosion. By now, you can already figure out that the desirable reaction for it is the D + T fusion. The design and engineering of such a device had been a challenge, and more advanced designs are still classified (secret) information. We can only give an oversimplified description here about its structure and working principles. In principle, a mixture of D, T and ^{6}Li heated to very high temperature and confined to a high density will start a chain fusion reaction, liberating an enormous amount of energy. The engineering of heating and confinement are tricky. In a thermonuclear bomb, the explosive process begins with the detonation of what is called the primary stage. This consists of a relatively small quantity of conventional explosives, its detonation brings together enough fissionable uranium to create a fission chain reaction, which in turn produces another explosion and a temperature of several million degrees. When the temperature of the mixture reaches 10,000,000 K, fusion reactions take place. In addition to the fission reaction of ^{235}U or ^{239}Pu or both, hydrogen bombs make use of the fusion reaction between deuterium and tritium:

$$^{2}D + {}^{3}T \rightarrow {}^{4}He + n + 17.6 \text{ MeV}$$

By incorporating ^{6}Li in the mixture, it absorbs the neutrons from the fission and fusion reactions, generating more T, which is an ingredient of the fuel mixture, by the reaction

$$n + {}^{6}Li \rightarrow T + {}^{4}He \quad (s = 942 \text{ b})$$

The more abundant isotope of Li may also be used, but its reaction with the neutron generates another neutron. However, its neutron capture cross section is much smaller.

$$n + {}^{7}Li \rightarrow T + {}^{4}He + n \ (s = 0.045 \text{ b})$$

The isotope ^{6}Li was more desirable because its cross section is 21,000 times larger than that of ^{7}Li. This illustrates the importance of techniques for isotope separation, which is much more difficult than chemical separation of elements. However, separation of ^{6}Li or ^{7}Li is easier than the separation of ^{235}U from natural uranium due to greater percentage of mass difference.

The neutrons are absorbed by uranium, causing more fission reaction. Thus, fusion bombs also produce radioactive fallout. When a mixture of fusion material is embedded in chemical explosives, the fusion reactions release many neutrons. These are called neutron bombs, which are smaller, releasing less energy than hydrogen bombs. Neutron bombs are designed to kill the operators (human) with neutron radiation, leaving buildings and machinery intact.

Hydrogen bombs can be thousands of times as powerful as fission bombs. The power of atomic bombs is expressed in kilotons of TNT. The explosive power of hydrogen bombs, is frequently given in megatons (1,000,000 tons). Bombs of more than 50 megatons have been tested. Strategic missiles usually range from 100 kilotons to 1.5 megatons.

Thermonuclear bombs can be made small enough (a few feet long) to fit in the warheads of intercontinental ballistic missiles that can travel almost halfway across the globe in 20 or 25 minutes. They have computerized guidance systems to land them within a few hundred yards of a designated target.

Tritium, T, is a beta emitter with a half life of 12.3 y. Therefore, fusion bombs requires a maintenance programme. Tritium reduction due to decay must be replenished. Keeping a stockpile of fusion bombs requires active organisation and effort.

Controlled Thermonuclear Fusion

Soon after the detonation of atomic bombs, controlled fission reactions were employed to produce electric energy. There are about 500 fission reactors running in the world today. In contrast, Edward Teller and other American scientists exploded the first hydrogen bomb at Enewetak atoll on Nov. 1, 1952, but controlled fusion reactions is not in sight yet.

- How can fusion chain reactions be controlled?

  What are the basic requirements for a controlled fusion reaction?

  Is there any fusion reactor in operation now, why or why not?

  What is the current status of controlled fusion research?

The fission and fusion reactions are fundamentally different. Fission is induced by thermal neutrons and each fission releases more than two neutrons. By properly controlling the number of active neutrons, the number of fission reactions can increase, decrease or be maintained at a constant. Thus, it is easy to have a controlled fission reactor. In fact, a controlled fission reaction was accomplished in 1942, and a fission explosion was detonated in 1945. The fission material has to be confined in a small volume long enough so that a lot of energy is released before the explosion breaks it apart.

Fusion is initiated by high temperatures. Since fusion reactions release energy, the plasma temperature increases causing more nuclei to fuse. Techniques of confinement had been developed during the Manhattan Project, and thus a thermonuclear explosion was achieved soon after World War II, in 1952. However, achieving controlled and sustained thermonuclear fusion to derive energy is a much more difficult task. So far, some milestones have been established, but deriving fusion energy for everyday usage is not yet a reality.

In parallel with the development of the hydrogen bomb, massive efforts towards achieving controlled thermonuclear fusion (CTF) were initiated in a number of countries. It is not easy to achieve this simple objective. Even the feasibility of controlled fusion power remains to be experimentally established. However, recent results are encouraging. Obviously, the deuteron-triton reaction:

$$D + T \rightarrow {}^{4}He + n + 17.6 \text{ MeV}$$

has received the most attention. This reaction releases a large amount of energy, and has the lowest ignition temperature, 40,000,000 K. So far, this reaction is the best choice. It has been the choice for thermonuclear bombs. The neutrons produced in these reactions are used to generate tritium in a tritium (T) breeding cycle:

$$n + {}^{6}Li \rightarrow T + {}^{4}He$$

$$n + {}^{7}Li \rightarrow T + {}^{4}He + n$$

The scheme for CTF includes a blanket of lithium around the fusion reactor.

Furthermore, a certain amount of neutron multiplication will be desirable. In some designs, niobium and molybdenum are used to blanket the walls for the (n, 2n) reactions. Tritium breeding ratios up to 1.3 appear possible. Neutrons carry off almost 80 percents of the fusion energy (14.1 MeV), and most fusion energy appears in the form of heat in the lithium blanket. Lithium has a low melting point (452 K), and if liquid metal is used, it can be circulated through heat exchange to generate steam for power generation.

Basic requirements for CTF of the above reaction are to:

1. achieve a temperature of 100,000,000 K,
2. confine a large number of nuclei in the plasma in a small volume for a length of time,

3. sustain the fusion
4. extract energy from the products after fusion.

To date, it is still a challenge to produce more energy than the energy used to heat the plasma and maintain a steady fusion rate.

Induction heating and laser heating have been used to achieve a high temperature in plasma, but these processes consume energy. Magnetic confinement and inertia confinement are well known techniques to maintain a large number of nuclei in a small volume for a period of time. An introduction has been given to magnetic confinement, but inertia confinement is often used when laser heating is used. Hot plasma loses energy mostly due to radiation of X-rays. Plasma leakage also causes energy loss. To achieve thermonuclear fusion, energy loss must be minimized and confinement techniques must be improved. Once fusion reactions started, the temperature may be maintained if the loss of radiation energy is limited. As mentioned above, most of the fusion energy (14.1 MeV) is carried out by the neutrons, which escape the plasma, and only 3.5 MeV is carried by the fused nucleus, which remains in the plasma. Transferring energy from neutrons to T or D nuclei in the plasma takes time, and the formed nuclei must be removed from the plasma to maintain a high efficiency of fusion.

Lawson, (1957) has developed a theoretical rule for maintaining the fusion reaction, and this is known as the Lawson criterion. It gives a minimum required confinement time, t_c, for the plasma to be energetically sustained. However, t_c depends on the plasma density, n, expressed as number of fusion nuclei per unit volume. The criterion requires that t_c times n be at least 10^{20} s m^{-3}. In other words, high temperature plasma with a high density of nuclei will lead to nuclear fusion.

Thermonuclear Fusion Research

Thermonuclear fusion research is very costly in manpower, equipment, and energy consumption. The European community, the Soviet Union, and the U.S. have large groups of scientists working on possible thermonuclear fusion reactions. With the industrial base broadened into Japan, China, and South East Asia, more efforts will be spent on nuclear fusion research worldwide.

- What is the current status of fusion research?

  What techniques are used for heating?

  What confinement techniques are available?

  Which one is the most promising?

  Has any machine generated more energy than it consumes?

Although plasma confinement by the magnetic bottle configuration was unstable, research goes on partly because of the desire to understand the instability. Magnetic bottles holding plasma with a high density of 10^{19} particles per m^3 have been achieved. This technique has been investigated with the intention to develop fusion energy by having many short-duration fusion reactors, each giving a small explosion. The explosions are often called shots. Continuous shots provide a large amount of energy, which can be collected or harvested using a connected apparatus in a series of operations.

Since the positive nuclei and the negative electrons move in a spiral fashion in a magnetic field, the obvious design is a ring-shaped magnetic field called a torus as pointed out earlier. Particles in the plasma have a closed path of motion. Despite the instability of the plasma in a simple torus, confinement by torus has been intensely investigated, because it has offered the best hope for success. Engineering design providing an additional toroidal magnetic field had provided excellent results. In 1968, the Soviet Union project reported a break through, achieving a temperature of 10,000,000 K in plasma enclosed in a torus they called Tokamak, (Rosenbluth and Rutherford, 1981). A project Tokamak Fusion Test Reactor (TFTR) raised the temperature to 100,000,000 K. Scientists in various fusion research centers went to the Soviet Union to see the Tokamak, and this technology has been used in the Joint European Torus (JET) project in England and the Princeton Large Torus (PLT) research.

The Tokamak at the I.V. Kurchatov Institute of the Atomic Energy in Moscow employs a closed magnetic trap. Temperature of 5,0000,000 K for deuterium and a density of 5×10^{19} particle/m^3 have been achieved. In 1996, a Tokamak-type device in Japan called JT 60U has generated 10 Mw of fusion power. However, it still consumes more energy than it produces. In the United States and Germany, an elaborated closed system called Stellarators originally proposed by Spitzer (1958) are being studied for an alternate method to the Tokamak confinement. The system employs a sophisticated strong magnetic field generated by superconductors outside the plasma region. It has received some success in 1991. Another idea called inertia confinement has been developed. This technique concentrated on LASER heating of frozen pellets, thus creating a series of fusion reactions, each similar to a small sun. This technique has achieved results only second to the Tokamak system. Scientific data such as tritium breeding, reactor economics, heat transfer from fusion to power generation, and mechanical response to heat and radiation are required before power reactor design can take place. These data have been measured at various laboratories such as the Lawrence Livermore National Laboratory, California, the Los Alamos National Laboratory etc (Cornwell and Pendergrass, 1985). All these projects are very expensive. Realising the difficulty, the United States and the Soviet Union agreed to cooperate in fusion research at the international level. To avoid tipping the balance of power, they also agreed to build a fusion reactor outside the two countries. Recently, the talk of a joint project with other countries such as Japan, Canada and the European Community for an International Thermonuclear Experimental Reactor has begun. At present, all these countries have fusion projects of their own, and they agreed to pool their data and work on common experiments. Although break even points have been reached in many research establishments, such as the JET, the Tokamak Fusion Test Reactor at Princeton, the Alcator A at MIT, and the fusion research Team in Japan, many problems remain to be solved before commercial fusion reactor starts to supply electrical energy. Some of the problems requiring further research are:

1. the vacuum wall blanket design, including heat transfer and stress analysis.
2. technology for handling and processing of tritium.
3. radiation damage to materials and suitability of structural materials for fusion reactors.
4. stability of plasma and magnetic systems required for fusion reactions.
5. energy transfer and technology required for converting fusion energy into electrical energy.
6. fueling the reactor, recovery of used fuel, and removal of waste.

Fusion and Electrolysis

Pons and Fleischmann electrolyzed a basic lithium oxide solution containing 0.1 mol of LiOD per litter. They used palladium electrodes and passed a current through it. During the electrolysis, deuterium, D_2, should be produced. They claimed that the deuterium produced might interact with the palladium metal they used as electrodes. As a result, palladium hydride was formed. In the hydride, deuterium nuclei are pushed closer than usual, and the palladium catalyzed the fusion. They showed some charts of g and b spectra, but the amounts of radiation were not significant. The amounts of energy produced were large, much more radiation should have been observed. Thus, doubts were raised about their claim.

- What phenomenon led Pons and Fleischmann to the claim of their discovery of cold fusion?

  What additional evidences will you look for in order to believe their claim?

  Is neutron, beta, gamma, or helium detected in their experiment?

At the time when fusion induced by electrolysis was in the news, another paper by Steven Jones and seven other coworkers of Brigham Young University was submitted for publication. Thus, non-scientists were often confused about the work carried out by the two Utah groups. Although both groups used electrolysis, their findings were rather different. Jones and his colleagues used 160 g D_2O solutions of various metal salts (0.1 g): $FeSO_4 \cdot 7H_2O$, $NiCl_2 \cdot 6H_2O$, $PdCl_2$, $CaCO_3$, $Li_2SO_4 \cdot H_2O$, $Na_2SO_4 \cdot 8H_2O$ and a small amount of AuCN. They adjusted the pH to be about 3 using nitric acid, and they used Ti-Au electrodes or Pd-Au pairs. The electrodes used for liberating D_2 molecules were Ti or Pd. They reported that "*H_2 bubbles were observed to form on the Pd foil only after several minutes of electrolysis, suggesting a rapid absorption of D in the foil; O_2 formed at the anode immediately*". They concluded that the fusion rate was approximately that predicted by muon-catalyzed fusion 10^{-20} per D_2 molecule per second. They detected very low levels of neutrons indicating very low fusion rates. Significance of the results was questioned, but Jones *et. al*, (1989) did not detect any excess energy generated. The controversies regarding fusion at low (room) temperatures remain and the problems are further complicated by competition for research funds and politics among scientists. The problems related to cold fusion are interesting. They not only involve scientific work, but also relate to social, political and economic problems.

Nuclear Power and Its Production

To provide the power for a dynamo-electric machine, or electric generator, nuclear power plants rely on the process of nuclear fission. In this process, the nucleus of a heavy element, such as uranium, splits when bombarded by a free neutron in a nuclear reactor.(1) The fission process for uranium atoms yields two smaller atoms, one to three free neutrons, plus an amount of energy. Because more free neutrons are released from a uranium fission event than are required to initiate the event, the reaction can become self sustaining—a chain reaction—under controlled conditions, thus producing a tremendous amount of energy.

In the vast majority of the world's nuclear power plants, heat energy generated by burning uranium fuel is collected in ordinary water and is carried away from the reactor's core either as steam in boiling water reactors or as superheated water in pressurized-water reactors. In a pressurized-water reactor, the superheated water in the primary cooling loop is used to transfer heat energy to a secondary loop for the creation of steam. In either a boiling-water or pressurized-water installation, steam under high pressure is the medium used to transfer the nuclear reactor's heat energy to a turbine that mechanically turns a dynamo- electric machine, or electric generator. Boiling-water and pressurized-water reactors are called light-water reactors, because they utilize ordinary water to transfer the heat energy from reactor to turbine in the electricity generation process. In other reactor designs, the heat energy is transferred by pressurized heavy water, gas, or another cooling substance.

Because the water used to remove heat from the core in a light-water reactor absorbs some of the free neutrons normally generated during operation of the reactor, the concentration of the naturally fissionable 235U isotope in uranium used to fuel light-water reactors must be increased above the level of natural uranium to assist in sustaining the nuclear chain reaction in the reactor core: the remainder of the uranium in the fuel is 238U. Increasing the concentration of 235U in nuclear fuel uranium above the level that occurs in natural uranium is accomplished through the process of enrichment, which is explained below.

The fuel core for a light-water nuclear power reactor can have up to 3,000 fuel assemblies. An assembly consists of a group of sealed fuel rods, each filled with UO2 pellets, held in place by end plates and supported by metal spacer-grids to brace the rods and maintain the proper distances between them. The fuel core can be thought of as a reservoir from which heat energy can be extracted through the nuclear chain reaction process. During the operation of the reactor, the concentration of 235U in the fuel is decreased as those atoms undergo nuclear fission to create heat energy. Some 238U atoms are converted to atoms of fissile 239Pu, some of which will, in turn, undergo fission and produce energy. The products created by the nuclear fission reactions are retained within the fuel pellets and these become neutron-absorbing products (called "poisons") that act to slow the rate of nuclear fission and heat production. As the reactor operation is continued, a point is reached at which the declining concentration of fissile nuclei in the fuel and the increasing concentration of poisons result in lower than optimal heat energy generation, and the reactor must be shut down temporarily and refueled.

The amount of energy in the reservoir of nuclear fuel is frequently expressed in terms of "full-power days," which is the number of 24-hour periods (days) a reactor is scheduled for operation at full power output for the generation of heat energy. The number of full power days in a reactor's operating cycle (between refueling outage times) is related to the amount of fissile 235U contained in the fuel assemblies at the beginning of the cycle. A higher percentage of 235U in the core at the beginning of a cycle will permit the reactor to be run for a greater number of full power days.

At the end of the operating cycle, the fuel in some of the assemblies is "spent," and it is discharged and replaced with new (fresh) fuel assemblies. The fraction of the reactor's fuel core replaced during refueling is typically one-fourth for a boiling-water reactor and one-third for a pressurized-water reactor.

The amount of energy extracted from nuclear fuel is called its "burn up," which is expressed in terms of the heat energy produced per initial unit of fuel weight. Burn up is commonly expressed as megawatt days thermal per metric ton of initial heavy metal.

Nuclear Fuel Cycle

The nuclear fuel cycle for typical light-water reactors. The cycle consists of "front end" steps that lead to the preparation of uranium for use as fuel for reactor operation and "back end" steps that are necessary to safely manage, prepare, and dispose of the highly radioactive spent nuclear fuel. Chemical processing of the spent fuel material to recover the remaining fractions of fissionable products, 235U and 239Pu, for use in fresh fuel assemblies is technically feasible. Reprocessing of spent commercial-reactor nuclear fuel is not permitted in the United States. The front end of the nuclear fuel cycle commonly is separated into the following steps.

Milling. Mined uranium ores normally are processed by grinding the ore materials to a uniform particle size and then treating the ore to extract the uranium by chemical leaching. The milling process commonly yields dry powder-form material consisting of natural uranium, "yellowcake," which is sold on the uranium market as U_3O_8.

Mining. Uranium ore can be extracted through conventional mining in open pit and underground methods similar to those used for mining other metals. *In situ* leach mining methods also are used to mine uranium in the United States. In this technology, uranium is leached from the in-place ore through an array of regularly spaced wells and is then recovered from the leach solution at a surface plant. Uranium ores in the United States typically range from about 0.05 to 0.3 percent uranium oxide (U_3O_8). Some uranium deposits developed in other countries are of higher grade and are also larger than deposits mined in the United States. Uranium is also present in very low grade amounts (50 to 200 parts per million) in some domestic phosphate-bearing deposits of marine origin. Because very large quantities of phosphate-bearing rock are mined for the production of wet-process phosphoric acid used in high analysis fertilizers and other phosphate chemicals, at some phosphate processing plants the uranium, although present in very low concentrations, can be economically recovered from the process stream.

Uranium conversion. Milled uranium oxide, U_3O_8, must be converted to uranium hexafluoride, UF6, which is the form required by most commercial uranium enrichment facilities currently in use. A solid at room temperature, UF6 can be changed to a gaseous form at moderately higher temperatures. The UF6 conversion product contains only natural, not enriched, uranium.

Fabrication. For use as nuclear fuel, enriched UF6 is converted into uranium dioxide (UO2) powder which is then processed into pellet form. The pellets are then fired in a high temperature sintering furnace to create hard, ceramic pellets of enriched uranium. The cylindrical pellets then undergo a grinding process to achieve a uniform pellet size. The pellets are stacked, according to each nuclear core's design specifications, into tubes of corrosion-resistant metal alloy. The tubes are sealed to contain the fuel pellets: these tubes are called fuel rods. The finished fuel rods are grouped in special fuel assemblies that are then used to build up the nuclear fuel core of a power reactor.

Enrichment. The concentration of the fissionable isotope, 235U (0.71 percent in natural uranium) is less than that required to sustain a nuclear chain reaction in light water reactor cores. Natural UF6 thus must be "enriched" in the fissionable isotope for it to be used as nuclear fuel. The different levels of enrichment required for a particular nuclear fuel application are specified by the customer: light-water reactor fuel normally is enriched up to about 4 percent 235U, but uranium enriched to lower concentrations also is required. Gaseous diffusion and gas centrifuge are the commonly used uranium enrichment technologies. The gaseous diffusion process consists of passing the natural UF6 gas feed under high pressure through a series of diffusion barriers (semiporous membranes) that permit passage of the lighter 235UF6 atoms at a faster rate than the heavier 238UF6 atoms. This differential treatment, applied across a large number of diffusion "stages," progressively raises the product stream concentration of 235U relative to 238U. In the gaseous diffusion technology, the separation achieved per diffusion stage is relatively low, and a large number of stages is required to achieve the desired level of isotope enrichment. Because this technology requires a large capital outlay for facilities and it consumes large amounts of electrical energy, it is relatively cost intensive. In the gas centrifuge process, the natural UF6 gas is spun at high speed in a series of cylinders. This acts to separate the 235UF6 and 238UF6 atoms based on their slightly different atomic masses. Gas centrifuge technology involves relatively high capital costs for the specialized equipment required, but its power costs are below those for the gaseous diffusion technology. New enrichment technologies currently being developed are the atomic vapor laser isotope separation (AVLIS) and the molecular laser isotope separation (MLIS). Each laser-based enrichment process can achieve higher initial enrichment (isotope separation) factors than the diffusion or centrifuge processes can achieve. Both AVLIS and MLIS will be capable of operating at high material throughput rates.

The back end of the cycle is divided into the following steps:

(i) *Reprocessing.* Spent fuel discharged from light-water reactors contains appreciable quantities of fissile (U-235, Pu-239), fertile (U-238), and other radioactive materials. These fissile and fertile materials can be chemically separated and recovered from the spent fuel. The recovered uranium and plutonium can, if economic and institutional conditions permit, be recycled for use as nuclear fuel. Currently, plants in Europe are reprocessing spent fuel from utilities in Europe and Japan.

(ii) *Interim Storage.* After its operating cycle, the reactor is shut down for refueling. The fuel discharged at that time (spent fuel) is stored either at the reactor site or, potentially, in a common facility away from reactor sites. If on-site pool storage capacity is exceeded, it may be desirable to store aged fuel in modular dry storage facilities known as Independent Spent Fuel Storage Installations (ISFSI) at the reactor site or at a facility away from the site. The spent fuel rods are usually stored in water, which provides both cooling (the spent fuel continues to generate heat as a result of residual radioactive decay) and shielding (to protect the environment from residual ionizing radiation).

(iii) *Waste Disposal.* A current concern in the nuclear power field is the safe disposal and isolation of either spent fuel from reactors or, if the reprocessing option is used, wastes from reprocessing plants. These materials must be isolated from the biosphere until the radioactivity contained in them has diminished to a safe level. Under the Nuclear Waste Policy Act of 1982, as amended, the Department of Energy has responsibility for the development of the waste disposal system for spent nuclear fuel and high-level radioactive waste. Current plans call for the ultimate disposal of the wastes in solid form in licensed deep, stable geologic structures.

Natural Units

The fundamental components of materials are called the natural units. In this study, we explore the material world. Unlike the elusive energy, a material has a mass and occupies a volume. Thus, we can easily recognize a piece of material. Driven by energy, materials undergo constant changes, generating all kinds of phenomena. Natural units participate in these changes, but they themselves remain the same. Unlike physical and chemical phenomena that involve large quantities of material, nuclear phenomena usually deal with interactions among natural units. Thus, we shall review the investigations on natural units before we introduce any nuclear phenomena.

- What are natural units?
- How many kinds of natural units are there?
- How do natural units make up the material world around us?
- Have we found the true natural units of the material world?

The spirit of science, including nuclear science, is to dig for the ultimate truth. Looking for the ultimate fundamental natural units of the material world started with philosophical approaches followed by scientific investigations. The digging has gone on for almost 5 thousand years, and the dig continues. Quickly, we review the philosophical approaches and experiments leading to the discovery of chemical elements, atoms, molecules, moles, and electrons in the past.

The search for natural units had a slow start, but the results grow exponentially. Following the discovery of radioactive decay, experiments have revealed not only the structure of atom but also smaller sub-atomic particles such as electrons, protons and neutrons. We accepted them as fundamental natural units of matter for a while, but further investigations revealed that protons and neutrons are composed of several particles. These will be discussed in future chapters.

Philosophical Approaches on the Fundamental Natural Unit of Materials

Ancient people did not have the energy concept, nor did they understand the material world. They did not have concepts to perceive deeper into the material world than what it appeared to their eyes. To them, wood logs burned and disappeared, leaving nothing but a little ash behind. Soil, rock, clay, and sand appeared to be indestructible by fire. Water came from the darkened sky and caused floods and mudslides that kill. However, some wise men asked fundamental questions and suggested that there are some fundamental components for the material world. Their suggestions were accepted on faith rather than on reason or comprehension. Philosophical approaches from Oriental and Hellenistic cultures on the fundamental components material are briefly described here. People in the orient believed in the dualism of *Yin* and *Yang* whereas Plato proposed the *primal substances*.

Natural Units of Dualism

With a history of more than 5000 years, the Chinese culture is wide spread, particularly in the orient.

What is the ancient Chinese view of the material world?

One of the Chinese classics called *I-Ching* dominated the ancient Chinese view of the material world. A translation of it is known as *Book of Change*, which was written during the *Chu* dynasty (1134-247 BC), in cryptic language, which is used by most fortunetellers as their folly. However, it also has its academic value as a classic. A dominant symbol in the book, the *Ultimate* (*Tai-chi*) and the trigrams, is shown here. The *Ultimate* consists of *Yin* and *Yang*, from which everything is derived. *Yang* is the cosmic male principle and *Yin* is the cosmic female principle.

Yin and *Yang* were the suggested natural units of all materials. *Yang* is represented by a solid line — whereas *Yin* by a broken line — -. Combining equal portions of *Yang* and *Yin* results in four things. *Fu-Hsi* showed the formation of eight things when they are divided in three parts each can be *Yin* or *Yang*, and he represented them by the tri-grams. When a whole has many parts of *Yin* and *Yang* many more things are generated. When a whole has many parts of *Yin* and *Yang* more things are generated. Thus, natural units *Yin* and *Yang* are the fundamental building blocks of materials. The dualism of *Yin-Yang* represents a view of the world that does not distinguish energy from matter. This view still affects the ways many Chinese think, eat, exercise, and live.

Plato's Natural Units

Plato and Aristotle left their postulates regarding the material world in writing. They had no concept of energy, and their postulates did not differentiate matter from energy, like the dualism of *Yin* and *Yang*. Their intellectual exercises qualitatively described the changing world, and they dominated the reasoning process for a long time.

What are the natural units proposed by Plato?

Plato (427-347 BC) postulated that combinations of four *primal substances: air, water, fire,* and *earth* made up all materials. Plato hinted at the fifth unobservable substance, *ether* or *quintessence*, but it was not widely known. Thus, he suggested four or five natural units.

In his *Timaeus,* Plato associated the regular polyhedra with the *primal substances:* cube with earth, tetrahedron with fire, octahedron with air, icosahedron with water, and dodecahedron with ether. From a geometrical viewpoint, there are five and only five regular polyhedra. He thought these mathematical forms are the ultimate representation of the world. This postulate connected geometrical results with the physical world, a system uniting the (mathematical) minds and the material world. Polyhedra have definite shapes. Size and quantity concepts were not present during his time.

Plato's student Aristotle (384-322 BC) examined the ideas of *primal substances* from the viewpoint of causes and effects. He believed in having a limited number of *primal materia*, but added four qualities: *hot, cold, wet,* and *dry*. In his mind, a combination of these qualities and *primal substances* caused all the phenomena of the world. The *primal materia* are connected to the *qualities* in Aristotle's world, as shown in a diagram here. Little further development was made regarding his *qualities* in terms of science.

Plato applied his knowledge in geometry to interpret the physical world. Aristotle (Sarton, 1970) took it a step further to explain the changes. These early developments illustrated human desire to know the natural units of the material world.

Plato's doctrine influenced scientific reasoning for almost two thousand years.

For example, Angelo Sala (1575-1640) considered fermentation as a regrouping of the *primal substance*. Daniel Sennert (1572-1627) postulated four kinds of *atoms* (a concept not of finite size) corresponding to the four primal substances. Few scholares challenged Plato's doctrine until it failed to explain the existence of various kinds of *air*.

During the period between the 5th and 3rd centuries BC, theories of nature in India and other parts of the world were based on conceptions of *materials* called fire, wind, water, earth, space, and vitalism (animated atoms). The dualism of love and hate also played a part. However, water, fire, wind and earth were considered as natural units in many cultures, perhaps originated from Plato.

Chemical Elements and Atoms

After the time of Plato and Aristotle, human activities became more and more complicated. Enterprises of alchemists, medicine men, and technologists flourished. Due to their wealth, these people had the leisure to pursue scientific reasoning. Slowly, people acquired the concept of quantity and size through commerce, before they acquired the energy concept.

Attitudes of philosophers began to change. Plato and Aristotle still had influence but they no longer dominated the academic circle. For example, Francis Bacon (1561-1626) introduced methods of fact collection and deduction. R. Descartes (1596-1650) introduced a principle of doubt, accepting nothing that cannot be clearly established to be true. Galileo (1564-1642) invented experimental methods, and other scholars soon use them.

The attitude change in the 16th century made a huge difference in the study of the material world. Technologies such as air pump were available to study air. Various gases were identified, and natural units *primal substances* is replaced by chemical elements

Gas Identification and Boyle's Concept of Chemical Elements

During the 16th century, sophisticated tools began to emerge. People studied jars of various gases. Results of these studies led scholars to reject Plato's *primal substances* as natural units. New concepts began to emerge, and the same fundamental question was asked again.

What are natural units of material?

How many kinds of units are there?

A Belgian scientist J. B. van Helmont (1580-1644) confirmed that gas generated from burning charcoal was the same as that generated from burning alcohol, but different from air. Thus, there is more than one kind of *air* (gas). The rich Englishman Robert Boyle (1627-1691) also experimented with gases and air. He noticed the different properties of various gases (*air*) produced in his experiments. The behavior of various gases convinced him that *fire, air, water,* and *earth* were not natural units of material. He published *The Sceptical Chymist* in 1661 to demolish the doctrine of *primal substances*.

In the 17th century, this was the way scholars wrote and spoke. He considered himself a chemist, and from a chemist's point of view, there must be some natural units based on which all materials are derived. In his mind, these natural units are fundamental substances called element that cannot be decomposed or change by whatever means. From these elements, all other materials in the world must have been made up or derived. He defined the term chemical element or element. An element is a substance that cannot be decomposed into anything simpler. Since elements are fundamental components of materials, they are natural units. Unfortunate for Boyle, he had not identified an element as an example. Oxygen was one of the few substance first identified as an element, and its discovery will be given shortly. By the way, Boyle also discovered the Boyle's law of gases, with the help of his assistant Robert Hooke (1635-1703). Since air pumps were available to them Boyle discovered the pressure-volume relationship of a certain amount of gases at constant temperature. On the social side, Boyle pushed for the formation of the Royal Society of England that had a long lasting effect in promoting scientific activities. He, a scientist, had made an impact on his society for the development of a field he dearly loved.

Discovery of Oxygen as an Element

Fire occurred frequently in ancient time that led Plato to consider it a *primal substance*. Questions remain, however.

- What is fire?

  What causes a fire and what really happens in a fire?

  How can phenomena related to fire be explained?

While Boyle experimented with gases, German scientist J.J. Becher (1635-1682) studied fire. Becher suggested that combustible material contained an inflammable *phlogiston*, which was released during burning. After release of phlogiston the material was no longer flammable. He recognized that when metals were burned, the products became heavier, but he could not explain. At the beginning of the 18th century, G.E. Stahl suggested *phlogiston* a substance with negative mass. When lost, the residue became heavier.

The *phlogiston* theory of fire was widely accepted for almost a hundred years until the air was intensely studied. Among many scholars that studied air, J. Priestley (1733-1804) found that *nitrous air* (actually nitrous oxide NO) reacted with portions of *air* to produce a water-soluble *air*. He mixed 2 parts *nitrous air* and 5 parts ordinary air in a marked vessel that was inverted on top of water. The reaction took place at room temperature, and the product (NO_2) dissolved in water. As a result, the water level raised to a level indicating $^1/_5$th (by volume) of the *air* was consumed. He called the consumed portion *good air*. Priestly went to Paris in October 1774, revealed his results to Lavoisier, who was very interested in the *good air*.

Independently C.W. Scheele and J. Priestly made some *good air* by heating various metal salts. Lavoisier later found that charcoal, diamond, sulfur, phosphorus, mercury, lead, and tin burned in the *good air*. Many products from the burning produced acids when dissolved in water. These results indicated to him that *air* was not a single substance. Furthermore, he named the *good air* oxygen[11] from the Greek words sxuz, (acid) and gqinomai, (I beget). He further concluded that oxygen was a chemical element as prescribed by Boyle, because it could not be decomposed into simpler things. He denounced the *phlogiston* theory of fire and Plato's doctrine. He formulated the burning of metal as follows.

Metal + Oxygen → *Calx*

If you think *Calx* a metal oxide, you are absolutely right.

Additional Chemical Elements

The production of Oxygen by Scheele and Priestly and its confirmation as an element by Lavoiser made a giant step forward in science, but more questions are raised.

What other chemical elements are there and how can they be identified?

How many kinds of chemical elements are there? The discovery of an element started a field of study. Lavoisier*, influenced by his teacher Jean E. Guettard's (1715-1786) who was interest in identification of minerals, seized the opportunity and dived into his research. As a result, he had identified the most number of chemical elements. His book, *Traite Elment de Chimie*, listed some elements as defined by Boyle (Sidgwick, 1950; McKie, 1952), but he left room by saying that some *bodies* may be decomposed to simpler elements in future research. It is interesting to note that he listed *light* and *caloric* as elements. Apparently, he had no concept of energy. He divided other *simple bodies* into metals and non-metals. We now know his *earthy substances* as compounds, but his metals and nonmetals are true chemical elements.

Dalton's Atomic Theory

The discovery of chemical elements by Lavoisier excited scientists at that time. This fresh field of study on chemical element attracted brilliant minds. One of the fundamental questions is.

What are natural basic units of chemical elements?

Other scholars examined Dalton's original theory as they reinterpreted natural changes based on the atomic model. Some are skeptical because the indivisible hard-sphere model could not explain the elasticity and thermal expansion of substances. However a chemical reaction can be viewed as a re-group of atoms. Elements and thus all materials consist of natural units called atoms.

Applying his atomic theory, Dalton further discovered the law of partial pressure of gases. In a container with many gases, the pressure due to a particular gas is called the partial pressure. He reasoned that the pressure of a fixed volume of gases was proportional to the number of atoms present. Thus the partial pressure was the same as the pressure if the container has no other gas. The law of multiple proportions – two elements combine with each other in some simple proportions is also to his credit, because they may combine in more than one way.

The Periodic Table of Chemical Elements

During the 17-18th centuries, more elements have been discovered, their melting points, densities, atomic weights, heat capacities, flame colors and other properties determined. The study of elements moved into a new phase.

- What is the correlation among properties of chemical elements?
- How can information on chemical elements be organised, classified and correlated for easy reference and access?

The effort to find correlation among chemical elements and to organise property information resulted in the discovery of periodic tables of chemical elements. These tables are tools for easy reference in our study of elements, and they are the foundation for experimental and theoretical chemistry.

As early as 1816, Dobereiner noticed some *triads* (three elements) with similar chemical properties. For example lithium, sodium, and potassium react with water to produce bases, so do calcium, strontium, and barium, but this group of bases is not very soluble. Hydrogen compounds of chlorine, bromine, and iodine are acidic. A few decades later, Dmitry I. Mendeleyev (1834 -1907) of Russia and Lothar Meyor (1830-1895) of Germany independently classified chemical elements. Mendeleyev made a bigger impact because he used his periodic tables to predict the existence of some elements unknown at that time. He arranged elements in his table in increasing order of atomic weights.

Atomic weights were derived from their relative amounts in reaction with oxygen, which was given an atomic weight of 16, so that the lightest element hydrogen has an atomic weight of 1.0. These were known as atomic weights in a chemical scale, which considered all atoms of an element identical. Today, the atomic mass is the mass compared to the mass of C, which is given as 12 exactly.

Periodic tables guided the search of new elements, served as a framework showing the relationship among elements, confirmed a substance as an element, and guided the synthesis of new elements (to be discussed in the Chapter on Nuclear Reactions). Periodic tables are tools as well as systems of chemical information classification.

Periodic tables are references to correlate chemical properties. For example, fluorine, chlorine, bromine, and iodine have very similar chemical properties, because they are in the same group. They react with most metals, especially with lithium, sodium, potassium, and rubidium forming binary salts with a stoichiometry similar to that of NaCl.

The similarity in chemical properties for each group is astonishing. Research to find the reason for this recurrence of chemical properties sizzled. Periodic tables changed the chemistry from practical work of analysis and identification to highly theoretical sciences.

Two most important elements in nuclear technology are hydrogen and uranium, as lightest and heaviest natural elements. Both were on Mendeleyev's periodic table. Uranium (U) was discovered in 1789 by Martin Heinrich Klaproth (1743-1817), a quantitative analytical chemist in Germany. He isolated a black powder from pitchblende and named it uranium after the planet Uranus. However, the black powder was actually uranium oxide. E.M. Peligot (1811-1890) reduced uranous chloride (UCl_4) to uranium metal in 1841. By the way, uranium is rather abundant on Earth.

Today, there are many forms of periodic table of chemical elements. The most common one is given on the cover of this booklet, because it is very useful in our discussion of nuclear technology. A real long form is given here, but the heavy synthetic elements are omitted. These long periodic tables arrange the elements according to their electronic configurations giving by quantum mechanical calculations.

Natural Units in Chemical Reactions

For some elements, atoms are natural units in chemical reactions. For them,

Dalton's atomic theory nicely explains chemical reactions. We are going to talk about atoms and ions first. For some chemical reactions, atoms and ions are the natural units.

However, many substances exist in a form more complicated than single atoms in their natural states of gas, liquid, or solid. Dalton did not realise this. Thus, he was frustrated for his failure to use atomic theory to explain some observations in chemical reactions. The challenge to explain these observations was a good opportunity for new theories to hypothesize natural units of substances at their natural states. This called for a radical thinking, and Avogado had brilliant idea He suggested molecules as the natural units. An additional natural unit called mole is derived from an artificial unit of atomic weight.

Atoms and Ions as Natural Units

Dalton's theory was a speculative approach by proposing that elements consist of particle-like atoms. His theory has to be scrutinised.

- What chemical evidences support the atomic theory?

  What chemical reactions can be explained by atomic theory?

At constant pressure, the volume of a gas is proportional to the quantity of gas. Certain metals reacted with acids to produce hydrogen gas. For a metal, the amount of hydrogen produced is proportional to the quantity of metal used. These observations support the atomic theory, at least in that all atoms of an element are the same.

Heating up elements iron and sulfur powders together results in a loss of the properties of both elements. The product was very different from either iron or sulfur. Another element hydrogen reacts with the element chlorine, and the product dissolves in water. The solution is acidic due to H^+ and Cl^- ions. Dalton's symbols for elements can be used to write the following equations to represent the reactions:

$$Fe + S = FeS$$
$$H + Cl = H^+ + Cl^-$$

Furthermore, some other chemical reactions can also be represented in a similar way. The interpretation is consistent with the idea that atoms are natural units in chemical reactions, but as you will see, if you are not already aware of, later that the explanation and representation are only partially true.

We shall discuss chemical reactions driven by electricity later. However, the fact that electricity induces chemical reactions suggested the presence of charge species called ions. The H^+ + Cl^- ions just mentioned serve as examples.

Molecule and Mole as Natural Units

According to Dalton's partial pressure theory, at the same temperature and pressure, equal volumes of gas contain the same number of atoms. Yet, 2 litters of hydrogen and 1 litter of oxygen reacted completely and produced two liters of water vapor when measured at the same temperature and pressure, observed Gay-Lussac. Dalton used H and O to represent elements hydrogen and oxygen respectively and the reaction equations were unsatisfactory to depict the volume relationship of the reaction.

$2\,H + O = 2\,HO$

$2\,H + O = H_2O$ (does not agree with volume measured)

$H + O = HO$ (does not agree with volume measured)

The atomic theory also failed to explain why 1 liter of some oxygen compounds such as nitrous oxide (NO) is lighter than 1 liter of oxygen gas at the same temperature and pressure.

- What is incorrect with Dalton's view of atoms?

  How do you explain these reactions and phenomena? (a challenge)

Dalton had the idea that at the same temperature and pressure, equal volumes of gases contain equal number of *particles* or atoms. Naturally, he expected oxygen to combine with another element to form a denser or heavier compound. He puzzled that carbonic oxide (actually carbon monoxide CO) was less dense than oxygen gas, but he did not pursue the matter further. Gay-Lussac also observed that when held at the same temperature and pressure, a volume of oxygen reacted completely with twice the volume of carbonic oxide. Solving these puzzles was a challenge, and Avogadro made a brilliant proposal for their explanations.

Amedeo Avogadro (1775-1856) accepted the hypothesis that at the same temperature and pressure, equal volumes of gases contain the same number of *particles*. He proposed, however, that these *natural units* are molecules rather than single atoms.

By considering the particles as molecules, Avogadro had an explanation for the whole-number volume relationship. He assumed hydrogen and oxygen molecules to have two atoms, and represented Gay-Lussac's experiments by these equations

$$1 \text{ vol. } O_2 + 2 \text{ vol. } H_2 \rightarrow 2 \text{ vol. } H_2O$$

and

$$2\,CO\,(g) + O_2\,(g) \rightarrow 2\,CO_2\,(g)$$

respectively. The coefficients agree with the volumes for gaseous substances at the same temperature and pressure. Avogadro's molecules are the natural units of substances as they appear in their states. Molecules are made up of atoms, bonded together via chemical bonds due to electric attraction. The diagrams here illustrate reactions at the molecular level. The introduction of molecules also introduced the concept of molecular weight, which is the sum of the atomic weights in the molecule. For convenience, an amount of any substance equal to the molecular weight in grams is called a mole. For example, one mole is 2.016 g H_2, 32.00 g O_2, 18.01 g H_2O, 28.00 g CO, 44.0 g CO_2 or 17.01 g] NH_3. Thus, mole is a natural unit derived from the artificial unit of gram in combination with atomic or molecular weight. Despite the different weights, one mole of any substance contains the same number of molecules, and this number is called Avogadro's number, 6.0221367×10^{23}, or 602213670000000000000000. This is a huge number! Such a number of atoms or subatomic particles is also called a mole. Study of quantitative relationships is called chemical stoichiometry, for example, 2 g of hydrogen will react with 16 g of oxygen. The mass relationships appear very strange, but in terms of natural units (atoms, molecules and moles) the relationships are logical and simple.

Natural Units in Electricity

Around 1800 Volta invented batteries. These devices are based on chemical or electrochemical reactions, which transfer electric charge through a wire or load. Other reactions also transfer electric charge but among atoms or groups of atoms. Driven by energy, transfer of electric charge regroups atoms to form new molecules. Both energy and electric charge are important factors in chemical reactions. At the turn of 19th century, people had the concept of quantities. Thus, they try to figure out the quantitative relationship between the illusive electricity and reactants or products. They not only quantify electricity they discovered the particles responsible for electric charges and named them electrons. This discovery led to a much better description of chemical reactions and other phenomena. Chemical reactions combine two or more of the 92 natural occurring elements on the periodic tables, making the millions of compounds in this world. In this part, we take a look on the early research on electricity leading to the discovery of electrons, and summarize the fundamental properties of electrons at the end.

The Coulomb Interaction — Electrostatic Force

Electrical attraction and repulsion causes many phenomena such as lightening and neutralization (chemical) reactions. The stories in the study of electricity are interesting, and the people involved had a rewarding career. Like other things, the beginning is always difficult, because electric charge, like energy, has no mass, size, shape or color. At the start, they asked these questions.

What are Electrostatic Charges?

Humans have noticed electricity for a long time, but the first detector was an electroscope invented by W. Gilbert (1544-1603), similar to the ones shown here. It consists of a pair of thin gold leaves suspended from an electrical conductor that leads to the outside of an insulated container. Touching the conductor with a charged rod causes the leaves to open up due to the repulsion of charges on the leaves. Using electroscopes, Charles du Fay (1698-1739) recognized two kinds of electricity. He rubbed a glass rod against a resin, and found them charged. He called the charge on glass *vitreous* and the charge on resin *resinous*. These charges are now called positive and negative. By the way, electroscope played an important role for the detection of radioactivity, and this will be explained later. For the study of electric forces, Charles-Augustin de Coulomb (1736-1806) invented a very sensitive torsion balance. The balance allowed him to measure very small forces exerted by tiny charged particles. He published his findings in 1785-89, and the result is now known as the Coulomb's law. According to this law, the attraction or repulsion force F is proportional to the product of the charges, q_1 q_2, and inversely proportional to the square of their distance of separation, r^2,

$$F = C\, q_1\, q_2\, r^{-2}.$$

Two different kinds of charge q_1 and q_2 give an attractive force (negative F) but a repulsive force (positive F) if q_1 and q_2 are the same kind. In his honor, the SI unit for charges is Coulomb, symbol C. The coulomb is defined as the quantity of electricity transported in one second by a current of one ampere. Two equal charges of 1 C separated by a distance of 1 m exert a force of 8.987551787×10^9 N. Thus, the Coulomb constant is 8.987551787×10^9 N m^2 C^{-2}. Of course, the constant depends on the units used. Forces due to electrostatic interaction are referred to as Coulomb interaction, together with gravitational interaction are two of the four fundamental forces of interaction. Coulomb's law and Newton's law of gravity have some similarities as shown in the boxes. The constants corresponding to SI units are given.

Electrochemical Reactions for Batteries

You already know that electric and chemical energies inter-convert into each other since the beginning of the universe. However, using electrochemical reactions to provide electric energy started only at the turn of 19th century. However, the invention of batteries was based on these reactions and batteries greatly helped scientific explorations, including the study of chemistry.

- How does electricity cause materials to change?

  Can chemical reactions produce electricity?

- What is the quantitative relationship between electric charge and quantities of chemicals reacted or produced?

Galvani, Luigi (1737-1798) studied medicine, but he taught at university and carried out research. He was fascinated by the electric effect on tissues and muscles. He discovered that electric charge caused muscles to twitch, and more interestingly two different pieces of metal touching muscles caused an electric current to flow.

Volta, Alessandro Giuseppe Antonio Anastasio (1745-1827) was Galvani's friend, and they share the interest in electricity. One day, Volta touched his tongue with two pieces of different metals and he experienced a bitter taste. Dissimilar metals touching the eye gave him a sensation of light. He tested the effects of different metals in the series: zinc, tin, lead, iron, copper, platinum, gold, silver, graphite, and charcoal. Later Volta found dissimilar metals placed in a salt solution caused electric charge to flow from one to another when connected. His applied his observations to make voltaic piles (batteries). He noticed the chemical reactions for electricity generation, and he connect electrochemical cells together to produce high voltages and these were known as voltaic piles. Today, they are called batteries.

Voltaic piles provided high voltage for scientific experiments, especially for chemical reactions caused by electric current. William Nicholson (1753-1815) observed bubbles forming on the surfaces of metals submerged in water when they are connected to a voltaic pile. Water was electrolyzed to produce hydrogen and oxygen. Application of voltaic piles in experiments allowed Humphry Davy (1778-1829) to observe hydrogen, metals, and metallic oxides depositing on the electrodes. Electric current.(flow of electric charge) decomposed his minerals into bases and acids. Since electricity causes chemical reactions, Davy applied electrolysis to obtain potassium and sodium metals from their hydroxides melts. Around the 1800s, voltaic piles, electrolysis, and electrochemistry were hot topics of research.

The Natural Unit — Faraday Constant

As mentioned earlier, electric charge has neither mass nor volume, and it was difficult to characterize. Yet, the fundamentals of electricity are worth investigating.

- What is the quantitative relationship between charge and reactants or products?

  What is responsible for electric charge?

Benjamin Franklin (1706-1790), who helped writing the Declaration of Independence, conducted electric experiments. In 1752, he showed that lightening was an example of electric conduction by a kite during a thunderstorm. Franklin enunciated a law stating that the (net) charge within an isolated system is a constant. Electric charge is always conserved. Franklin considered that electricity is associated with some sort of fluid that is present in all materials. He also called the two types of electricity positive and minus charges.

Faraday (1791-1860) discovered chemical reactions caused by electricity, and he further investigated electrolysis. For example, he set up electrolytic cells similar to the one shown in the diagram. He measured the amount of charge flew through the cells and the quantities of products deposited on the electrodes. This allowed him to study quantitative relationship between masses deposited on electrodes and charges. The mass of an element deposited on an electrode by electrolysis (of a solution containing a compound of that element) depends on atomic weight and the total electric charge that passes through the solution. "*There is a certain absolute quantity of the electric charge associated with each atom of matter*", he wrote.

Faraday laid the foundation for the quantitative relationship in electrochemical reactions. The relationship is now well established. Passing a charge of 96485.309 coulomb through the cell will deposit 108 g (1 mole) of silver (Ag), or 1.008 g (also 1 mole) of hydrogen in an electrolysis experiment. This amount of charge is called the Faraday constant F in his honor.

The Electron Particle with a unit charge

Using electrolysis experiments, Faraday showed quantitative relationship between charges and reactants or products, but the reasons for electrolytic reaction and origin of electric charge were unknown.

- Why does charge cause chemical reactions? What is the identity and origin of charge?
- Are electric charges associated with any particle in matter?

  What are these particles?

With battery power and air pumps available to William Crookes (1832-1919), he attached high a voltage to two electrodes in a partially evacuated tube as shown here. He saw an invisible beam of rays emitted from the cathode traveling in a straight line to the anode, and called them cathode rays. He made the rays visible by placing a fluorescence screen on its path. In 1879, Crookes found strong indications that the cathode rays are negatively charged particles.

In 1897 J.J. Thomson: deflected the cathode rays using electric and magnetic fields. He concluded that cathode rays are composed of corpuscles and called them electrons. He further suggested that electrons are present in all kinds of metals. He determined the charge to mass ($^{e}/_{m}$) ratio for electrons. After having studied the mobility of positively charged ions in solution during electrolysis, he concluded that the amount of charge on the electron is the same as that on the hydrogen ion, H^+, is, but of different nature. He guessed the mass of the electron to be about $^1/_{2000}$ of that of a hydrogen atom, in agreement with the value known today $^1/_{1836}$.

Thomson's discovery indicated that electrons were part of atoms, and that they could be separated from atoms, contradicting Dalton's theory that atoms were indivisible.

By the turn of the 19th century, Thomson's discovery of electrons was widely known. For his achievement he received the Nobel Prize for physics in 1906, and knighthood in 1908.

Reasoned that an electron must carry the smallest amount of charge, Robert A. Millikan (1868 - 1953) designed experiments tried to measure the smallest amount of charge attached to an oil drop. He measured the speeds of oil drops in a special chamber, and calculated their sizes from the known gravitational field and air resistance. He then turned on the electric field of his oil drop chamber, and measured their new speeds. Some droplets abruptly changed their speeds by certain constants, showing the capture or loss of an electron. The force F experienced by the oil drop should be equal to the product of the charge q and the field E, $F = q E$. He determined the smallest amount of charge from the sudden change of speed, and suggested that as the charge of an electron. Some of the oil drops have charges as multiples of the smallest charge.

His measurements and calculations established the charge carried by an electron within 0.6 percent of the current value. Later measurements of the Avogadro's number and the Faraday constant F gave the charge on an electron of 4.80324×10^{-10} esu or $1.602189(5) \times 10^{-19}$ Coulomb. Millikan's experiment showed that the smallest natural unit of charge is that of an electron, and for it he was honored with a Nobel Prize for physics in 1923.

The discovery of the electron enables us to explain and write the chemical equations for electrolysis of water as shown in the box here. The battery (or DC power supply) *pumps* the electrons from one electrode to another forcing chemical reactions to take place at the electrodes. Thus, hydrogen and oxygen gases are produced at the electrodes.

Properties of Electrons

J.J. Thomson and Millikan have determined the fundamental properties of electrons. Due to change of standards and refined measurements, these values change slightly over the years. Up to date properties of electron are summarized in a Table here.

The rest mass and charge are basic properties of electron, and these are given in various units. The charge of an electron is the natural unit of charge. Additional properties are spin and magnetic moment.

In 1896, the Dutch physicist P. Zeeman (1865-1943), shared physics Nobel Prize with H.A. Lorentz in (1902-??) observed the broadening of the sodium D-lines when the source is placed in a magnetic field. The theoretical explanation is provided almost 30 years later by quantum mechanics. George E. Uhlenbeck (1900-1988) and Samuel A. Goudsmit (1902-1978) proposed that the electrons might have an intrinsic angular momentum or spin. In other words, the electron has two (quantum) states which are represented by the quantum number $s = +½$ or $-½$ (In general the electron is said to have a spin of ½). The intrinsic magnetic moment associated with the angular momentum (spin) for the electron is 9.284770×10^{-24} J/tesla or 1.0011596522 Bohr magnetons. These are some of the fundamental natural constants.

The spin of electron contributes to its intrinsic magnetic moment. The magnetic moment in tern contributes to the magnetic properties of many materials. As a result, electron spin plays an important role in electron-electron interactions.

Today, sophisticated cathode ray tubes are built for X-ray generation, TV tubes, computer monitors, electron microscopes, *etc*. In these instruments, electrons are forced to leave the surface of a hot filament and accelerated by high voltages in the grid to form electron beams. By varying the intensity of the beam and deflected by electric and electromagnetic fields, electron beams create various images on the TV screens and computer monitors. The images tell stories and communicate ideas. The inventions of television and computer have changed the way we live. In the 1990s, our lives are dependent on electronic equipment.

REFERENCES

Atkinson, R. E., and Houtermans, F.G. (1929) *Z. Physic.* 54, 656-665.

Bethe, H.A. (1967), *Energy production in stars*, in *Nobel Lectures in Physics*, published for the Nobel Foundation in 1972 by Elsevier.

Bromberg, J.L., (1982) *Fusion Ä science, politics and invention of a new energy source*, MIT Press.

Cornwell, J.B. and Pendergrass, J.H., (1985), *Fusion Technology*, 8, 1861.

Fleischmann, M. and Pons, S. (1989) *J. Electroanal. Chem.*, 261, 301-308.

Frank-Kamenetskii, D.A. (1972), *Plasma - the forth state of matter*, Translated from Russian by J. Norwood Jr. Plenum Press.

Huizenga, J.R., (1992) *Cold fusion, the scientific fiasco of the century*, Univ. of Rochester Press.

Jones, S., Palmer, E.P., Czirr, J.B., Decker, D.L., Jensen, G.L., Thorne, J.M., Taylor, S.F., and Rafelski, J., (1989), *Nature* 338, 737

Lawson, J.D. (1957), *Proc. Phys. Soc.* (London) B70, 6

Libowitz, G.G., (1965), *The solid-state chemistry of binary metal hydrides*, Benjamin.

Mallove, E.F., (1992), *Fire from ice*, John Wiley and Sons.

Pagel, B., (1965), *New Scientist*, 8 April 1965.

Peat, F.D., (1989), *Cold fusion - the making of a scientific controversy*, Comtemporary Books.

Post, R.F., (1970), *Ann. Rev. Nucl. Sci.* 20, 509.

Rose, D.J. and Clark, M. Jr. (1961), *Plasmas and controlled fusion*, MIT Press, 1961 (QC718.R6.,

Rosenbluth, M.N. and Rutherford, P.H., (1981), *Tokamak plasma stability*, in "*Fusion* Vol. 1.part A", edited by E. Teller, Academic Press.

Bibliography

Achinstein, P. (1968), *Concepts of Science*, Baltimore: John Hopkins University Press, Chs. 7 & 8.

Achinstein, P. (1992), 'Inference to the Best Explanation: Or, Who Will Win the Mill-Whewell Debate', *HistPhilSci* 23, p. 349.

Achinstein, P. and O. Hannaway (1985), *Observation, Experiment and Hypothesis in Modern Physical Science*, Cambridge Mass: MIT Press.

Adler, J.E. (1990), 'Conservativism and Tacit Confirmation' *Mind*, pp. 559-70.

Agar, M., Hobbs, J. (1982), Interpreting discourse. In *Discourse Processes,* 5 pp. 1-32.

Albrecht, T.R., Akamine, S., Carver, T.E., and Quate, C.F. (1990) Microfabrication of cantilever styli for the atomic force microscope. J. Vac. Sci. Technol. A 8(4), 3386-3396

Albrecht, T.R., Grütter, P., Horne, D., and Rugar, D. (1991) Frequency modulation detection using high-Q cantilevers for enhanced force microscope sensitivity. J. Appl. Phys. 69(2), 668-673

Allara, D.L. (1996), Nanoscale structures engineered by molecular self-assembly of functionalized monolayers. In *Nanofabrication and biosystems*, ed. Hoch et al. Birge, R.R. 1995. Protein based computers. *Sci. Am.* (Mar.):90-95.

Allen, J. F., Hendler, J., and Tate, A., editors (1990a), *Readings in Planning*. Morgan Kaufmann, San Mateo, California.

American Board of Criminalistics (ABC), GKE Examination Philosophy. http://www.criminalistics.com/ABC/abc002.htm

American Wind Energy Association. (1992), *Wind Energy for Sustainable Development.* Washington, DC: American Wind Energy Association; 22 pp.

American Wind Energy Association. (1993), *Small Wind Energy Systems Applications Guide.* Washington, DC: American Wind Energy Association; 8 pp.

American Wind Energy Association. (1994), *American Wind Energy Association's 1994 Membership Directory*. Washington, DC: American Wind Energy Association; 42 pp.

Angel, R (1980), *Relativity: The Theory and its Philosophy*, Oxford: Pergamon Press, ch.5.

Anne Eisberg, "Benign Viruses Shine on the Silicon Assembly Line," New York Times, February 12, 2004.

Antonio Regalado (2004), "Nanotechnology Patents Surge as Companies Vie to Stake Claim," Wall Street Journal, June 18, p. 1.

Araj, K.J.; Fisher, T.A.; Kronenburg, J.C. (National Research Council, Washington, DC), (1991), *Assessment of Research Needs for Wind Turbine Rotor Materials Technology*. DOE/AL/58181-T1. Washington, DC: U.S. Department of Energy; 108 pp. (NTIS no. DE92018563),

Arlazarov, V.L., Adelson-Velsky, G. M., and Donskoy, M. V. (1975), Some methods of controlling the tree search in chess programmes. *Artificial Intelligence*, 6(4):361-371.

Armin Rahn (1993), *The Basics of Soldering, 1st ed.*, John Wiley and Sons (New York).

Arons, Arnold. (1993). Guiding insight and inquiry in the introductory physics laboratory. *Physics Teacher*. 31(5), p. 278-82.

Aronson, J., Harré, R. & Way, E. C. (1993), *Realism Rescued*, London: Duckworth.

Ashbaugh, D. R., *Quantitative-Qualitative Friction Ridge Analysis: An Introduction to Basic and Advanced Ridgeology*, CRC Press, Boca Raton, FL, 2000.

Ashwill, T.D.; Leonard, T.M. (1986), *Developments in Blade Shape Design for a Darrieus Vertical Axis Wind Turbine*. SAND-86-1263. Albuquerque, NM: Sandia National Laboratories; 21 pp.

Auld, B.A. (1973), *Acoustic Fields and Waves in Solids*, John Wiley & Sons (New York).

Ayer, A. J. (1972), *Probability and Evidence*, London: Macmillan, Pt. I.

Ayer, A.J. (1957), 'The Conception of Probability as a Logical Relation' in *Observation and Interpretation: Proceedings of the Ninth Symposium of the Colston Research Society*, S. Körner (ed.), Butterworth: London.

Babbie, E. (1989), *The Practice of Social Research*, 5th ed. Belmont, CA: Wadsworth.

Babcock, B.A.; Conover, K.E. (R. Lynette & Associates, Inc.), (1994), "Design of Cost-Effective Towers for an Advanced Wind Turbine." *Wind Energy 1994*. SED-Vol. 15. New York, NY: American Society of Mechanical Engineers; pp. 261-268.

Bachand, G. and C.D. Montemagno (2000), Constructing organic/inorganic NEMS devices powered by biomolecular motors. *Journal of Biomedical Microdevices.* 2, No.3, 179-184.

Bagno, E. and Eylon, B.-S. (1997). From problem solving to a knowledge structure; an example from the domain of electromagnetism. *Am. J. Physics* 65(8): 726-736.

Ball, P. (2002), 'Tutorial: Natural Strategies for the Molecular Engineer', *Nanotechnology*, 13, 15-28.

Barnard, J.C. (1990), *An Evaluation of Three Models Designed for Siting Wind Turbines in Areas of Complex Terrain.* PNL-7357. Richland, WA: Pacific Northwest Laboratory; 50 pp. (NTIS no. DE90013625),

Barnard, J.C.; Wegley, H.L. (1987), *A Preliminary Evaluation of the Performance of Wind Tunnel and Numerical Modeling Simulations of the Wind.* PNL-6105. Richland, WA: Pacific Northwest Laboratory; 20 pp. (NTIS no. DE87004634),

Barnard, J.C.; Wegley, H.L.; Hiester, T.R. (1985), *Improving the Performance of Mass- Consistent Numerical Models Using Optimization Techniques.* PNL-5566. Richland, WA: Pacific Northwest Laboratory; 33 pp. (NTIS no. DE86001224), An article based on this report appeared in the Journal of Climate and Applied Meteorology (26:6); pp. 675-686.

Barnwell, R.W.; Hussaini, M.Y. (1992), *Natural Laminar Flow and Laminar Flow Control.* New York, NY: Springer-Verlag; 411 pp.

Barrangán Muñoz, J. M. (1997), Medio ambiente y desarrollo en las áreas litorales: Guia práctica para la planificación y gestión integradas. Barcelona, OIKOS-TAU, 160 pp.

Bauccio, M. (ed.) (1993), *ASM Metals Reference Book, 3rd ed.*, ASM International (Materials Park, OH).

Bauccio, M. (ed.) (1994), *ASM Engineering Materials Reference Book, 2nd ed.*, ASM International (Materials Park, OH).

Becker, M.F., J.R. Brock, H. Cai, N. Chaudhary, D. Henneke, L. Hilsz, J.W. Keto, J. Lee, W.T. Nichols, and H.D. Glicksman (1997), Nanoparticles generated by laser ablation. In *Proc. of the Joint NSF-NIST Conf. on Nanoparticles.*

Beer, F.P., Johnston, Jr., E.R. (1981), *Mechanics of Materials*, McGraw-Hill, Inc. (New York).

Ben-David, Joseph and Randall Collins (1966), "Social Factors in the Origins of a

New Science: the Case of Psychology." *American Sociological Review*. Vol. 31 pp 451-465.

Ben-Menahem, Y. (1990), 'The Inference to the Best Explanation' *Erkenntnis*, pp. 319-344

Bergström, L. (1984), 'Underdetermination and Realism', *Erkenntnis* 21, pp.349-365.

Berkowitz, A.E., J.R. Mitchell, M.J. Carey, A.P. Young, S. Zhang, F.E. Spada, F.T. Parker, A. Hutten, and G. Thomas (1992), *Phys. Rev. Lett*. 68:3745.

Berliner, H. J. and Ebeling, C. (1989), Pattern knowledge and search: The SUPREM architecture. *Artificial Intelligence*, 38(2):161-198.

Bigelow, J. & Pargetter, R. (1990), *Science and Necessity*, Cambridge: CUP, Ch. 5.

Black, M. (1966), 'Notes on the Paradoxes of Confirmation', in Hintikka, J & Suppes, P (eds.), *Aspects of Inductive Logic*, Dordrecht: Reidel.

Blackburn, B. (1993), 'Truth, Realism and the Regulation of Theory', in *Essays on Quasi-Realism*, Oxford: OUP.

Bohr, N (1949), 'Discussions with Einstein on Epistemological Problems in Atomic Physics', in Schillpp, P A (ed.), *Albert Einstein: Philosopher-Scientist*, La Salle: Open Court.

Boyd, R. (1979), 'Metaphor and Theory Change: What is "Metaphor" a Metaphor For?', in A. Ortony (ed.), *Metaphor and Thought*, Cambridge: CUP.

Boyd, R. (1980), 'Scientific Realism and Naturalistic Epistemology', *PSA* Vol. 2, pp. 613-662.

Boyd, R. (1989), 'What Realism Implies and What It Does Not', *Dialectica* 43.

Boyd, R. (1990), 'Realism, Approximate Truth, and Philosophical Method' in C. Wade Savage (ed.), *Scientific Theories, MinnStud*.

Brady, G.S., et al. (ed.) (1997), *Materials Handbook, 14th ed.*, McGraw-Hill (New York).

Braithwaite, R. (1953), *Scientific Explanation*, Cambridge: CUP, Ch. 2.

Braithwaite, R.B (1959), *Scientific Explanation*, Cambridge: Cambridge University Press, ch.9.

Brandes, E.A., et al. (ed.) (1992), *Smithells Metals Reference Book, 7th ed.*, Butterworth-Heinemann (UK).

Bransford, John D., Brown, Ann L., and Cocking, Rodney R. (editors). (1999). *How People Learn: Brain, Mind, Experience, and School*. Washington, DC: National Academy Press.

Bremner, A.A. (1990), 'Holism a Century Ago: The Elaboration of Duhem's Thesis', *Synthese*, 83, 3, pp.325-336.

Bridgman, P.W. (1960), *The Logic of Modern Physics*, New York: Macmillan.

Brink, C. (1989), 'Verisimilitude: Views and Reviews' , *History and Philosophy of Logic* 10, pp. 181-201.

Bromwich, M, R., *et al.*, *USDOJ/OIG Special Report, The FBI Laboratory: An Investigation into Laboratory Practices and Alleged Misconduct in Explosives-Related and Other Cases*,1997. http://www.usdoj.gov/oig/fbilab1/fbil1toc.htm

Brown, H.I. (1987), 'Naturalizing Observation' in Nersessian, N.J. (ed.), *The Process of Science*, Dordrecht: Kluwer.

Brown, H.I. (1987), *Observation and Objectivity*, New York: OUP.

Brown, J.R. (1982), 'The Miracle of Science', *PhilQuart* 32, pp. 232-44.

Brown, J.R. (1985), 'Explaining the Success of Science' *Ratio* 27, p. 49.

Brown, J.R. (1986), 'Thought Experiments Since the Scientific Revolution', *ISPS* 1, pp. 1-15.

Brown, J.R. (1988), 'The Experimenter's Social Circle', *Philosophy of Social Sciences* 18, pp. 101-6.

Brown, J.R. (1989), *The Rational and the Social*, London: Routledge, Ch. 4.

Brown, J.R. (1991), *The Laboratory of the Mind: Thought Experiments in the Natural Sciences*, London: Routledge.

Browne, Lois; Blackburn, Edward. (1999). Teaching introductory organic chemistry. A problem solving and collaborative-learning approach. *Journal of Chemical Education*. 76(8), p. 1104-07.

Buck, J.W.; Renne, D.S. (1985), *Observations of Wake Characteristics at the Goodnoe Hills MOD-2 Array*. PNL-5565. Richland, WA: Pacific Northwest Laboratory; 77 pp. (NTIS no. DE85018282),

Budinski, K.G., Budinski, M.K. (ed.) (1999), *Engineering Materials: Properties and Selection, 6th ed.*, Prentice Hall (Upper Saddle River, NJ).

Bunge, M. (1963), *The Myth of Simplicity*, Englewood Cliffs: Prentice-Hall.

Bunge, M. (1973), *Method, Model and Matter*, Dordrecht: D.Reidel, Chs. V, VI, VII.

Bush, V. (1946), *Endless Horizons,* Washington. DC: Public Affairs Press.

Butterfield, C.P.; Musial, W.P.; Scott, G.N.; Simms, D.A. (1992), *NREL Combined Experiment Final Report-Phase II (draft),* NREL/TP-442-4807. Golden, CO: National Renewable Energy Laboratory; 32 pp. (plus appendices).

Caglayan, Sehvar. (1994). Effectiveness of an active method in teaching physiology. *Am. J. Physiol.* 267 (*Adv. Physiol. Educ.* 12): S81-86

Caldwell, N.H.M., Breton, B.C., & Holburn, D.M. (1997), Towards the Intelligent SEM. Electron Microscopy and Analysis, (Institute of Physics Conference Series Number 153), pp 53-56, Institute of Physics Publishing.(1997).

Caldwell, N.H.M., Breton, B.C., & Holburn, D.M. XpertEze (1997), A Knowledge-Based Approach to Scanning Electron Microscopy. Applications and Innovations in Expert Systems V (pp 127-140), SGES Publications. Callinicos, A. (1989), *Against Postmodernism.* Cambridge: Polity Press.

Callister, W.D. (1985), *Materials Science and Engineering: An Introduction*, John Wiley (New York).

Campbell, N.R. (1957), *Foundations of Science: The Philosophy of Theory and Experiment*, New York: Dover.

Canon, John. (1997). The constructivist learning environment survey may help halt student exodus from college science courses. *Journal of College Science Teaching*. 27(1), p. 67-71.

Capobianco, M., (1998), Discussion Document on EIA&ICZM. Tecnomare Report 636A0753-RAP-D100-004.

Caprio, M. (1994). Easing into constructivism. *Journal of College Science Teaching*. 23(6), p. 210-212.

Carnap, R. (1950), *Logical Foundations of Probability*, Chicago: UCP, Ch. 1, 2, 7.

Carnap, R. (1952), *The Continuum of Inductive Methods*, Chicago: UCP.

Carnap, R. (1956), 'The Methodological Character of Theoretical Concepts' in Feigl & Scriven (eds.), *MinnStud*, Minneapolis: University of Minnesota Press, pp.38-76.

Carnap, R. (1962), *The Logical Foundations of Probability*, London, Routledge.

Carnap, R. (1963), 'Carnap's Intellectual Autobiography' in Schillpp, P (ed.), *The Philosophy of Rudolf Carnap*, (The Library of Living Philosophers), La Salle: Open Court.

Carnap, R. (1967), *The Logical Structure of the World*, London: Routledge.

Carnap, R. (1968), 'Inductive Logic and Inductive Intuition' in Lakatos, I. (ed.), *The Problem of Inductive Logic*, Amsterdam: North-Holland, pp.258-314.

Carrier, M. (1991), 'What is Wrong with the Miracle Argument?', *HistPhilSci* 22, pp. 23-36.

Carroll, J.S. (2001), 'Social Science Research Methods for Assessing Societal Implications of Nanotechnology', in: M.C. Roco and W.S. Bainbridge (eds.), *Societal Implications of Nanoscience and Nanotechnology*, Dordrecht: Kluwer, pp. 188-192.

Carroll, J.W. (1994), *Laws of Nature*, Cambridge: Cambridge University Press.

Carsley, J.E., R. Shaik, W.W. Milligan, and E.C. Aifantis (1997), In *Chemistry and physics of nanostructures and related non-equilibrium materials*. ed. E. Ma, B. Fultz, R. Shull, J. Morral, and P. Nash. Warrendale, PA: TMS.

Carter, G.F., Paul, D.E. (ed.) (1991), *Materials Science and Engineering*, ASM International (materials Park, OH).

Carter, R.W.G., (1988), Coastal environments. An introduction to the physical, ecological and cultural systems of coastlines. Academic Press, London.

Cartwright, N. (1980), *How the Laws of Physics Lie*, Oxford: Clarendon Press, Ch. 3.

Cartwright, N. (1983), 'When Explanation Leads to Inference' *How the Laws of Physics Lie*, Oxford: OUP, Essay 5.

Cartwright, N. (1983), *How the Laws of Physics Lie*, Oxford: Clarendon Press, Ch. 1, 4

Cartwright, N. (1989), *Nature's Capacities and their Measurement*, Oxford: Clarendon Press.

Cartwright, N. (1994), 'Fundamentalism vs. the Patchwork of Laws' *PASS* 93.

Casey, Jr., H.C., and M.B. Panish (1978), *Heterostructure Lasers, Part A: Fundamental Principles*. (Academic Press. New York), pp. 1-9.

Cencini, C. and Varani, L., (1989), Degradation of coastal dune systems through anthropogenic action. In Coastlines of Italy. Ed. Paolo Fabri. American Society of Civil Engineers, New York, 55-69.

Cervantes M., (1998), Public/Private Partnerships in Science and Technology: An Overview. STI Review No. 23, OECD, Paris, 12 pp.

Cherif, Abour; Adams, Gerald;Cannon, Charles. (1997). Nonconventional methods in teaching matter, atoms, molecules and the periodic table for nonmajor students. *American Biology Teacher*. 59(7), p. 428-38.

Chi, Michelene T. H. et al. (1989). Self-explanations: how students study and use examples in learning to solve problems. *Cognitive Science* 13: 145-182.

Chihara, C. (1987), 'Some Problems for Bayesian Confirmation Theory', *BJPS* 38, pp.551-560.

Childs, S.E.; Hughes, P.S.; Saeed, A. (Atlantic Orient Corporation), (1993), "Development of a Dynamic Brake Model." *Windpower '93 Conference; July 12-16, 1993; San Francisco, California.* Washington, DC: American Wind Energy Association; pp. 366-373.

Christianson, Roger; Fisher, Kathleen. (1999). Comparison of student learning about diffusiona nd osmosis in constructivist and traditional classrooms. *International Journal of Science Education*. 21(6), p. 687-98.

Churchland, P. (1979), *Scientific Realism and the Plasticity of Mind*, Cambridge: CUP (Chs. 1 &2).

Churchland, P. (1985), 'The Ontological Status of Observables: in Praise of Superempirical Virtues', in Churchland, P. & Hooker, C. (eds.), *Images of Science*, Chicago: UCP.

Churchland, P. M. (1979), *Scientific Realism and the Plasticity of Mind*, Cambridge: CUP.

Churchland, P.M. (1990), 'On the Nature of Theories: A Neurocomputational Perspective', *MinnStud* XIV, Minneapolis: Univ. Minnesota Press, pp. 59-101.

Clark, R.N. (1985), "Wind-diesel Hybrid System for Water Pumping." *Proceedings of the Wind Energy Exposition and National Conference of the American Wind Energy Association.* Washington, DC: American Wind Energy Association; pp. 221–225.

Clark, R.N. (1989), "Stand-alone Wind Electric Water Pumping." *Proceedings of the European Wind Energy Conference, EWEC '89.* Amsterdam: Elsevier; pp. 650–654.

Clark, R.N.; Davis, R.G. (1991), "Performance Changes Caused by Rotor Blade Surface Debris." *Windpower '91 Conference; September 24-27, 1991; Palm Springs, California.* Washington, DC: American Wind Energy Association; pp. 470-474.

Cohen, L.J. (1970), *The Implications of Induction*, London: Methuen

Cohen, L.J. (1980), 'What Has Science to Do With Truth?' *Synthese* 45, p. 489.

Cole, D. (1984), 'Thought and Thought Experiments', *Philosophical Studies*, 45, pp.431-444.

Coleman, C. (Northern Power Systems), (1990), *Wind Hybrid System Research and Testing: Telecommunications Systems.* SERI/STR-257-3686. Golden, CO: Solar Energy Research Institute.

Coleman, C; McNiff, B.P. (Nothern Power Systems), (1990), *Dynamic Response Testing of the North Wind 100 Wind Turbine.* SERI/STR-257-3686. Golden, CO: Solar Energy Research Institute.

Commission of the European Communities (CEC), DGXVII, (1996), European Energy to 2020: a Scenario Approach. 120 pp.

Connell, J.R. (1988), "A Primer of Turbulence at the Wind Turbine Rotor." *Solar Energy* (41:3); pp. 281–293. New York, NY: Pergamon Press.

Connell, J.R. (Colorado State University); Morris, V.R. (1991), *Calibration Data for Improved Correction of UVW Propeller Anemometers.* PNL-7824. Richland, WA: Pacific Northwest Laboratory.

Connell, J.R.; George, R.L. (1987), "Accurate Correlation of Wind Turbine Response With Wind Speed Using a New Characterization of Turbulent Wind." *Journal of Solar Energy Engineering;* Vol. 109. New York, NY: American Society of Mechanical Engineers.

Connell, J.R.; Morris, V.R. (1990), "Turbulent Wind at the Equatorial Segment of an Operating Darrieus Wind Turbine Blade." *Ninth ASME Wind Energy Symposium.* SED-Vol. 9. New York, NY: American Society of Mechanical Engineers.

Connell, J.R.; Powell, D.C. (1989), *User's Guide for Personal Computer Model of Turbulence at a Wind Turbine Rotor.* PNL-6959. Richland, WA: Pacific Northwest Laboratory.

Cornman, J.W. (1972), 'Craig's theorem, Ramsey-sentences and Scientific Instrumentalism', *Synthese*, 25, pp.82-126.

Costanza, R. and Ruth, M., (1997), Using Dynamic Modeling to Scope Environmental Problems and Build Consensus. Environmental Management,Vol. 22, pp. 183-196.

Craig, W. (1956), 'Replacement of Auxiliary Expressions', *PhilRev*, 65, pp.38-55.

Cramer, G.; Kleinkauf, W.; Schott, T. (1994), "Wind/Diesel/Battery Systems-Applications, Experience, Economy of Different System Configurations and Sizes." *Wind Engineering* (17:5); p. 228. Brentwood, England: Multi-Science Publishing Company, LTD.

Crow, M. & Sarewitz, D. (2001), 'Nanotechnology and Societal Transformation', in: M.C. Roco and W.S. Bainbridge (eds.), *Societal Implications of Nanoscience and Nanotechnology*, Dordrecht: Kluwer, pp. 45-54.

Cummisky, D. (1992), 'Reference Failure and Scientific Realism: A response to the meta-induction', *BJPS*, 43, 1, 21-40.

Davis, J.R. (ed.) (1992), *ASM Materials Engineering Dictionary*, ASM International (materials Park, OH).

Day, T. & Kincaid, H. (1994), 'Putting Inference to the Best Explanation in its Place' *Synthese* 98.

de Costa, N.C.A., & S. French (1990), 'The Model-Theoretic Approach in the Philosophy of Science', *PhilSci* 57, pp. 248-265.

de Finetti, B. (1969), 'Initial Probabilities: A Prerequisite for any Valid Induction', *Synthese* 20.

de Finetti, B. (1974), *Theory of Probability*, New York: Wiley.

DeGarmo, E.P., et al. (ed.) (1997), *Materials and Processes in Manufacturing, 8th ed.*, Prentice Hall (Upper Saddle River, NJ).

Devitt, M. (1979), 'Against Incommensurability', *Australasian JPhil*, 57, pp. 29-50.

Devitt, M. (1991), 'Aberrations of the Realist Debate' *Philosophical Studies* 61.

Devitt, M. (1991), *Realism and Truth*, Oxford: Blackwell.

Dieter, G.E. (1993), *Mechanical Metallurgy*, McGraw-Hill (New York).

Domininghaus, H. (1993), *Plastics for Engineers: Materials, Properties, Applications*, Hanser Publishers (Munich).

Doran, J.C. (1981), *Inherent Uncertainties in Meteorological Parameters for Wind Turbine Design*. PNL-SA-736. Richland, WA: Pacific Northwest Laboratory; 18 pp.

Doran, J.C.; Packard, K.R. (1982), *Comparison of Model and Observations of the Wake of a MOD-0A Wind Turbine*. PNL-4433. Richland, WA: Pacific Northwest Laboratory; 43 pp.

Downes, S. (1992), 'The Importance of Models in Theorizing: A Deflationary Semantic View', *PSA* Vol 1, p. 142.

Doyle, J. and Patil, R. S. (1991), Two theses of knowledge representation: language restrictions, taxonomic classification, and the utility of representation services. *Artificial Intelligence*, 48(3):261-297.

Dretske, F. I (1977), 'Laws of Nature', *Philosophy of Science*, 44, pp. 248-68.

Duhem, P. (1954), *The Aim and Structure of Physical Theory*, Princeton: PUP, Ch. 4.

Duhem, P. (1962), *The Aim and Structure of Physical Theory*, New York: Atheneum, Ch. 6-7.

Duhem, P. (1969), *To Save the Phenomena*, Chicago: UCP.

Dupuy, J. P. (2004), 'Complexity and Uncertainty: A Prudential Approach To Nanotechnology', in: European Commission (Community Health and Consumer Protection): *Nanotechnologies: A Preliminary Risk Analysis on the Basis of a Workshop, Brussels, 1-2 March 2004*, pp. 71-93 (www.europa.eu.int/comm/health/ ph_risk/documents/ev_20040301_en.pdf).

Earman, J (ed.), (1983), *Testing Scientific Theories, MinnStud* Vol 10, Minneapolis: University of Minnesota Press.

Earman, J. (1992), *Bayes or Bust? A Critical Examination of Bayesian Confirmation Theory*, Cambridge, Mass: MIT Press, esp. Ch. 1-3, 6.

Earman, J. (1993), 'Underdetermination, Realism and Reason', in French, Uehling & Wettstein (eds.), *Midwest Studies in Philosophy* Vol. 18, Notre Dame: University of Notre Dame Press, pp.19-38.

Ebert-May, Diane; Brewer, Carol; Allred, Slyvester. (1997). Innovation in large lectures – teaching for active learning. *Bioscience*. 47(9), p. 601-07.

Eggleston, D.M.; Starcher, K.L. (1990), "A Comparative Study of the Aerodynamics of Several Wind Turbines Using Flow Visualization." *Journal of Solar Energy Engineering*. Vol. 112. New York, NY: American Society of Mechanical Engineers; pp. 301-309.

Eggleston, D.M.; Stoddard, F.S. (1987), *Wind Turbine Engineering Design*. New York, NY: Van Nostrand Reinhold Company, Inc.; 352 pp.

Elkins, C. (1979), 'Science Fiction versus Futurology: Dramatic Versus Rational Models', *Science Fiction Studies*, 6, 20-31.

Elliott, D.L. (1991), "Status of Wake and Array Loss Research." *Windpower '91 Conference; September 24-27, 1991; Palm Springs, California*. Washington, DC: American Wind Energy Association.

Elliott, D.L.; Barnard, J.C. (1990), "Effects of Trees on Wind Flow Variability and Turbulence." *Journal of Solar Energy Engineering;* Vol. 112. New York, NY: American Society of Mechanical Engineers.

Elliott, D.L.; Buck, J.W.; Barnard, J.C. (1988), *An Examination of Wake Effects and Power Production for a Group of Large Wind Turbines*. PNL-6528. Richland, WA: Pacific Northwest Laboratory; 58 pp. (NTIS no. DE88009906),

Elliott, D.L.; Cadogan, J.B. (1990), "Effects of Wind Shear and Turbulence on Wind Turbine Power Curves." *European Community Wind Energy Conference and Exhibition; September 10-14, 1990; Madrid, Spain*. Bedford, England: H.S. Stephens & Associates.

Ellis, B. & Bowman, J. (1967), 'Conventionality in Distant Simultaneity', *PhilSci* 34, pp. 116-136.

Ellis, B. (1968), *Basic Concepts of Measurement*, Cambridge: Cambridge University Press, Ch. XI

Ellsberg, D. "Risk, Ambiguity and the Savage Axioms." Quarterly Journal of Economics 75 (1961): 643-669.

Eppler, R. (1990), *Airfoil Design and Data.* New York, NY: Springer-Verlag; 562 pp.

Evans, D.F.; Wennerstrom, H. (1999), *The Colloidal Domain: Where Physics, Chemistry, Biology and Technology Meet*, John Wiley & Sons, New York.

Evans, J.H. (2002), *Playing God? Human Genetic Engineering and the Rationalization of Public Bioethical Debate*, Chicago University Press, Chicago.

F. Liu, R. Sha and N.C. Seeman (1999),, Modifying the Surface Features of Two-Dimensional DNA Crystals, Journal of the American Chemical Society. 121, 917-922.

Falmagne, J.C. (1966), 'A Probabilistic Theory of Extensive Measurement', *Synthese*, 16, pp. 125-69.

Faraday, Michael (1855), *Experimental Researches in Electricity*, London.

Fardoun, A.A.; Fuchs, E.F. (University of Colorado); Carlin, P.W. (1993), "A Variable-Speed, Direct- Drive Transmission Wind Power Plant." *Windpower '93 Conference; July 12-16, 1993; San Francisco, California.* Washington, DC: American Wind Energy Association; pp. 134-141.

Farrell, John; Moog, Richard; Spencer, James (1999). A guided inquiry general chemistry course. *Journal of Chemical Education.* 76(4), p. 570-74.

Faulds, H., On the skin furrows of the hand, *Nature*, 22, 605, 1880.

Fennema, O. R. (1973) in *Low-Temperature Preservation of Foods and Living Matter*, eds. Fennema, O. R., Powrie, W. D. & Marth, E. H. (Dekker, New York), pp. 436-475.

Feyerabend, P. (1965), 'On the 'Meaning' of Scientific Terms', *JPhil* 62, pp. 266-74.

Field, H. (1973), 'Theory Change and the Indeterminacy of Reference', *JPhil* 70, pp. 462-81.

Fine, A. (1975), 'How to Compare Theories: Reference and Change', *Nous*, 9, pp. 17-32.

Fine, A. (1984), 'The Natural Ontological Attitude', in J. Leplin (ed.), *Scientific Realism*, Berkeley: Univ. California Press. [Repr. in his *The Shaky Game* (1986).]

Fine, A. (1986), 'Unnatural Attitudes: Realist and Instrumentalist Attachments to Science', *Mind* 95, pp. 149-79.

Fine, A. (1991), 'Piecemeal Realism', *Philosophical Studies* 6, pp. 79-96.

Fodor, J. (1984), 'Observation Reconsidered', *PhilSci* 51, pp. 23-43.

Foley, R. (1983), 'Epistemic Conservativism' *Philosophical Studies* 43.

Franklin, A. (1987), *The Neglect of Experiment*, Cambridge: Cambridge University Press.

Friedler, Yael and others. (1992). Problem-solving inquiry-oriented biology tasks integrating practical laboratory and computer. *Journal of Computers in Mathematics and Science Teaching*. 32(5), p. 347-57.

Friedman, M. (1972), 'Grünbaum on the Conventionality of Geometry'; *Synthese*, 24, pp.219-235.

Frost, W.; Long, B.H.; Turner, R.E. (1978), *Engineering Handbook on the Atmospheric Environmental Guidelines for Use in Wind Turbine Generator Development*. NASA-TP-1359. Huntsville, AL: National Aeronautics and Space Administration, Marshall Space Flight Center; 372 pp.

Galison, P. (1987), *How Experiments End*, Chicago: UCP.

Galison, P. (1988), 'Philosophy in the Laboratory', *JPhil* LXXXV, pp. 525-7.

Gallet, Christian. (1998). Problem solving teaching in the chemistry laboratory: Leaving the cooks.... *Journal of Chemical Education*. 75(1), p. 72-77.

Galton, F., Personal Identification and Description II, *Nature*, 38, 201-202, 1888.

Gardner, Marjorie, Greeno, James G., Reif, Frederick, Schoenfeld, Alan H., diSessa, Andrea, and Stage, Elizabeth (editors). (1990). *Toward a Scientific Practice of Science Education*. Hllsdale, NJ: Lawrence Erlbaum Associates.

Georges, T.M., Harlan, J.A., Leben, R. R., Lematta, R. A., (1998), A Test of Ocean Surface-Current Mapping with Over-the-Horizon Radar. IEEE Trans. Geosci. Remote Sensing, vol. 36(1), pp. 101-110

Gere J.M., Timoshenko, S.P. (1984), *Mechanics of Materials, 2nd ed.*, Brooks/Cole (Monterey, CA).

Giedymin, J (1982), *Science and Convention*, Oxford: Pergamon Press, esp. Ch. 1, 3, 4.

Giere, R. (1984), *Understanding Scientific Reasoning* (2nd Ed.), NY: Holt, Rinehart and Winston.

Giere, R. N. (1988), *Explaining Science*, Chicago: UCP.

Giere, R.N. (1973), 'Objective Single Case Probabilities and the Foundations of Statistics' in *Logic, Methodology and the Philosophy of Science* IV, P. Suppes & L. Henkin (eds.).

Giere, R.N. (1980), 'Causal Systems and Statistical Hypotheses', in Cohen, L J & Hesse, M (eds.), *Applications of Inductive Logic*, Oxford: Clarendon Press.

Gillies, D.A. (1972), 'Operationalism', *Synthese*, 25, pp.1-24.

Gillies, D.A. (1973), *An Objective Theory of Probability*, London: Methuen & Co.

Gipe, P. (1993), *Wind Power for Home & Business.* Post Mills, VT: Chelsea Green Publishing Company; 413 pp.

Glynn, Shawn M., Yeany, Russell, H., and Britton, Bruce K. (editors). (1991). *The Psychology of Learning Science.* Hillsdale, NJ: Lawrence Erlbaum Associates.

Grant, G. (1990). 'Transcendence through detournement in William Gibson's *Neuromancer,' Science-fiction Studies.* 17(5).

Hacking, I. (1967), 'Slightly More Realistic Personal Probability', *Philosophy of Science*, 34, pp. 311-25

Hacking, I. (1983), *Representing and Intervening: Introductory Topics in the Philosophy of Natural Science*, Cambridge University Press, Cambridge, UK.

Hacking, I. (1992), 'Do Thought Experiments have a life of their own?' *PSA* 2, pp.302-308.

Hacking, I. (1992), 'The Self-Vindication of the Laboratory Sciences', in: A. Pickering (ed.), *Science as Practice and Culture*, University of Chicago Press, Chicago, pp. 29-64.

Hake, Richard. (1998). Interactive-engagement versus tradition methods: a six-thousand-student survey of mechanics test data for introductory physics courses. *Am. J. Physics* 66(1): 64-74.

Hanfling, O., ed. (1981), *Essential Readings in Logical Positivism*, Oxford: Blackwell.

Hansen, A.C. (1992), *Yaw Dynamics of Horizontal Axis Wind Turbines: Final Report.* NREL/TP-442- 4822. Golden, CO: National Renewable Energy Laboratory; 175 pp. (NTIS no. DE92001245),

Hansen, A.C. (University of Utah), (1987), *The TEETER Program-Analysis and User's Guide.* SERI/STR-217-3050. Golden, CO: Solar Energy Research Institute.

Hansen, A.C.; Butterfield, C.P. (1993), "Aerodynamics of Horizontal-Axis Wind Turbines." *Annual Review of Fluid Mechanics.* Vol. 25, pp. 115-149.

Hanson, N. R. (1958), *Patterns of Discovery*, Cambridge: CUP.

Harré R. (1981), *Great Scientific Experiments*, Oxford: Phaidon.

Harré, R (1970), *The Principles of Scientific Thinking*, London: Macmillan, Ch. 4.

Harré, R. (1976), 'The Constructive Role of Models', in Collins, L (ed.), *The Use of Models in the Social Sciences*, London: Tavistock.

Harré, R. (1986), *Varieties of Realism*, Oxford: Blackwell.

Harré, R. (1993), *Laws of Nature*, London: Duckworth.

Harvey, P.D. (ed.) (1982), *Engineer Properties of Steel*, American Society for Metals (Metals Park, OH).

Hayward, R.C.; Saville, D. A.; Aksay, I.A. (2000), "Electrophoretic assembly of colloidal crystals with optically tunable micropatterns", *Nature*, 404, 56-59.

Healy M.G. and Doody, J.P. (eds), (1995), Directions in European coastal management. EUCC, Samara Publishing, Cardigan.

Heelan, P.A. (1988), 'Experiment and Theory', *JPhil*, LXXXV, pp. 515-24.

Held, R., T. Heinzel, A.P. Studerus, K. Ensslin, and M. Holland (1997), Semiconductor quantum point contact fabricated by lithography with an atomic force microscope. *Appl. Phys. Lett.* 71:2689-91.

Hellemans, A. (1998), X-rays find new ways to shine. *Science* 277:1214-15.

Hellman, G. (1983), 'Realist Principles', *PhilSci* 50, pp. 227-49.

Helmreich, S. (1998), *Silicon Second Nature: Culturing Artificial Life in a Digital World*, University of California Press, Berkeley, CA.

Hempel, C. (1965), 'Aspects of Scientific Explanation' in *Aspects of Scientific Explanation*, NY: Free Press, Sec. 6.

Hempel, C. (1965), 'The Theoretician's Dilemma' in his *Aspects of Scientific Explanation*, New York: Free Press [orig. in *MinnStud* Vol 2]

Hempel, C. (1972), 'Formulation and Formalization of Scientific Theories: A Summary Abstract' in F. Suppe (ed.), *The Structure of Scientific Theories*, Urbana: Univ. of Illinois Press.

Hendry, F. R. (1995), 'Realism and Progress: Why Scientists Should be Realists' in Fellows, R. (ed.), *Philosophy and Technology*, Cambridge: CUP.

Herrera, J.I.; Reddoch, T.W. (Electrotek Concepts, Inc.), (1988), *Analysis of the Electrical Harmonic Characteristics of a Westinghouse Variable Speed*

Generating System for Wind Turbine Applications. SERI/STR-217-3133. Golden, CO: Solar Energy Research Institute.

Hesse, M. (1952), 'Operational definition and analysis in physical theory', *BJPS* pp.281-94.

Hesse, M. (1966), *Models and Analogies in Science*, Notre Dame: University of Notre Dame Press.

Hesse, M. (1967), 'Simplicity' in *The Encyclopedia of Philosophy* (ed.), Paul Edwards, N.Y. Macmillan.

Hesse, M. (1970), 'Is there an independent observation language?', in Colodny (ed.), *The Nature and Function of Scientific Theories*, Pittsburgh: Univ. of Pittsburgh Press.

Hesse, M. (1974), *The Structure of Scientific Inference*, Basingstoke: Macmillan, Chs. 1,2.

Higgins, R.J. (1997), An economical process for manufacturing of nano-sized powders based on microemulsion-mediated synthesis. In *Proc. of the Joint NSF-NIST Conf. on Nanoparticles.*

Hill, Douglas. (1985). Inquiry experiences for elementary education. *Journal of College Science Teaching*. 14(4), p. 349-51.

Hintikka, J. & Suppes, P. (eds.), (1966), *Aspects of Inductive Logic*, Dordrecht: Reidel.

Hoefer, C. & Rosenberg, A. (1994), 'Empirical Equivalence and Systems of the World', *PhilSci* 61, 4, pp.592-607.

Holman, E.W. (1974), 'Extensive Measurement without an Order Relation', *Philosophy of Science* 41, pp. 361-73

Hong, H. "Effects of Mathematics Learning Through Children's Literature on Math Achievement and Dispositional Outcomes." *Early Childhood Research Quarterly* (11), 4 (1996): 477-494.

Hooker, C. A. & Churchland, P. M., eds. (1985), *Images of Science*, Chicago: UCP (esp. papers by Boyd, Churchland, Ellis, Hooker, Musgrave, and van Fraassen).

Horowitz, T. & Massey, G.J. (eds.), (1991), *Thought Experiments in Science and Philosophy*, Lanham: Rowman & Littlefield.

Horwich, P. (1982), 'How to Choose between Empirically Indistinguishable Theories', *JPhil* 79, pp.61-77.

Horwich, P. (1982), 'Three Forms of Realism', *Synthese* 52, pp.181-201.

Horwich, P. (1982), *Probability and Evidence*, Cambridge: CUP, esp. Ch. 1-4.

Horwich, P. (1986), 'A defence of conventionalism' in McDonald, G. (ed.), *Fact, Science and Morality*, Oxford: Blackwell, pp.163-187.

Howard, J. (2001), Mechanics of Motor Proteins and the Cytoskeleton. Sinauer Associates, Sunderland, Massachusetts.

Howland, H. (1962), Structural, hydraulic and economic aspects of leaf venation and shape ', in: E.E. Bernard & M.R. Kare (eds.), *Biological prototypes and synthetic systems*, Plenum Press, New York, vol. 1, pp. 183-192.

Howson, C. & P. Urbach (1989), *Scientific Reasoning: The Bayesian Approach*, La Salle: Illinois: Open Court.

Hull, D., M. Forbes & K. Okruhlike (1992), *PSA* 2 Part IX, *Thought Experiments: The Theoretician's Laboratory*, papers by Brown, Gooding, Nersessian and Hacking.

Humphreys, P. (1993), 'Seven Theses on Thought Experiments', in Earman, J. et. al. (eds.), *Philosophical Problems of the Internal and External World*, Pittsburgh: University of Pittsburgh Press.

James, G.H., III; Carne, T.G.; Lauffer, J.P. (1993), *The Natural Excitation Technique (NExT) for Modal Parameter Extraction from Operating Wind Turbines.* SAND-92-1666. Albuquerque, NM: Sandia National Laboratories; 46 pp. (NTIS no. DE93010611),

Jardine, N. (1986), *The Fortunes of Inquiry*, Oxford: Clarendon Press.

Jeffrey, R. (1970), 'Dracula Meets Wolfman: Acceptance vs. Partial Belief', Induction, Acceptance, and Rational Belief, M. Swain (ed.), D. Reidel: Dordrecht.

Jeffrey, R. (1983—2nd ed.), *The Logic of Decision*, Chicago: Univ. of Chicago Press.

Jeffrey, R. (1992), *Probability and the Art of Judgment*, Cambridge: CUP, Ch. 4. [Reprinted from *Observation, Experiment, and Hypothesis in Modern Physical Science*, P. Achinstein & O. Hannaway (eds.), Boston: MIT Press.]

Jeffreys, H. (1939), *Theory of Probability*, Oxford: Clarendon Press.

Jensen, Murray S. and Finley, Fred N. (1996). Changes in students' understanding of evolution resulting from different curricular and instructional strategies. *J. res. Science Teaching* 33(8): 879-900.

Jones, Loretta and others. (1997). Preparing preservice chemistry teachers for constructivist classrooms through use of authentic activities. *Journal of Chemical Education*. 74(7), p. 787-88.

Kallery, Maria. "Making the Most of Questions and Ideas in the Early Years." *Primary Science Review* 61 (Jan./Feb. 2000): 18-19.

Kanger, S. (1972), 'Measurement; An Essay in the Philosophy of Science', *Theoria*, XXXVIII, pp. 1-44.

Kaplan, W.A. (1999), *Modern Plastics Encyclopedia '99*, McGraw-Hill (New York).

Kear, B., and G. Skandan (1998), Nanostructural bulk materials: Synthesis, processing, properties and performance. *In R&D status and trends*, ed.

Kear, B.H., R.K. Sadangi, and S.C. Liao (1997), Synthesis of WC/Co/diamond nanocomposites. In *Proc. of the Joint NSF-NIST Conf. on Nanoparticles*.

Kebel, T.E. (1996), 'Anti-Foundationalism and the Vienna Circle's Revolution in Philosophy", *BJPS* 47.

Kelley, N.D. (1989), *An Initial Look at the Dynamics of the Microscale Flow Field Within a Large Windfarm in Response to Variations in the Natural Inflow.* SERI/TP-257-3591. Golden, CO: National Renewable Energy Laboratory; 5 pp.

Kelley, N.D.; Wright, A.D. (1991), *A Comparison of Predicted and Observed Turbulent Wind Fields Present in Natural and Internal Wind Park Environments.* NREL/TP-257-4508. Golden, CO: National Renewable Energy Laboratory; 9 pp.

Kennedy, P. (1987), *The Rise and Fall of the Great Powers, Economic Change and Military Conflict from 1500 to 2000*,Random House.

Keokebakker, P. and Peet, G., (1987), Coastal Zone Planning and Management in the Netherlands. Coastal Management 15: 121-134.

Kerns, Tim. (1996). Should we use cooperative learning in college chemistry? *Journal of College Science Teaching*. 25(6), p. 435-38.

Keynes, J.M. (1957), *A Treatise on Probability*, New York: Macmillan

Keyssar, Alexander (1986), *Out of Work: The First Century of Unemployment in Massachusetts* (Cambridge: Cambridge University Press).

King, Alison. (1995). Inquiring minds really do want to know: Using questioning to teach critical thinking. *Teaching of Psychology*. 22(1), p. 13-17.

Kitcher, P. (1978), 'Theories, Theorists & Theoretical Change', *PhilRev*, 87, pp. 519-547.

Kitcher, P. (1993), *The Advancement of Science*, Oxford: OUP, Ch.5.

Klimas, P.C. (1984), *Tailored Airfoils for Vertical Axis Wind Turbines.* SAND-84-1062. Albuquerque, NM: Sandia National Laboratories; 12 pp.

Knight, A.E. and Molloy, J.E. (2000), "Muscle, myosin and single molecules". In "Molecular Motors" *Essays In Biochemistry* Vol. 35 (S. J. Higgins and G. Banting, Eds.), London, Portland Press.

Knight, F.H. Risk (1933), Uncertainty and Profit. London School of Economics and Political Science, London.

Knorr-Cetina, K. (1992), 'The Couch, the Cathedral, and the Laboratory: On the Relationship between Experiment and Laboratory in Science', in: A. Pickering (ed.), *Science as Practice and Culture*, University of Chicago Press, Chicago, pp. 113-138.

Knorr-Cetina, K. (1999), *Epistemic Cultures: How the Sciences Make Knowledge*, Harvard University Press, Cambridge, MA.

Koch, C.C. (1998), Bulk behavior. In *R&D status and trends*, ed. Siegel et al. 32 *Evelyn L. Hu and David T. Shaw.*

Kocurek, D. (1987), *Lifting Surface Performance Analysis for Horizontal Axis Wind Turbines.* SERI/STR-217-3163. Golden, CO: Solar Energy Research Institute; 319 pp. (NTIS no. DE87001176),

Kolata, G. (1995), "A Vat of DNA May Become a Vast Computer of the Future," N.Y. Times, 11 p. C1. *Large, much cited spread at the top of the front page of the Tuesday "Science Times" Section that first brought the Adelman discovery to wide public attention. Beautiful illustrations that explain how it works.*

Koyré, A (1968), 'Galileo's Treatise *De Motu*: The Use and Abuse of Imaginary Experiment', in *Metaphysics and Measurement*, London: Capman and Hall.

Krantz, D.H., Luce, R.D., Suppes, P. & Tversky, A. (1971), *Foundations of Measurement*, New York: Academic Press.

Kripke, S. (1972), *Naming and Necessity*, Oxford: Blackwell, esp. Lecture III.

Kristeva, J. (1990). 'Psychoanalysis and the polis'. in Gayle L. Ormiston and Alan D. Schrift (eds), *Transforming the Hermenentic Context.* Albany. NY: SUNY Press.

Kruskal, M., and N. Zabusky (1965), Interaction of solitons in a collisionless plasma and the recurrence of initial states, *Phys. Rev. Lett.* Vol. 15. 240-243.

Kuhn, T S (1977), 'A Function for Thought Experiments', reprinted in *The Essential Tension,* Chicago: UCP.

Kuhn, T. (1970), *The Structure of Scientific Revolutions* (second edition), University of Chicago Press, Chicago.

Kuhn, T. (1977), 'Objectivity, Value Judgement, and Theory Choice' in *The Essential Tension*, Chicago: UCP.

Kuhn, T. S. (1982), 'Commensurability, Comparability, Communicability', *PSA* Vol. 2, pp. 669-688.

Kuhn, T.S. (1990), 'Dubbing and Redubbing: The Vulnerability of Rigid Designation' in Wade Savage (ed.), *MinnStud XIV*, Minneapolis: University of Minnesota Press, pp. 298-318.

Kuhn, T.S. (1996), *The Structure of Scientific Revolutions*, University of Chicago Press, Chicago.

Kuipers, T.A.F. (1982), 'Approaching Descriptive and Theoretical Truth', *Erkenntnis* 18, pp. 343-378.

Kukla, A. (1994), 'Non-empirical theoretical virtues and the argument for underdetermination', *Erkenntnis* 41, 2, pp.157-170.

Kukla, A. (1994), 'Scientific Realism, Scientific Practice and the Natural Ontological Attitude', *BJPS* 45, pp.955-975.

Kumar, A., and G.M. Whitesides (1993). Features of gold having micrometer to centimeter dimensions can be formed through a combination of staming with an elastomeric stamp and an alkanethiol ink followed by chemical etching. *App. Phys. Lett.* 63:2002-2004.

Kupferman, R.,M. Kawaguchi, and M.M. Denn (2000), Emergence of structure in models of liquid crystalline polymers with elastic coupling. *J. Non-Newtonian Fluid Mech.* Vol. 91. 255-271.

Kuusi, O. (1999), "Epistemic value analysis – a tool to evaluate technological options." Personal communication.

Kuusi, O. and M. Meyer (2001), "Technology Generalizations and Leitbilds – The Anticipation of Technological Opportunities." Manuscript submitted to *Technological Forecasting and Social Change*.

Kyberg, H.E. (1968), *Philosophy of Science: A formal Approach*, New York: Macmillan, Ch. 3.

Kyberg, H.E. (1970), *Probability and Inductive Logic*, London: Collier-Macmillan.

Kyburg, H. and H.E. Smokler (1964), *Studies in Subjective Probability*, New York: J.Wiley. [esp paper by de Finetti]

Kyburg, H.E. (1984), *Theory and Measurement*, Cambridge: Cambridge University Press.

Kyprianidou-Leodidou, T., W. Caseri, and V. Suter (1994), *J. Phys. Chem*. 98:8992. Kung, H.H., and E.I. Ko, 1996. *Chem. Eng. J*. 64:203.

L.J. Lauhon, W. Ho, "Inducing and observing the abstraction of a single hydrogen atom in bimolecular reaction with a scanning tunneling microscope," *J. Phys. Chem*. 105 (2000): 3987-3992.

L.S. Penrose, "Self-Reproducing Machines," *Scientific American* 200(June 1959):105-114.

Laibinis, P.E.; Hickman, J.J.; Wrighton, M.S.; Whitesides, G.M. 1989. "Orthogonal Self-Assembled Monolayers - Alkanethiols On Gold and Alkane Carboxylic-Acids On Alumina", *Science*, 245, 845- 847.

Lakatos, I. (1968), 'Changes in the Problem of Inductive Logic' in Lakatos (ed.), *The Problem of Inductive Logic*, Amsterdam: North-Holland, pp.315-417.

Lake, David A. (1999). Peer tutoring improves student performance in an advanced physiology course. *Am. J. Physiol*. 276 (*Adv. Physiol. Educ*. 21): S86-92.

Latour, B. & Woolgar, S. (1979), *Laboratory Life*, Los Angeles: Sage.

Laudan, L. (1965), 'On the Impossibility of Crucial Falsifying Experiments: Grünbaum on 'The Duhemian Argument" *PhilSci* 32, pp.295-299.

Laudan, L. (1977), *Progress and its Problems*, Berkeley: Univ. of California Press.

Laudan, L. (1984), 'A Refutation of Convergent Realism' in Leplin, J., ed.

Laudan, L. (1990), 'Demystifying Underdetermination' in Wade-Savage (ed.), *MinnStud*, Vol. 14, Minneapolis: University of Minnesota Press, pp. 267-297.

Lauw, H.K.; Weigand, C.H.; Marckx, D.A. (Electronic Power Conditioning, Inc.), (1993), *Variable-Speed Wind System Design: Final Report*. DOE/BP/99893-TI. Washington, DC: U.S. Department of Energy; 81 pp. (NTIS no. DE94006210),

Lave, L.B., A. Russell, F.C. McMichael, and C.T. Hendrickson. 1996. Environmental implications of battery powered vehicles as a transitional technology: ozone vs. lead. *Environmental Science & Technology*. p. 30.

Lave, L.B., C.T. Hendrickson, and F. C. McMichael (1995b), Environmental implications of electric cars. *Science*. vol. 268: 992-5. May 19.

Lawler, E. L. and Wood, D. E. (1966), Branch-and-bound methods: A survey. *Operations Research*, 14(4):699-719.

Lax, Peter D. (1989), The flowering of applied mathematics in America. in *A Century of Mathematics in America-Part II*. edited by Peter Duren et al. American Mathematical Society. 455-466.

Laymon, R. (1985), 'Idealizations and the Testing of Theories by Experimentation', in *Observation Experiment and Hypothesis in Modern Physical Science*, P. Achinstein and O. Hannaway (eds), Cambridge Mass: The MIT Press.

Layton, E.T., Jr. (1971), 'Mirror-Image Twins: The Communities of Science and Technology in 19th-Century America', *Technology and Culture*, 12, 562-580.

Lazzaro, Joseph J. Opening doors for the disabled: adaptive technology lets blind, deaf, and motor-disabled personal computer users lead more productive lives. Byte, v. 15, August 1990: 258, 260, 262-264, 266, 268.

Le Cun, Y., Jackel, L. D., Boser, B., and Denker, J. S. (1989), Handwritten digit recognition: applications of neural network chips and automatic learning. *IEEE Communications Magazine*, 27(11):41-46.

Leafe, R., Pethick, J., Townend, I., (1998), Realising the benefits of shoreline management. The Geographical Journal, 164(3), 282–290.

Leah Eisenberg, "Medicating Death Row Inmates So They Qualify for Execution," AMA Case in Health Law, Vol. 6, No. 9, September 2004.

Leatherdale, W.H. (1974), *The Role of Analogy, Model and Metaphor in Science*, NY: A.M. Elsevier.

Ledley, Fred D. (1995), "Nonviral Gene Therapy: The promise of Genes as Pharmaceutical Products," *Human Gene Therapy*, Vol. 6, Sep, pp 1129-1144.

Lee, E., (1993), Coastal Planning and Management: A Review prepared by Rendel Geotechnics for the Department of the Environment. London, HMSO.

Lee, M.S. and M. Head-Gordon (2000), Absolute and relative energies from polarized atomic orbital selfconsistent field calculations and a second order convergence with size and composition of the secondary basis. *Computers and Chemistry*. Vol. 24(3,4), 295-301.

Lee, S.C. (1998), The nanobiological strategy for construction of nanodevices. *Biological molecules in nanotechnology: the convergence of biotechnology, polymer chemistry and materials science*, edited by S.C. Lee and L. Savage (IBC Press, Southborough, MA).

Leeds, S. (1994), 'Constructive Empiricism', *Synthese 101*, pp. 187-221.

Lefebvre, C., (1998), Coastal Conservation Policy - Guidelines for Political Actors, Programme 1998-2000. IUCN, Commission on Environmental, Economic and Social Policy Coastal Working Group.

Leonard, William. (1988). Overcoming obstacles in teaching large-enrollment lab courses. *American Biology Teacher*. 50(1), p 23-28.

Leplin, J. & Laudan, L. (1991), 'Empirical Equivalence and Underdetermination', *JPhil*, 88, pp.449-472.

Leplin, J. (1988), 'Is Essentialism Unscientific?', *PhilSci*, 55, pp. 493-510.

Leplin, J., ed. (1984), *Realism*, Berkeley: University of California Press, (esp. papers by McMullin, Boyd, Fine, & Leplin).

Levine, M. (1979), 'On Theory-Change and Meaning-Change', *PhilSci*, 46, pp. 407-424.

Lewis, D. (1970), 'How to Define Theoretical Terms', *JPhil* 67 pp. 427-46 [Repr. in his *Philosophical Papers* Vol. 1, Oxford: OUP (1987)]

Lewis, D. (1980), 'A Subjectivist's Guide to Objective Chance" in R. Jeffrey (ed.), *Studies in Inductive Logic and Probability* Vol. II, Berkeley: Univ. of California Press [repr. in Lewis' *Philosophical Papers* Vol 2].

Lide, D.R. (ed.) (1995), *CRC Hankbook of Chemistry and Physics, 76th ed.*, CRC Press (Boca Raton, FL).

Lind, Karen K. *Exploring Science in Early Childhood Education: A Developmental Approach*. 3rd. edition. Delmar, 2000.

Lind, Karen K. *Science in Early Childhood: Developing and Acquiring Fundamental Concepts and Skills*.

Lin-Easton, P.C. (2001), 'It's Time for Environmentalists to Think Small ' Real Small: A Call for the Involvement of Environmental Lawyers in Developing Precautionary Policies Molecular Nanotechnology', *Georgetown International Law Review*, 14, 106-134.

Lissaman, P.B.S.; Gyatt, G.W.; Zalay, A.D. (AeroVironment), (1982), *Numeric Modeling Sensitivity Analysis of the Performance of Wind Turbine Arrays*. PNL-4183. Richland, WA: Pacific Northwest Laboratory; 97 pp.

Logue, J. (1995), *Projective Probability*, Oxford: Clarendon Press.

Lord, Thomas. (1994). Using constructivism to enhance student learning in college biology. *Journal of College Science Teaching*. 23(6), p. 346-48.

Lord, Thomas. (1998). Cooperative learning that really works in biology teaching:

Using constructivist-based activities to challenge student teams. *American Biology Teacher*. 60(8), p. 580-88.

Lucas, J.R. (1970), *The Concept of Probability*, Oxford: Clarendon Press, Chs. II & IV.

Lucas, J.R. (1984), *Space, Time and Causality*, Oxford: Clarendon Press, Ch. VI.

Luckenbach, S.A. (1972), *Probabilities, Problems and Paradoxes*, Vencino CA: Dickenson.

Lundsager; Norgaard. (1988), *The 55/30 kW Experimental Wind/Diesel System at Riso National Laboratory*. RISO-M-2717. Roskilde, Denmark: Riso National Laboratory.

Mach, E (1976), 'On Thought Experiments', in *Knowledge and Error*, Dordrecht: Reidel.

Mach, E. (1960), *The Science of Mechanics*, La Salle: Open Court, pp. 588ff.

Mackie, J. L. (1963), 'The Paradox of Confirmation', *BJPS* 13, pp. 265-77.

Mackie, J.L. (1973), *Truth, Probability and Paradox*, Oxford: Clarendon Press, Ch. 5

Maher, P. (1993), *Betting on Theories*, Cambridge: CUP.

Mahmood U., C.H. Tung, A. Bogdanov Jr., R. Weissleder (1999), "Near-Infrared Optical Imaging of Protease Activity for Tumor Detection." *Radiology*, Dec; 213 (3): pp. 866-870.

Manwell, J.F.; McGowan, J.G. (1992), "Advances in Quasi-Steady Wind Diesel System Simulation." *11th ASME Wind Energy Symposium*. New York, NY: American Society of Mechanical Engineers.

Mao, C., W. Sun, and N.C. Seeman (1997), Construction of Borromean rings from DNA. *Nature* 386:137-138.

Margalit, Malka (1990), Effective technology integration for disabled children: the family perspective. New York: Springer-Verlag, 226p.

Marine Institute, (1996), Towards a Marine Policy for Ireland. Dublin, Marine Institute.

Marino, M. (1988), 'Inference to the Best Explanation: van Fraassen and the Case of the 'Fifth Force", *ISPS* 3, p. 35.

Mark A. Reed, James M. Tour (2000), "Computing with molecules," *Scientific American* 282, 86-93.

Marshall, K. (2004), 'Atomizing the Risk Technology', in: N.K. Hayles (ed.), *Nanoculture: Implications of the New Technoscience*, Bristol, UK: Intellect Books, pp. 147-160.

Martin, T.P., U. Naher, H. Schaber, U. Zimmerman (1993), *Phys. Rev. Lett.* 70:3079. Prigogine, I., and S. Rice. 1988. *Advances in chemical physics*, Vol. 70, Parts 1 & 2. New York: J. Wiley.

Martuccelli, A.M., (1994), Il Ruolo della Bonifica per la Costruzione, Tutela e Gestione del Territorio Polesano. in "Il Fiume e la sua Terra. Tutela e Gestione del Territorio a Quarant'Anni dall'Alluvione del Polesine. 1951-1991. Atti del Convegno di Studi.

Marty, A. (2003), *Testimony before the Committee on Science on HR 766, Nanotechnology Research and Development Act*, US Government Printing Office, Washington, DC.

Marvin, Carolyn (1987), *When Old Technologies Were New: Thinking about Electric Communication in the Late Nineteenth Century.* New York: Oxford University Press.

Matsuki, H., K. Ofuji, N. Chubachi, and S. Nitta (1996), Signal transmission for implantable medical devices using figure-of-eight coils. *IEEE Trans. on Magnetics.,* vol. 32. No.5, 5121. September.

Matsuki, H.,Y. Yamakata, N. Chubachi, S. Nitta, and H. Hashimoto (1996), Transcutaneous DC-DC converter for totally implantable artificial heart using synchronous rectifier. *IEEE Trans. On Magnetics*. 32, No.5, 5118 (Sept.).

Maxwell, G. (1962), 'The Ontological Status of Theoretical Entities', in *MinnStud*, vol. 3, Minneapolis: University of Minnesota Press.

Mayer, S. (2002), 'From genetic modification to nanotechnology: the dangers of 'sound science", in: T. Gilland (ed.), *Science: Can We Trust the Experts?*, London: Hodder and Stoughton, pp. 1-15.

Mayer, T. and d'Ozouville, L. (1998), Europe's Grand Challenge: Achieving Coordination in Marine Research. Ocean Challenge 8(2): 20-21.

Mazlish, B. (1967), 'The fourth discontinuity'. *Technology and Culture,* 8(1),

McAllister, J. W. (1993), 'Scientific Realism and the Criteria for Theory Choice', *Erkenntnis* 38, 2, pp.203-222.

McCaffery, L. (1991), 'Introduction: the desert of the real', in L. McCaffery *(ed.), Storming the Reality Studio,* Durham, NC: Duke University Press.

McCain, L. (2002), 'Informing Technology Policy Decisions: the US Human Genome Project's Ethical, Legal and Social Implications Programmes as a Critical Case', *Technology and Society*, 24, 111-132.

McCammon, J. A., Gelin, B. R. & Karplus, M. (1977), *Nature (London)* 267, 585-590.

McConnell, H.M. (1996), Light-addressable potentiometric sensor: Applications to drug discovery. In *Nanofabrication and biosystems*, ed. Hoch et a–*ms*, Plenum Press, New York, vol. 1., p. 393-97.

McGuiness, B. ed. (1985), *Moritz Schlick*, Dordrecht: D.Reidel. [Reprint from Syntheses, Vol. 64, No. 5.]

McKenna, T. (1991), *The Archaic Revival,* New York: HarperCollins.

Mcknight, C, Dillon, A., Richardson, J. (1991), *Hypertext in Context.* Cambridge: Cambridge University Press.

McMullin, E. (1984), 'The Goals of Natural Science', *Proceedings of the American Philosophical Association* 58.

McMullin, E. (1987), 'Explanatory Success and the Truth of Theory' in N. Rescher (ed.), *Scientific Inquiry in Philosophical Perspective*, N.Y.: Univ. Press of America.

McMullin, E. (1991), 'Comment: Selective Anti-Realism' *Philosophical Studies* 61.

McNerney, G.M. (U.S. Windpower, Inc.), (1994), "U.S. Windpower 56-100 Stall Operation." *Wind Energy 1994.* SED-Vol. 15. New York, NY: American Society of Mechanical Engineers; pp. 219-226.

McNiff, B.P. (Second Wind Inc.); Musial, W.P. (National Renewable Energy Laboratory); Errichello, R. (GEARTECH), (September 1990), "Gear Fatigue Life Variations for Different Wind Turbine Braking Strategies." *Windpower '90 Conference; September 24-28, 1990; Washington, DC.* Washington, DC: American Wind Energy Association; 10 pp. (NTIS no. DE91002161),

Mellor, D H (ed.), (1980), *Science, Belief and Behaviour*, Cambridge: Cambridge University Press, (esp. essays 5 & 6).

Mellor, D.H. (1971), *The Matter of Chance*, Cambridge: Cambridge University Press.

Merrifield, D.B., (1990), Industrial competitiveness. In: Technology Commercialization and Competitiveness. IC2 Institute, University of Texas, Austin, TX, pp. 23–29.

Merz, J.L. (1986), "The Optoelectronics Joint Research Laboratory: Light Shed on Cooperative Research in Japan." *Scientific Bulletin* (ONR Far East Office), 11 (4), 1-30 (Oct.-Dec.).

Meyer, G. and Amer, N.M. (1988), Novel optical approach to atomic force microscopy. Appl. Phys. Lett. 53(12), 1045-1047

Meyer, G. and Amer, N.M. (1990), Simultaneous measurement of lateral and normal forces with an optical-beam-deflection atomic force microscope. Appl. Phys. Lett. 57(20), 2089-2091

Meyer, M. & Kuusi, O. (2004), 'Nanotechnology: Generalizations in an Interdisciplinary Field of Science and Technology', *Hyle: International Journal for Philosophy of Chemistry*, 10(2), 153-168.

Meyer, M. (2000), "Patent citations in a novel field of technology: What can they tell about interactions of emerging communities of science and technology?" *Scientometrics*, XXXXIIX (2), 151-178.

Meyer, M. (2000a), 'Does science push technology? Patents citing scientific literature', *Research Policy*, 29, 409-434.

Meyer, M. (2000b), 'Patent citations in a novel field of technology: What can they tell about interactions of emerging communities of science and technology?', *Scientometrics*, 48, 151-178.

Meyer, M. (2000b), "Hurdles on the way to growth. Commercializing novel technologies." Institute of Strategy and International Business, Working Papers Series 2000/1, Helsinki University of Technology, Espoo.

Meyer, M. (2001a), 'Patent citations in a novel field of technology: An exploration of nano-science and nano-technology', *Scientometrics*, 51, 163-183.

Meyer, M. (2001b), 'Socio-economic Research on Nanoscale Science and Technology: A European Overview and Illustration', in: M.C. Roco and W.S. Bainbridge (eds.), *Societal Implications of Nanoscience and Nanotechnology*, Dordrecht: Kluwer, pp. 217-241.

Meyer, M. and O. Persson (1998), "Nanotechnology - interdisciplinarity, patterns of collaboration and differences in application," *Scientometrics*, XXXXII (2), 195-205.

Meyer, M.; Persson, O. (1998), 'Nanotechnology ' interdisciplinarity, patterns of collaboration and differences in application', *Scientometrics*, 47, 195'205.

Micelli, Pat (1994), 'Censorship: it's OK to read about Homolka on computer, lawyers say'. *McGill Daily*. 24 January: 3.

Michael Page, Donald W. Brenner, "Hydrogen abstraction from a diamond surface: *Ab initio* quantum chemical study using constrained isobutane as a model," *J. Am. Chem. Soc.* 113(1991):3270-3274.

Michalet, X. and D. Bensimon (1988), Vesicles of toroidal topology: observed morphology and shape transformations. *J. Phys. II*. France. Vol. 5. 598.

Miksanek, T. (2001), 'Microscopic Doctors and Molecular Black Bag: Science Fiction's Prescription for Nanotechnology and Medicine', *Literature and Medicine*, 21(1), 55-70.

Milburn, C. (2002), 'Nanotechnology in the Age of Posthuman Engineering: Science Fiction as Science', *Configurations*, 10, 261-295.

Milburn, C. (2004), 'Nano/Splatter: Disintegrating the Postbiological Body', *New Literary History*, 35 (in print).

Miley, S.J. (Texas A&M University), (1982), *A Catalog of Low Reynolds Number Airfoil Data for Wind Turbine Applications.* RFP-3387. Golden, CO: Rocky Flats Wind Energy Research Center; 615 pp. (NTIS no. DE82021712),

Miller, A.H.; Dugundji, J.; Martinez-Sanchez, M.; Gohard, J.; Chung, S.; Wendell, J. (1978), *Wind Energy Conversion, Volume 1: Methods for Design Analysis of Horizontal Axis Wind Turbines*. COO-4131-Tq (Vol. I), Washington, DC: U.S. Department of Energy; 253 pp.

Miller, T. L. (1993). Demonstration-exploration-discussion: teaching chemistry with discovery and creativity. *J. Chem. Education*, 70(3): 187-189.

Minsky, M. (1995), 'Virtual Molecular Reality', in: Krummenacker, M. & Lewis, J. (eds.), *Prospects in Nanotechology. Proceedings of the 1st general conference on nanotechnology: developments, applications, and opportunities, November 11-14, 1992, Palo-Alto*, John Wiley & Sons, New York, pp. 187-205.

Minsky, M. L. (1954), *Neural nets and the brain-model problem*. PhD thesis, Princeton University.

Minstrell, Jim and van Zee, Emily. (2000). *Inquiring into Inquiry Learning and Teaching in Science*. Washington, DC: American Association for the Advancement of Science.

Mintzes, Joel L., Wandersee, James H., and Novak, Joseph D. (1997). *Teaching Science for Understanding: A Human Constructivist View*. San Diego, CA: Academic Press.

Misa, T. (1988), 'How Machines Make History and How Historians (and Others) Help Them to Do So', *Science, Technology, and Human Values*, 13, 308-331.

Mishra, R.S., and A.K. Mukherjee (1997), Oral presentation at TMS meeting, Indianapolis, Indiana, 16-18 September 1997, to be published in proceedings of Symp. "Mechanical Behavior of Bulk Nano-Materials."

Mishra, R.S., R.Z. Valiev, and A.K. Mukherjee (1997), *Nanostructured Materials* 9:473. Morris, D.G., and M.A. Morris. 1991. *Acta Metall. Mater*. 39:1763-1779.

Mitsch, W.J. and Gosselink, J.G. (1993), Wetlands. 2nd Edition. New York: Van Nostrand Reinhold.

Mitsch, W.J. and Reeder, B.C., (1991), Modelling nutrient retention of a freshwater coastal wetland: estimating the roles of primary productivity, sedimentation, resuspension and hydrology. Ecological Modelling 54, pp. 151-187.

Mnyusiwalla, A.; Abdallah, S.D.; Singer, P.A. (2003), 'Mind the gap: science and ethics in nanotechnology', *Nanotechnology*, 14, R9-R13.

Mnyusiwalla, A.; Daar, A.S.; Singer, P.A. (2003), '"Mind the Gap": Science and Ethics in Nanotechnology, *Nanotechnology*, 14, R9-R13.

Mody, C.C.M. (2000), "A New Way of Flying': Différance, Rhetoric, and the Autogiro in Interwar Aviation', *Social Studies of Science*, 30, 513-543.

Mody, C.C.M. (2001), 'A Little Dirt Never Hurt Anyone: Knowledge-Making and Contamination in Materials Science', *Social Studies of Science*, 31, 7-36.

Mody, C.C.M. (2004), 'How Probe Microscopists Became Nanotechnologists', in: D. Baird, A. Nordmann & J. Schummer (eds.), *Discovering the Nanoscale*, Amsterdam: IOS Press, pp. 119-133.

Molloy, J.E. (1998), "Optical chopsticks: Digital synthesis of multiple optical traps." *Meth. Cell Biol.* 55: 205-216.

Moore, G., Brief History of Fingerprint Identification, http://onin.com/fp/fphistory.html.

Morelli, Ralph. (1990). The student as knowledge engineer: A constructivist model for science education. *Journal of Computing in Higher Education*. 10(5), p. 78-102.

Morris, V.R.; Barnard, J.C.; Wendell, L.L.; Tomich, S.D. (1992), "Comparison of Anemometers for Turbulence Characterization." *Windpower '92 Conference; October 19-23, 1992; Seattle, Washington.* Washington, DC: American Wind Energy Association; 9 pp. (NTIS no. DE93004336),

Mortensen, C. (1983), 'Relevance and Verisimilitude' *Synthese* 55 (1983), p. 353-364.

Muhlholzer, F. (1995), 'Science without Reference?', *Erkenntnis*, 42, 2, pp. 203-222.

Mui, C. and C. Musgrave (2004), "The Hydroxylation and Oxidation of the Ge(100)-2x1 Surface by Hydrogen Peroxide and Water," *Langmuir,* 20, 7604-7609.

Mui, C., Y. Widjaja, J. Kang, and C. Musgrave (2004), "Surface Reaction Mechanisms for Atomic Layer Deposition of Silicon Nitride," *Surface Science,* 557, 159-170.

Mullard, S., (1995), Towards an EU Strategy for Integrated Coastal Zone Management. European Environmental Law Review 4(January): 16-20.

Mumford, Lewis (1934), *Technics and Civilization* (New York: Harcourt Brace Jovanovich, Inc.), National Science and Technology Council (NSTC), 1999. *Nanotechnology: Shaping the World Atom by Atom* (Washington, DC: September; available at http://www.nano.gov).

Munn Sanchez, E. (2004), 'The Expert's Role in Nanoscience and Technology', in: D. Baird, A. Nordmann & J. Schummer (eds.), *Discovering the Nanoscale*, Amsterdam: IOS Press, pp. 257-266.

Nagel, E. (1961), *The Structure of Science*, London: Routledge, Ch. 4.

Nagel, E., Bromberger, S. & Grünbaum, A. (eds.), (1971), *Observation and Theory in Science*, Baltimore & London: John Hopkins Press.

Narens, L. (1974), 'Measurement without Archimedean Axioms', *Philosophy of Science* 41, pp. 374-93.

Nerlich, G. (1976), *The Shape of Space,* Cambridge: CUP, Chs. 5-9.

Nersessian, N. (1984), *Faraday to Einstein: Constructing Meaning in Scientific Theories*, Dordrecht: Martinus Nijhoff.

Nersessian, N. J. (1987), 'A cognitive-historical approach to meaning in scientific theories', in Nersessian (ed.), *The Process of Science*, Dordrecht: Martinus Nijhoff.

Neurath, Carnap & Morris, eds. (1955), *International Encyclopedia of Unified Science*, Vols. I & 2, Chicago: UCP (esp. articles by Neurath and Carnap).

Neurath, O. (1973), *Empiricism and Sociology*, (Vienna Circle Collection, Vol. 1), Dordrecht: D.Reidel, esp. Ch. 9.

Newton-Smith W.H. (1987), 'Realism and Inference to the Best Explanation', *Fundamenta Scientiae* 7.

Newton-Smith, W. H. (1978), 'The Underdetermination of Theory by Data', *Proceedings of the Aristotelian Society*, Supp. vol. 52, pp. 71-91.

Newton-Smith, W. H. (1981), *The Rationality of Science*, London: Routledge, ch.2.

Neyman, A. (1985), Bounded complexity justifies cooperation in the finitely repeated prisoners' dilemma. *Economics Letters*, 19:227-229.

Niaz, Mansoor. (1995). Cognitive conflict as a teaching strategy in solving chemistry problems: A dialectic-constructivist perspective. *Journal of Research in Science Teaching*. 32(9), p. 959-70.

Nierenberg, R. (Altamont Energy Corporation), (1989), *Free-Flow Variability on the Jess and Souza Ranches, Altamont Pass.* SERI/STR-217-3404. Golden, CO: Solar Energy Research Institute.

Niiniluoto, I. (1989), 'Corroboration, Verisimilitude, and the Success of Science', in *Imre Lakatos and Theories of Scientific Change*, K. Gavroglu, Y. Goudaroulis and P. Nicolacopoulos, Dordrecht: Kluwer.

Niven, D. ed. (1890), *The Scientific Papers of James Clerk Maxwell*, Cambridge: CUP.

Nola, R. (1980), 'Fixing the Reference of Theoretical Terms', *PhilSci.*, 47, pp. 505-531.

Nola, R., ed. (1988), *Relativism and Realism in Science*, Dordrecht: Kluwer.

Noll, R.B.; Ham, N.D. (Aerospace Systems, Inc.), (1983), *Dynamic Stall of Small Wind Systems.* RFP-3523. Golden, CO: National Renewable Energy Laboratory. (NTIS no. DE84006827),

Nordmann, A. (2003), 'Shaping the World Atom by Atom: Eine nanowissenschaftliche WeltBildanalyse', in: A. Grunwald (ed.), *Technikgestaltung zwischen Wunsch und Wirklichkeit*, Berlin: Springer, pp. 191-199.

Norton, P.B., Esposito, J.J. (officers) (1994), *Encyclopaedia Britannica, 15th ed.*, Encyclopaedia Britannica, Inc. (Chicago).

O'Donnell, Angela M. and King, Alison. (1999). *Cognitive Perspectives on Peer Learning.* Mahwah, NJ: Lawrence Erlbaum Associates.

Oberg, E., Jones, F.D., Horton, H.L., Ryffel, H.H. (1996), *Machinery's Handbook, 25th ed.*, Industrial Press (New York).

Oddie, G. (1978), 'Verisimilitude and Distance in Logical Space' *Acta Philosophica Fennica* 30.

Olsen, S., Tobey, J., et al., (1998), A Common Framework for Learning from ICM Experience. Ocean and Coastal Management 37(2): 155-174.

Olsen, S.B. (1993), A Learning Approach to the Management of Coastal Ecosystems: Lessons from High and Low Income Nations. Preparing to Meet the Challenges for the 21st Century: Proceedings of the World Coast Conference, Netherlands.

Olson, C. F. (1994), Time and space efficient pose clustering. In *Proceedings of the IEEE Conference on Computer Vision and Pattern Recognition*, pages 251-258, Seattle, Washington.

Ong, W.J. (1977), *Interfaces of the Word,* Ithaca, NY: Cornell University Press.

Ostowari, C.; Niak, D. (Texas A&M University), (1985), *Post-Stall Wind Tunnel Data for NACA 44xx Series Airfoil Sections.* SERI/STR-217-2559. Golden, CO: Solar Energy Research Institute.

Pap, A. (1963), *An Introduction to the Philosophy of Science*, London: Eyre and Spottiswode, Ch. 11.

Papineau, D. (1979), *Theory and Meaning*, Oxford: OUP, chs.1, 2 & 5.

Papineau, D. (1987), *Reality and Representation*, Oxford: Blackwell.

Papineau, D., ed. (1996), *Philosophy of Science*, Oxford: OUP.

Parker, S.P. (1994), *Dictionary of Science and Technical Terms, 5th ed.*, McGraw-Hill (New York).

Pearce, G. & Maynard, P. (eds.), (1973), *Conceptual Change*, Dordrecht: Reidel.

Pease R.F.W. and Nixon W.C. "Microformation of filaments," in First International Conference on Electron and Ion Beam Science and Technology, Toronto, May 1964, edited by R. A. Bakish (Wiley, New York) pp 22~230. (1965).

Peirce, C. S. (1870), Description of a notation for the logic of relatives, resulting from an amplification of the conceptions of Boole's calculus of logic. *Memoirs of the American academy of arts and sciences*, 9:317-378.

Peng, X.G.; Manna, L.; Yang, W.D.; Wickham, J.; Scher, E.; Kadavanich, A.; Alivisatos, A.P. (2000), "Shape control of CdSe nanocrystals", *Nature*, 404, 59-61.

Penley, C. (1989), *The Future of an Illusion: Film, Feminism and Psychoanalysis,* Minneapolis: University of Minnesota Press.

Penning-Rowsell E., Green, C.H., Thompson, P.M., et al., (1992), The Eco-nomics of Coastal Management - A Manual of Benefit Assessment Techniques. Belhaven Press, London, UK, 140 pp.

Penning-Rowsell, E., (1997), A General Model for Promoting the Integration of National Natural Resources Management. GeoJournal, v. 43, n. 3, pp. 247-262.

Pereira, F. C. N. and Shieber, S. M. (1987), *Prolog and Natural-Language Analysis*. Center for the Study of Language and Information (CSLI), Stanford, California.

Perkal M., C. Marks, M.I. Lorber, W.H. Marks (1992), "A three-year experience with serum anodal trypsinogen as a biochemical marker for rejection in pancreatic allografts. False positives, tissue biopsy, comparison with other markers, and diagnostic strategies." *Transplantation*, Feb; 53 (2): pp. 415-419.

Poe, E.A., The Purloined Letter, *The Gift*, 1845. (also at http://pambytes.com/poe/poe.html)

Poincaré, H (1952), *Science and Hypothesis*, New York: Dover Publications, esp. Ch. 3-5.

Poincaré, H. (1958), *The Value of Science*, New York: Dover Publications, Pt. 3.

Popper, K (1959), *The Logic of Scientific Discovery*, London: Hutchinson, (Appendix XI).

Popper, K. & Miller, D.W.(1983), 'A Proof of the Impossibility of Inductive Probability', *Nature* 302, pp.687-688.

Popper, K. (1959), *The Logic of Scientific Discovery*, London: Hutchinson and Co., Ch. VII.

Psillos, S. (1995), "Is Structural Realism the Best of Both Worlds?" *Dialectica* 49.

Psillos, S. (1996), 'On van Fraassen's Critique of Absolute Reasoning', *PhilQuart* 46.

Putnam, H (1975), 'The Refutation of Conventionalism', in *Mind, Language and Reality: Philosophical Papers Vol. 2*, Cambridge: CUP.

Putnam, H. (1963), 'Probability and Confirmation' reprinted in *Mathematics, Matter and Method: Collected Papers Vol. 1* (2nd ed.), Cambridge: CUP, pp.293-304.

Putnam, H. (1965), 'Craig's Theorem', repr. in *Mathematics, Matter and Method, Philosophical Papers Vol. 1*, Cambridge: CUP, pp.228-236.

Putnam, H. (1975), 'What Theories are Not', in *Mathematics, Matter and Method, Collected Papers Vol. 1*, Cambridge: CUP.

Putnam, H. (1978), *Meaning and the Moral Sciences*, London: Routledge, Lectures I-VI.

Putnam, H. (1979), *Mind, Language and Reality, Philosophical Papers Vol. 2*, Cambridge: CUP, (esp. 'The Meaning of Meaning').

Putnam, H. (1982), *Reason, Truth and History*, Cambridge: CUP.

Putnam, H. (1987), *The Many Faces of Realism*, La Salle: Open Court.

Rahn, Armin (1993), *The Basics of Soldering, 1st ed.*, John Wiley and Sons (New York).

Ralph C. Merkle (1995), "Design-ahead for nanotechnology," in Markus Krummenacker, James Lewis, eds., *Prospects in Nanotechnology: Toward Molecular Manufacturing*, John Wiley & Sons, New York, pp. 23-52.

Rautalahti-Miettinen, E., Ripatti, V., (1998), Development of Environmental Awareness. Workshop on Education and Learning for Sustainable Consumption, 14-15 September, OECD Headquarters, Paris.

Reese, W.L. (1980), *Dictionary of Philosophy and Religion*, Atlantic Highlands, NJ: Humanities Press.

Reichenbach, H. (1949), *The Theory of Probability*, Berkeley: University of California Press.

Reichenbach, H. (1950), *The Philosophy of Space and Time*, New York: Dover Publications, esp. Ch. 1.

Reichenbach, H. (1959), *The Rise of Scientific Philosophy*, Berkeley: U. of California Press.

Reichenbach, H. (1961), *Experience and Prediction*, Chicago: UCP, Ch. 5.

Reid, Brian (1993), 'Usenet readership summary report'. Palo Alto. CA: Network Measurement Project at the DEC Western Research Laboratory (March),

Reigeluth, Charles M. (1999). *Instructional-Design Theories and Models: A New Paradigm of Instructional Theory*, Volume 11. Mahwah, NJ: Lawrence Erlbaum Associates.

Reinhardt, C. (2004), 'Chemistry in a Physical Mode: Molecular Spectroscopy and the Emergence of NMR', *Annals of Science*, 61, 1-32.

Resnik, D.B. (1992), 'Convergent Realism and Approximate Truth', *PSA* Vol. 1, pp. 421-434.

Richardson, R. (1986), 'Models and Scientific Explanation' *Philosophica* 37, p. 59.

Robinson, B.H. and N.C. Seeman (1987), The Design of a Biochip: A Self-Assembling Molecular-Scale Memory Device. Protein Engineering 1, 295-300.

Roco, M.C. (2003a), 'Broader Societal Issues of Nanotechnology', *Journal of Nanoparticle Research*, 5, 181-189.

Roco, M.C. (2003b), 'Coherence and Divergence in Megatrends in Science and Engineering', in: M.C. Roco, W.S. Bainbridge (eds.), *Converging Technologies for Improving Human Performance*, Kluwer, Dordrecht, pp. 79-96.

Roco, M.C. (2003c), 'Fundamentally New Manufacturing Processes and Products', in: Roco, M.C. & Bainbridge, W.S., (eds.), *Converging Technologies for Improving Human Performance*, Kluwer, Dordrecht, pp. 300-302.

Roco, M.C. (2004), 'The Emergence and Policy Implications of Converging Technologies' [www.nsf.gov/home/crssprgm/nano/nbic_roco_04_0422_@aaas_57sl.pdf, accessed May 2004].

Rogers, A; Manwell, J.F.; Stein, W.M.; McGowan, J.G. (1993), "Experimental Testing of Variable Speed Diesels for Wind/Diesel System Applications." *Wind Energy-1993*; January 31-February 4, 1993; Houston, Texas. SED-Vol. 14. New York, NY: American Society of Mechanical Engineers.

Rohrbach, C.; Wainauski, H.; Worabel, R. (1977), *Experimental and Analytical Research on the Aerodynamics of Wind Driven Turbines.* COO-2615-T2. Windsor Locks, CT: Hamilton Standard; 250 pp.

Roper, J.E. (1982), 'Models and Lawlikeness', *Synthese* 52, pp. 313-323.

Rorty, R. (1965), Mind-body identity, privacy, and categories. *Review of Metaphysics*, 19(1):24-54.

Rose, D.J. and Clark, M. Jr. (1961), *Plasmas and controlled fusion*, MIT Press, 1961 (QC718.R6.,

Rosenkrantz, R.D. (1981), *Foundations and Applications of Inductive Probability*, Atascadero, California: Ridgeview Press.

Roth, Wolff-Michael; Roychoudhury, Anita. (1993). Using vee and concept maps in collaborative settings: Elementary education majors construct meaning in physical science courses. *School Science and Mathematics*. 93(5), p. 237-44.

Salmon, M.H. et. al. (1992), *Introduction to the Philosophy of Science*, N.J.: Prentice-Hall, pp. 66-100.

Salmon, N. U. (1982), *Reference and Essence*, Oxford: Blackwell, esp. Ch. 4.

Salmon, W.C. (1966), *The Foundations of Scientific Inference*, USA: University of Pittsburgh

Salmon, W.C. (1969), 'The Conventionality of Simultaneity', *PhilSci* 36, pp.44-63.

Sankey, H. (1994), *The Incommensurability Thesis*, Avebury: Brookfield.

Savage, L.J. (1971), 'Probability in Science: A Personalistic Account' in *Logic, Methodology and Philosophy of Science* IV, P. Suppes & L. Henkin (eds.).

Savage, L.J. (1972), *Foundations of Statistics*, New York: Dover

Schluter, L.L. (Sandia National Laboratories); Vachon, W.A. (W.A. Vachon & Associates), (1993) "Optimizing Wind Turbine Control System Parameters." *Windpower '93 Conference; July 12-16, 1993; San Francisco, California.* Washington, DC: American Wind Energy Association; pp. 175-182.

Sharpe, Pamela J. "Thinking About Thinking: A Study of the Adult's Role in Providing for the Development of Number Awareness in Children." *Early Child Development & Care* 144 (1998): 79-89.

Shymansky, James. (1992). Using constructivist ideas to teach science teachers about constructivist ideas, or teachers are students too! *Journal of Science Teacher Education*. 3(2), p. 53-57.

Simms, D.A. (National Renewable Energy Laboratory); Cousineau, K.L. (Zond Systems, Inc.), (June 1993), *A Rugged, Low-Cost, Advanced Data Acquisition System for Field Test Projects.* NREL/TP-442-5617. Golden, CO: National Renewable Energy Laboratory; 11 pp. (NTIS no. DE93010041),

Simms, D.A.; Butterfield, C.P. (1990), *A Low-Cost PC-Based Telemetry Data-Reduction System.* SERI/TP-257-3737. Golden, CO: Solar Energy Research Institute.

Simon, R.L. (1988), *Wake Effects in a Fayette 85-IIS Wind Turbine Array.* SERI/STR-217-3186A. Golden, CO: Solar Energy Research Institute; 75 pp.

Simon, R.L.; Matson, D.F.; Fuchs, J.M. (Fayette Manufacturing Corporation), (1987), *Wake Effects in a Fayette 95-IIS Wind Turbine Array.* SERI/STR-217-3186. Golden, CO: Solar Energy Research Institute.

Smith, J.C. (1991), *Studies on the Modeling of Wind-Diesel Systems and the Integration of Wind Energy Systems with Electric Utilities.* Prepared for DOE/EC Wind Energy Assessment and Review, Albuquerque, New Mexico, June 4, 1991, by Electrotek Concepts, Inc.

Snider, G.L.; Orlov, A.O.; Amlani, I.; Zuo, X.; Bernstein, G.H.; Lent, C.S.; Merz, J.L.; Porod, W. (1999), "Quantum-dot cellular automata: Review and recent experiments (invited)", *Journal of Applied Physics*, 85, 4283-4285.

Snow, A.L.; Heberling, C.F., III; Van Bibber, L.E. (Westinghouse Electric Corporation), (1989), *The Dynamic Response of a Westinghouse 600-kW Wind Turbine.* SERI/STR-217-3405. Golden, CO: Solar Energy Research Institute; 185 pp.

Snow, E.S., P.M. Campbell, and F.K. Perkins (1997), Nanofabrication with proximal probes. *Proceedings of the IEEE* 85:601-11.

Soong, R.K.; Bachand, G.D.; Neves, H.P.; Olkhovets, A.G.; Craighead, H.G.; Montemagno, C.D. (2000), "Powering an inorganic nanodevice with a biomolecular motor", *Science*, 290, 1555-1558.

Sorensen, J. and McCreary, S., (1990), Institutional Arrangements for Managing Coastal Resources and Environments, National Park Service, US Department of the Interior & USAID.

Sorensen, J., (1993), The International Proliferation of Integrated Coastal Zone Management Efforts. Ocean and Coastal Management 21(1-3): 45-80.

Souers, P.C., (1986), *Hydrogen properties for fusion energy*, Univ. of California Press. (QC 791.S68)

Spera, D.A. (1994), *Wind Turbine Technology: Fundamental Concepts of Wind Turbine*

Engineering. No. 100368. Fairfield, NJ: American Society of Mechanical Engineers; 700 pp.

Spiegelhalter, D. J. (1986), Probabilistic reasoning in predictive expert systems. In Kanal, L. N. and Lemmer, J. F., editors, *Uncertainty in Artificial Intelligence*, pages 47-67. Elsevier/North-Holland, Amsterdam, London, New York.

Spigel, Lynn (1992), *Make Room for TV: Television and the Family Ideal in Postwar America,* Chicago: University of Chicago Press.

Spirtes, P., Glymour, C., and Scheines, R. (1993), *Causation, prediction, and search*. Springer-Verlag, Berlin.

Stoddard, F.S. (1990), "Wind Turbine Blade Technology: A Decade of Lessons Learned." *Energy and the Environment, Into the 1990's, 1st World Renewable Energy Congress*. Edited by A.A.M. Sayigh. Vol. 3. Tarrytown, NY: Pergamon Press; 1570 pp.

Stoddard, F.S.; Porter, B.K. (1986), *Wind Turbine Aerodynamics Research Needs Assessment*. DOE/ER/30075-H1. Washington, DC: U.S. Department of Energy; 126 pp. (NTIS no. DE86009189),

Swartz, N (1985), *The Concept of Physical Law*, Cambridge: Cambridge University Press.

Swenson, Loyd S. (1979), *The Etherial Aether: A History of the Michelson-Morley Experiment*, Texas: Austin.

Tangler, J.L. (1987), *Status of the Special-Purpose Airfoil Families*. SERI/TP-217-3264. Golden, CO: Solar Energy Research Institute; 9 pp. (NTIS no. DE88001124/XSP),

Tangler, J.L.; Smith, B.; Jager, D. (1992), *SERI Advanced Wind Turbine Blades*. NREL/TP-257-4492. Golden, CO: National Renewable Energy Laboratory. (NTIS no. DE92001216),

Tangler, J.L.; Smith, B.; Kelley, N.D.; Jager, D. (1992), *Measured and Predicted Rotor Performance for the SERI Advanced Wind Turbine Blades*. NREL/TP-257-4594. Golden, CO: National Renewable Energy Laboratory. (NTIS no. DE92001215),

Thagard, P. (1978), 'The Best Explanation: Criteria for Theory Choice' *JPhil* 75.

Thresher, R.W.; Holley, W.E.; Wright, A.D. (January 1989), *Prediction of Stochastic Blade Responses Using a Filtered Noise Turbulence Model in the FLAP Code*. SERI/TP-217-3413. Golden, CO: Solar Energy Research Institute.

Thresher, R.W.; Wright, A.D. (January 1989), *Prediction of Stochastic Blade Responses*

Using Measured Wind-Speed Data as Input to the FLAP Code. SERI/TP-217-3394. Golden, CO: Solar Energy Research Institute.

Tohru Watanabe, Watanabefv@aol.com, Department of Robotics, Ritsumeikan University; personal communication to Tihamer Toth-Fejel, 21 March 2000.

Tol, R., Klein, R., Jansen, H., Verbruggen, H., (1996), Some economic considerations on the importance of pro-active integrated coastal zone management. Ocean & Coastal Management, v. 32, n. 1, pp.39-55.

Torrey, D.A. (Rensselaer Polytechnic Institute); Childs, S.E. (Atlantic Orient Corporation), (1993), "Development of Variable-Reluctance Wind Generators." *Windpower '93 Conference; July 12-16, 1993; San Francisco, California.* Washington, DC: American Wind Energy Association; pp. 258-265.

Towns, Marcy H. and Grant, Edward R. (1997). "I believe I will go out of this class actually knowing something": cooperative learning activities in physical chemistry. *J. res. Science Teaching* 34(8): 819-835.

Tres, P.A. (1998), *Designing Plastic Parts for Assembly, 3rd ed.*, Hanser Publishers (Munich).

Trombulak, Stephen. (1995), Merging inquiry-based learning with near-peer teaching. *Bioscience.* 45(6), p. 412-16.

Vachon, W.A. (1990), "Smart Control Algorithms for Operation of Variable-Speed Wind Turbines." *Proceedings of the Ninth ASME Wind Energy Symposium.* Edited by D.E. Berg. SED- Vol. 9. New York, NY: American Society of Mechanical Engineers.

Vallega, A., (1992), Sea Management - A Theoretical Approach. Elsevier Applied Science, pp. 259.

Vallentyne, P. (1988), 'Explicating Lawhood', *PhilSci* 55, pp.598-613.

Van Dijk, H., (1994), Integrated Management and Conservation of Dutch Coastal Areas: Premises, Practice and Solutions. Marine Pollution Bulletin 29(6-12): 609-616.

van Vlack, L.H. (1985), *Elements of Materials Science and Engineering*, Addison-Wesley (Reading, MA).

Veers, P.S. (1988), *Three-Dimensional Wind Simulation.* SAND-88-0152. Albuquerque, NM: Sandia National Laboratories; 36 pp. (NTIS no. DE88009793/XSP),

Velleman, Ruth A. (1990), New technology. In Velleman, Ruth A., Meeting the needs of people with disabilities: a guide for librarians, educators, and other service professionals. Phoenix, AZ: Oryx Press, p. 79-94.

Venturi, R.. Brown, D. S., Izenour, S. (1972), *Learning from Las Vegas.* Cambridge, Mass.: The MIT Press.

Verbeek, A. & Callaert, J. (2002), *Detailed analysis of the science-technology interaction in the field of Nanotechnology*, Luxemburg: Office for Official Publications of the European Communities (Linking Science to Technology. Bibliographic References in Patents, Vol. 9) [79 pp.].

Vickers, J. (1988), *Chance and Structure*, Oxford: Clarendon Press

Viterna, L.A.; Janetzke, D.C. (1982), *Theoretical and Experimental Power from Large Horizontal-Axis Wind Turbines.* DOE/NASA/20320-41. Cleveland, OH: NASA Lewis Research Center; 18 pp. (NTIS no. DE83001099),

Walker, S.N.; Wade, J.E. (Pacific Wind Energy), (1988), *Effects of Precipitation on Wind Turbine Performance.* SERI/STR-217-3287. Golden, CO: Solar Energy Research Institute.

Warren, D. H. D. (1983), An abstract Prolog instruction set. Technical Note 309, SRI International, Menlo Park, California.

Watts, Mike; Jofili, Zelia; Bezerra, Risonilta. (1997). A case for critical constructivism and critical thinking in science education. *Research in Science Education.* 27(2), p. 309-22.

Weber, T.L.; Wilson, R.E.; Walker, S.N. (Oregon State University), (1991), *Analysis of a Teetered, Variable-Speed Rotor.* SERI/TP-267-4366. Golden, CO: Solar Energy Research Institute; 59 pp. (NTIS no. DE91002190),

Wehrey, M.; Redmond, I.; Anderson, C.; Jamieson, R. (Southern California Edison, James Howden and Company Ltd.), (1988), *Dynamic Response of a 330-kW Horizontal Axis Wind Turbine Generator.* SERI/STR-217-3203. Golden, CO: Solar Energy Research Institute; 103 pp.

Wendell, L.L; Barnard, J.C.; Morris, V.R.; Gower, G.L. (1993), "Turbulence Characteristics in Complex Terrain." *Wind Energy 1993.* SED-Vol. 14, Wind Energy. New York, NY: The American Society of Mechanical Engineers.

West, Leo H. T. and Pines, A. Leon (editors). (1985). *Cognitive Structure and Conceptual Change.* Orlando, FL: Academic Press.

Wilder, K., H.T. Soh, T. Soh, A. Atalar, and C.F. Quate (1997), Hybrid atomic force/ scanning tunneling lithography. *J. of Vacuum Science & Technol. B.* 15:1811-17.

Wilensky, R. (1983), *Planning and Understanding*. Addison-Wesley.

Wilkes, K W (1988), *Real People: Personal Identity Without Thought Experiments*, Oxford: Oxford University Press.

Williamson, J. (1989), 'Every virus tells a story: the meaning of HIV and AIDS', in E. Carter and S. Watney (eds), *Taking Liberties: AIDS and Cultural Politics*, London: Serpents Tail/ ICA.

Wilson, M. (1980), 'The Observational Uniqueness of Some Theories', *JPhil* 77, pp.208-33.

Wilson, R.E.; Walker, S.N. (1985), "State-of-the-Art in Wind Turbine Aerodynamics." *Windpower '85 Conference; August 27, 1985; San Francisco, California.* Alexandria, VA: American Wind Energy Association; pp. 30-38.

Wright, A.D.; Buhl, M.L., Jr.; Thresher, R.W. (1988), *FLAP Code Development and Validation.* SERI/TR-217-3125. Golden, CO: Solar Energy Research Institute.

Wright, A.D.; Butterfield, C.P. (1991), *The NREL Teetering Hub Rotor Code: Final Results and Conclusions.* NREL/TP-257-4517. Golden, CO: National Renewable Energy Laboratory; 10 pp. (NTIS no. DE92001187),

Wyatt, A. (Research Triangle Institute), (1992), *Wind Electric Pumping Systems: Sizing and Cost Estimation.* Washington, DC: American Wind Energy Association; 39 pp.

Zamora, J.H., Procedure glitch undoes drug cases. *San Francisco Examiner*, 2000. http://www.examiner.com/000209/0209chemists.html

Zeilik, Michael, Schau, Candace, and Mattern, Nancy. (1999). Conceptual astronomy. II. replicating conceptual gains, probing attitude changes across three semesters. *Am. J. Physics* 67(10): 923-927.

Index

C

D

E

F

G

H

T

U

V

W

X

Z